OCEAN WAVE DYNAMICS
FOR COASTAL AND
MARINE STRUCTURES

ADVANCED SERIES ON OCEAN ENGINEERING

ISSN 1793-074X

Editor-in-Chief
Philip L-F Liu (*School of Civil and Environmental Engineering,
Cornell Univ., Hollister Hall Ithaca, NY*)

Published

*For the complete list of titles in this series, go to https://www.worldscientific.com/series/asoe

Advanced Series on Ocean Engineering — Volume 52

OCEAN WAVE DYNAMICS FOR COASTAL AND MARINE STRUCTURES

Vallam Sundar

Indian Institute of Technology Madras, India

World Scientific

NEW JERSEY · LONDON · SINGAPORE · BEIJING · SHANGHAI · HONG KONG · TAIPEI · CHENNAI · TOKYO

Published by

World Scientific Publishing Co. Pte. Ltd.

5 Toh Tuck Link, Singapore 596224

USA office: 27 Warren Street, Suite 401-402, Hackensack, NJ 07601

UK office: 57 Shelton Street, Covent Garden, London WC2H 9HE

Library of Congress Cataloging-in-Publication Data
Names: Sundar, V. (Vallam), author.
Title: Ocean wave dynamics for coastal and marine structures /
    Vallam Sundar, Indian Institute of Technology, Madras, India.
Description: Hackensack, NJ : World Scientific, [2021] | Series: Advanced series on
    ocean engineering, 1793-074X | Includes bibliographical references and index.
Identifiers: LCCN 2021009799 | ISBN 9789811236662 (hardcover) |
    ISBN 9789811236679 (ebook for institution) |
    ISBN 9789811236686 (ebook ebook for individuals)
Subjects: LCSH: Shore protection. | Harbors--Protection. | Offshore structures--Dynamics. |
    Surface waves (Oceanography)--Mathematical models. | Ocean waves--Mathematical models.
Classification: LCC TC330 .S86 2021 | DDC 627/.042--dc23
LC record available at https://lccn.loc.gov/2021009799

British Library Cataloguing-in-Publication Data
A catalogue record for this book is available from the British Library.

For any available supplementary material, please visit
https://www.worldscientific.com/worldscibooks/10.1142/12268#t=suppl

Desk Editors: Anthony Alexander/Yu Shan Tay

Typeset by Stallion Press
Email: enquiries@stallionpress.com

Dedicated to my beloved parents
Prof. Vallam Venkataswami &
Smt. Vallam Sarojini

Preface

The behavior of surface ocean waves was well understood by representing them mathematically as regular wave forms. The studies then focused on random waves with unidirection, and by early 1980's, the focus then shifted to a better representation by considering their associated directions, which leads to closer observation of waves in the ocean. Due to an ever increase in exploration and exploitation of ocean resources, maritime trade, and Ocean energy, a variety of structures are constructed, and new concepts have been emerging. All these developments necessitate a comprehensive and in-depth knowledge of the behavior of ocean waves in the offshore as well as its characteristics while propagating toward the coast, which include the phenomena of wave generation, propagation, deformation, and its effects on marine structures. This book is published to serve the graduate students and researchers working in ocean, coastal, and harbor engineering to understand mechanics of ocean waves. This book is well drafted such that the understanding of each chapter can be greatly enhanced by reading its preceding chapter. It encompasses the fundamentals of the subject with sufficient description and illustrations.

This book can serve as a textbook for the students specializing in the field of ocean engineering related topics. It offers the application of theoretical formulas to practical relevance through solving worked out examples and is also expected to be of great help to the professionals in major ports, harbors, offshore and marine industries, consultancy agencies, etc.

The *Introduction* chapter describes the different terminologies pertaining to the study of wave mechanics and provides an insight to the different types of structures in the ocean front, physical, and ocean environmental parameters and their impact on structures.

Prior to the concept of wave mechanics, it is mandatory to have a strong foundation on basics in fluid mechanics. Hence, the chapter on *Fluid Mechanics* is included which provides a brief elucidation about the basic definitions and discussions on the topics of greater importance.

A sizeable share of this book emphasizes on the *Ocean Wave Mechanics* and the types of *deformation* the waves undergo, while propagating from offshore to coastal waters. A notable account on small amplitude wave theory followed by profound explanation and inferences of dynamic properties of standing and progressive waves is included. The mathematical and empirical relations derived for the resultants of wave action are also discussed.

A detailed account of *Finite Amplitude Wave Theory* is discussed with mathematical and statistical approaches highlighting the established theories such as Stroke's, Cnoidal, and Solitary wave theories which are important in understanding the behavior of waves in shallow water conditions.

Having understood the linearity of periodic/specific wave forms, mathematical statistics tools are employed to decode the parameters of random waves. The book then discusses the *Characteristics of Random Waves*, under the sixth chapter, through both time- and frequency-domain analyses. After realizing the importance of multi-directional waves, researchers in mid-1990s focused a great deal attention on forces and motion response characteristics on structures in three-dimensional waves, the details of which are covered herein.

This book would fail to be wholesome and complete if the impacts of *Wave-Induced Forces on Structures* are not discussed, and thus the various force regimes and evaluation wave loads on structures are discussed through several worked-out examples.

Most numerical predictions are either verified through physical modeling and a few problems still bank mainly on the experimental investigations, such as the evaluation of the stability of rubble mound breakwaters. The importance of dimensional analysis through examples on wave structure interaction problems and case studies is discussed in detail in the chapter *Physical Modeling*.

Last but not the least is the chapter on *Laboratory Generation of Waves*, which provides elaborately the need for generating different types of waves and interaction with structures. A list of possible experiments to be carried out by students as a part of the curriculum is presented.

The references subscribed in this book are vast and extensive, yet they would furnish more exhaustive topics pertaining to the field of study. Several worked-out examples are provided that could facilitate students easy enhanced learning. The appendices enclosed along with the chapters ensure a comprehensive knowledge of the subject.

About the Author

Prof. Vallam Sundar is a civil engineer and received his PhD from IIT Madras. He has served as a faculty at IIT Madras since 1981 and currently a Professor Emeritus in the Department of Ocean Engineering. Prof. Sundar has supervised 27 PhD, 14 MS and 12 MTech theses. He owns to his credit about 550 publications in conferences and journals — highest by an Indian in the field of Ocean Engineering. He has retired from service in June 2018 and has rejoined as Professor Emeritus from 1 July 2018. He is a member of the editorial board for about 10 international journals. He was awarded an honorary doctorate award by University of Wuppertal, Germany, and a lifetime achievement award by Indian Society of Hydraulics. Apart from this, he has been awarded 10 more. The International Association for Hydro-environment Engineering and Research (IAHR) has elected him as Chairman of its Asia-Pacific Division in 2006. His solutions to number of coastal erosion problems have led to fruitful results. His reports on tsunami mitigation master plan for the two maritime states, Tamil Nadu and Kerala, remain basic document for planning and implementation. For three years, from 2014, he has served as a member of the panel of five international experts to review the Coastal Adaptation Study for Singapore. During 2016–2018, he was also a member of the Board of Governors of IIT Madras. He was instrumental in establishment of a National Technology Center for Ports, Waterways and Coasts.

Acknowledgments

I should like to place on record my sincere thanks to all colleagues Prof. S.A. Sannasiraj, and Dr. V. Sriram for their suggestions and comments. The lecture material has been classroom tested and the support extended by a number of students in particular, Ms. Sukanya, R., are gratefully acknowledged.

Am thankful to my wife Mrs. Poornima, daughters Dr. V.S. Suriya Priya and Ms. V.S. Banu Priya for their excellent support in making the completion of this book a success.

Contents

Chapter 1

Introduction

1.1 General

It is believed that oceans constitute about 70% of the surface of the earth, and about 97% of water content belongs to the ocean. It is believed that about 1% is fresh water, and 2–3% in glaciers and ice caps. The ocean is rich in abundance of natural living and non–living resources. In order to explore and exploit the vast non-living resources like oil, gas, and minerals from beneath the ocean floor, a variety of structures are needed to be designed and installed. The perennial instability of shorelines leading to coastal erosion along certain stretches of the coast, natural coastal hazards like storm surge, tsunami, etc., necessitates protection measures which are mostly structures, like seawalls, groins, offshore detached breakwaters, etc. constructed along or near the coast. One major development activity along the coast is the development of harbors by constructing breakwaters. The major environmental parameters that need to be considered in the design of all the different types of structures are the wind, waves, tides, and currents. Among these, although, the loads due to waves dominate and dictate the design of structures in the marine environment, characteristics of other parameters are also essential.

Further, the continuous depletion of the conventional energy resources, rapid research towards *alternative or renewable energy*, a term used for an energy source that is an alternative to using fossil fuels is in progress. Alternative energy resources are the renewable energy resources that are naturally available and can be naturally replenished. This energy cannot be exhausted and is constantly renewed unlike energy generated from fossil fuels. Apart from the solar and wind energy, the energy from the ocean comes from several sources like waves, tides, and currents. In addition, the gradient between salient, temperature, and density apart from tides and waves generates ocean currents that can be effectively utilized for extraction of energy. For designing efficient energy extraction devices from these natural resources, a critical knowledge on the characteristics and the basic physics of waves, tides, and currents are essential. Hence, this chapter would discuss the important basic aspects of these parameters.

1.2   Structures in the Marine Environment

The structures in the marine environment can broadly be classified as coastal structures, port, and harbor structures and offshore structures which are briefly discussed in this section.

1.2.1   *Coastal structures*

1.2.1.1   *General*

The types of coastal structures are (a) breakwaters, (b) seawall and bulkheads, (c) revetments, and (d) groins. In addition, we have intake structures for drawing seawater continuously for power plants, aquaculture, etc. and outfalls for discharging effluents into the marine environment.

1.2.1.2   *Seawalls*

Seawalls and bulkheads are structures placed parallel or nearly parallel to the shoreline to separate the land area from water area. The primary purpose of a bulkhead is to retain or prevent sliding of the land, with a secondary purpose of affording protection to the backshore against damage by wave action. Seawalls are used primarily to protect areas in the rear of the beach against severe attack of waves and storms. They are necessarily massive and expensive and should be constructed only where the adjoining shore is highly developed and storm attack is severe. Seawall is in essence, a retaining wall which, in addition to earth pressure from landside, is acted upon by the impact force of the waves. A typical cross-section of a seawall is shown in Fig. 1.1.

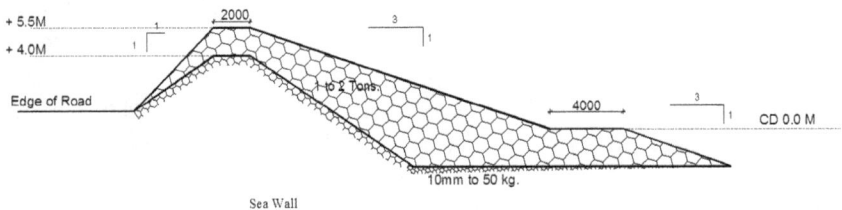

Fig. 1.1   Typical cross-section of a seawall.

1.2.1.3 *Groins*

A groin is usually perpendicular to the shore, extending from a point land-ward of possible shoreline recession into the water; a sufficient distance to stabilize the shoreline at a desirable location. Groins may be classified as permeable or impermeable, high or low, and fixed or adjustable. They may be constructed using timber, steel, stone, concrete, or other materi-als, or combinations thereof. Groins are built in order to halt or reduce shoreline erosion by means of controlling the rate of alongshore or littoral drift of beach material. They are designed to trap the maximum volume of material. Functional design is the determination of length, spacing, height, alignment, and type of groin which will halt or reduce beach erosion to an acceptable degree. The groins should extend beyond the zone of breaking of waves and distance between them should be 1.5–3 times its length. An aerial view of a groin field protecting Paravoor ($8°49'52.7''$N; $76°38'9.9''$E) along the Paravoor, Kerala coast and Chennai, Tamil Nadu coast are shown in Fig. 1.2.

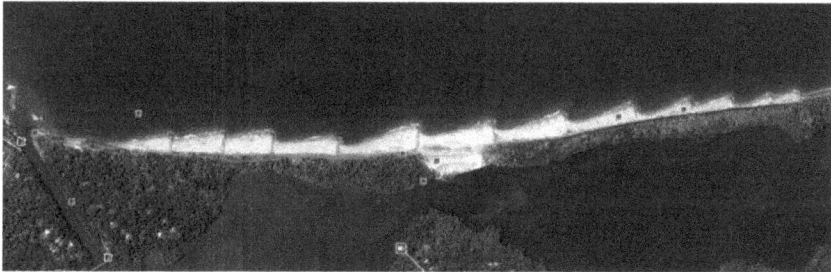

(a)

(b)

Fig. 1.2   (a) A view of the coast of Paravoor, Kerala showing the effect of groin field. (b) A view of the groin field protecting coast of Royapuram, North of Chennai harbor.

1.2.1.4 *Breakwaters*

Harbors are classified broadly as natural and artificial harbors. Artificial harbors are formed by constructing usually a pair of breakwaters. They protect the coast or a harbor from waves, thereby preventing their destructive influence on the coast or within the area of harbor. Isolated offshore breakwaters, usually of mound type, are aligned parallel to the shoreline to serve as a shore-protection measure. Offshore breakwaters are expensive and should be used after good model tests as sometimes they bring in precisely the opposite effects as against what is desired. From the shape of the breakwaters they are classified as vertical wall type, sloping mound, and composite type consisting of a vertical wall type kept over sloping mount type (Fig. 1.3).

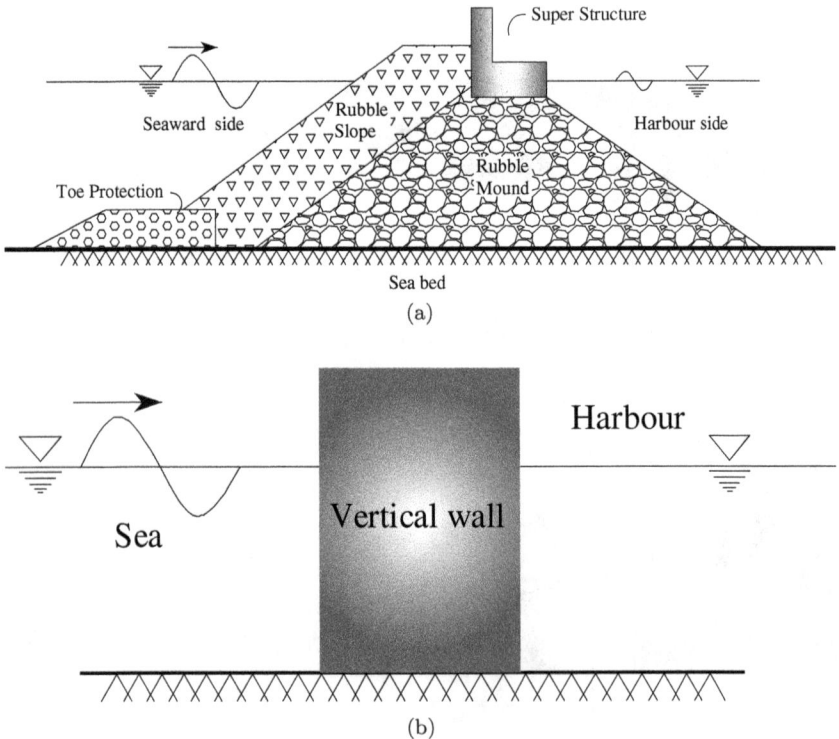

Fig. 1.3   (a) Rubble mound breakwater. (b) Vertical breakwater. (c) Typical cross-section of horizontal composite breakwaters.

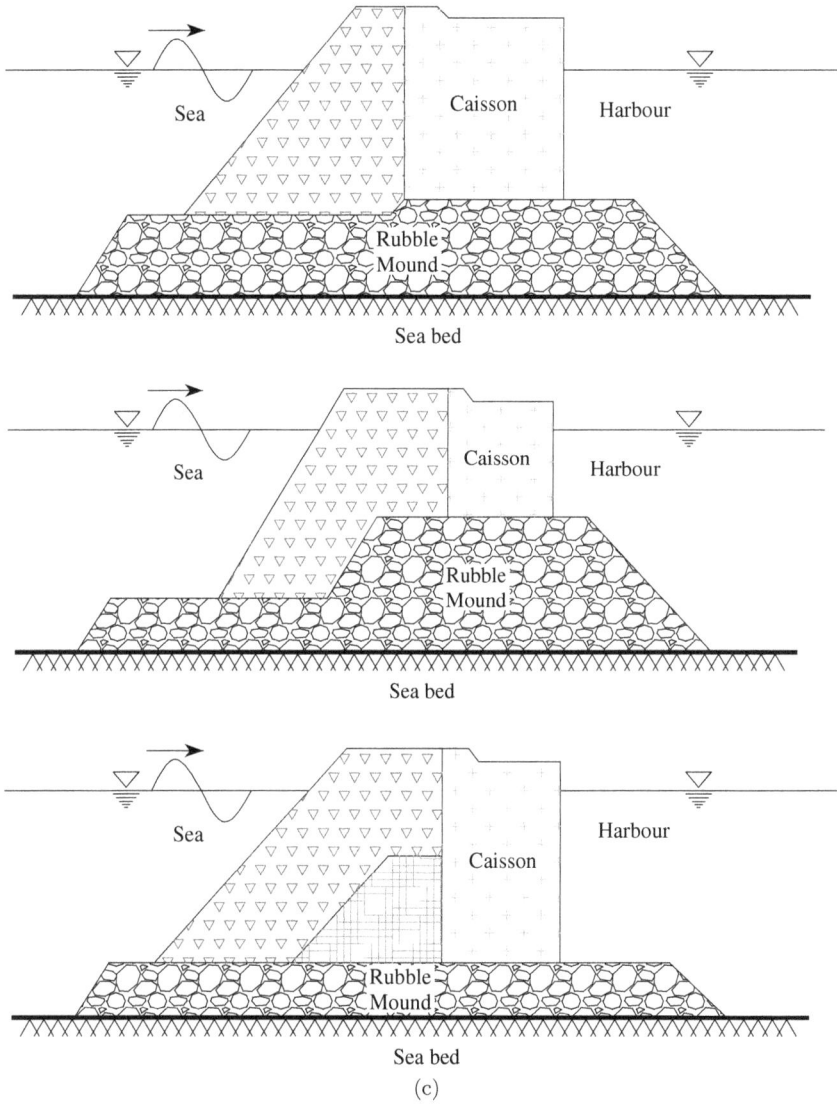

Fig. 1.3 (*Continued*)

1.2.1.5 *Revetment*

To improve the stability of slope of the shorelines, Riprap Revetment provides another method of shore-protection measure. A typical cross-section or Riprap Revetment is shown in Fig. 1.4. For additional

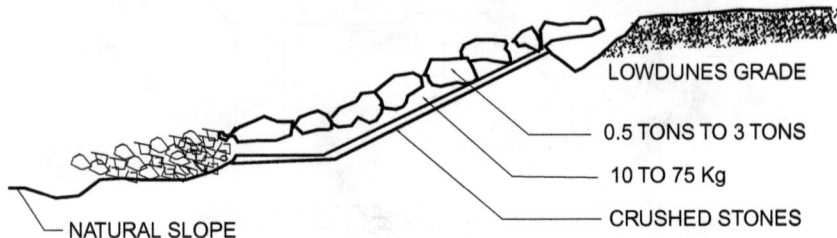

Fig. 1.4   Typical cross-section of a revetment.

details on coastal structures, the reader is recommended to refer to Coastal Engineering Manual (2006), Horikawa (1978), and Sorenson (1978).

1.2.2   *Port structures*

1.2.2.1   *General*

The most important structure to form a harbor, the main component of a port, is also an important structure that falls under the category of coastal structure that has been briefly discussed earlier. A port or a marine terminal is intended to provide facilities for transshipment of cargo and movement of passengers. The structures constructed as facilities for ports or harbors are quays or wharves, jetties, piers, dolphins are denoted by a general name "Docks".

A *wharf* is a structure supported on piles and constructed along a river or a sea that facilitates safe berthing of vessels.

A *quay* is generally erected along the river bank or shoreline, where it is ideal for the purpose of docking. Quays are very much similar in function to a wharf, except for its platforms are long enough and is fairly above the water surface which can sustain loading and unloading of vessels and its supporting component is made up of a solid vertical wall touching the sea/river bed.

A docking platform handling cargo is constructed for handling a specific type of cargo like cement or oil or grains, etc. Hence the type of docking platform depends on the type of cargo to be handled. The main factors that need to be considered in their design are direction of wind and waves, size of ships to be berthed and soil conditions. The total number of wharves for a particular location is determined based on the mooring population and its capacity. The quays are constructed as sheet pile wall, open piles structure, or as gravity structure.

A *pier* projects/overhangs from shore to the sea, above the water level such as to enable disembarkation of passengers at deeper waters further away from the shore, also along the length of pier smaller boats can be berthed. The piers are of three types: (a) piers consist mainly of reclaimed land and bordered by quays, (b) piled piers, and (c) floating piers. The first type is similar to quays already discussed. Piled piers consist of a deck supported on piles. The deck is of concrete and the piles may of steel or reinforced concrete. To take effectively, the horizontal loads, a few piles may be kept inclined which are called battered piles. Floating pile is a type of structures used where soil conditions are too poor for a fixed structure or water depths are excessive to make (fixed structure uneconomical). It consists of a floating pontoon, an anchoring system, and an access bridge connected to an abutment at the shore. Pile structure are often widely spaced between each other, where as wharves are narrowly spaced and they are commonly used for loading and unloading of vessel or embarking and disembarking of passengers in pleasure boats.

A schematic plan view of a marina formed by a pair of breakwaters, berthing structure as well as finger piers for berthing small boats is shown in Fig. 1.5.

Jetties are structures projecting into the water. Jetties may have berths on two sides and about land over their full width. A closed jetty consists of solid wall up to the water bed, thereby obstructing the flow of currents. They can also be supported on piles (open jetty). The characteristics of a jetty are that it consists of a number of individual structures each of which support special type of loads.

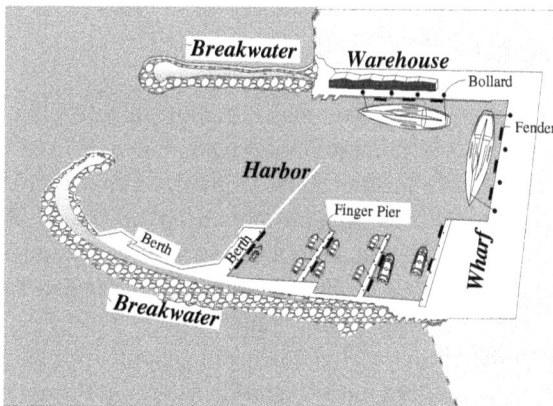

Fig. 1.5   A schematic plan view of a marina.

A *mole* is a large jetty constructed in view to form an artificial harbor. *Groins* are aimed to increase/advance sand build up in a beach. In order to prevent siltation at mouth, jetties in pairs are commonly flanked at either sides of the river as it enters a sea or lake.

A *sheet pile wall* consists of a vertical wall to retain earth. The wall is laterally supported by the rods at level near the top. In the sheet pile with relieving platform, the weight of the backfill or part thereof, immediately behind the wall is carried by piles directly to a firm stratum, whereby, the active earth pressure on the sheep pile will be reduced and the wall relieved.

Dolphins are structures at the entrance of a locked basin or alongside a pier or a wharf. Their main purpose is to absorb the impact force of the ships and also to provide mooring facility. Dolphins are broadly classified in two types, namely, breasting type and mooring type. Breasting dolphins are provided in front of the seaface of the berth. The bollards or the mooring posts are housed on them. As sometimes, the mooring dolphins are insufficient to hold the vessel in position against currents acting away from the berth; mooring dolphins are provided as counter measures. The mooring dolphins pickup the pull from hawsers, the breasting dolphins support fenders which absorb berthing impacts and onshore wind loads on the moored vessel and loading platforms support special loading or unloading equipment.

A *trestle* is a structure connecting the main berthing pier to the shore or shore-connected facilities. They are supported on piles which need designed for forces due to waves and currents. For additional details on ports and harbor structures, it is suggested to refer Brunn (1981).

1.2.3   *Offshore structures*

1.2.3.1   *General*

The diversities of factors involved makes it difficult to have a classification of offshore platforms. The number of parameters include functional aspects, construction and installation methods, geometric forms, etc. However, they can be broadly classified as bottom supported units and floating units. Jacket structure, gravity based, compliant structures and guyed towers are a few bottom supported units. Semi-submersibles, tension Leg platforms, SPAR, and FPSO are a few floating units.

Based on its application, the platforms in general are categorized exploration and exploitation/production platforms. Exploration platforms are employed for searching for oil reserves in the ocean and these are the drill ships, semi-submersibles, and jack-up platforms. The exploratory rigs can move from one location to another in search of oil.

1.2.3.2 Exploration platforms

The *drill ships* are ship shaped and self-propelled. The ship motions and thruster or anchor capacity limit the operational weather conditions. The *semi-submersibles* may be with four, six, or eight legged. Legs connected by buoyancy chambers. The primary buoyant members are positioned below SWL and hence unaffected by waves. They are difficult to tow and expensive particularly beyond 450 m. The *Jack up rigs* are towed to any desired location in the sea. They float on their hull. The platforms are initially lowered to the sea floor using pneumatic jacks, hydraulic jacks, or electric rack, and pinion drives. To achieve the required clearance, the hull is lifted above the sea surface. It can operate up to about 200 m. Once rigged stable, these platforms can continue operations in high seas.

1.2.3.3 Production platforms

The production of oil and gas from these platforms can last for usually 20–30 years. The sea bed-base structure usually operates in less than 500 m and may operate even up to 800 m. The floating structures can operate in water depths greater than 800 m.

The *jacket platform* of steel on a pile foundation is the most common kind of offshore structure which exists worldwide as the oldest type. These constitute framed open trusses which are driven into the seabed through piles. The jacket platform can support two to three decks. All the platforms in Bombay High, India are jacket type. The use of these platforms is generally limited to a water depth of about 150–180 m in the North Sea environment. The horizontal and vertical loads that effected due to the action of wind waves and weight of deck, machinery for drilling, or production operation, are transmitted to the ocean floor/seabed through the piles by lateral thrust and skin friction. When four main piles are insufficient to transmit, the horizontal and vertical loads, additional piles called the skirt piles, are sometimes driven. These supplementary skirt piles, are connected to the jacket tower, with skirt sleeves extending to one or two panel lengths in the tower.

The *gravity platform* structures are offshore structures that are placed on the seafloor and held in place by their weight. These structures do not require piles or anchors. Moreover, the huge bottom section is quite suited for production and storage of oil. Since gravity base structures require large volume and high weight, concrete has been the most common material for gravity structures. The type of foundation is the pivotal parameter to decide the type of platform. Considerable horizontal load and moments are experienced by the offshore structures at the mud level due to the action of wave loads. The gravity platform resists moment and horizontal load by non-uniform distribution of soil up thrust and shear resistance of the soil, respectively. Gravity-type structures are preferred when the soil surface condition at the site can withstand heavy loads. Construction of these structures can be done either on shore or in sheltered waters. They are then towed to offshore location and positioned by controlled flooding with water in a shortspan of time. Typical views jack-up rig, jacket platform, gravity platform and semi-submersible are projected in Fig. 1.6.

An *articulated platform* is column attached with buoyancy tanks at one and two locations along its depth. It is fixed through a universal swivel joint to a plate resting on the seabed. The column with buoyancy tanks are usually of truss-type. Buoyancy tanks can be attached at convenient levels to the truss-type columns. The advantages of articulated platforms include lowered cost of construction, lesser siting problems and an ecofriendly loading method.

A *guyed tower* consists of a slender steel space frame and the vertical forces on the platform are taken by a foundation base. The base can partially penetrate the seabed. The support structure is held upright by several guy lines that run to clump weights on the ocean floor. The horizontal loads from waves and currents are transferred to the clump weights through guy lines. These effects cause them to be lifted from the seafloor under certain operational loads. The weight and the foundation constraints of conventional offshore structures make them scarcely sought-after forms, for deep waters usually more than 300 m depth. Guyed towers and the tension leg platforms are preferable forms of design for such high-water depths.

The *tension leg platform* (*TLP*) is a one which is buoyant and fixed to the sea floor through vertical cables under tension. The mode of operation is such that loading or unloading would not cause any heave, as vertical loads are taken by anchor cables. The key advantage of TLP is that there is no considerable increase in cost with rise in water depth.

The *Spar* usually a cylindrical caisson moored to the seabed and is adopted for drilling, production, as well as for storage. The operations are

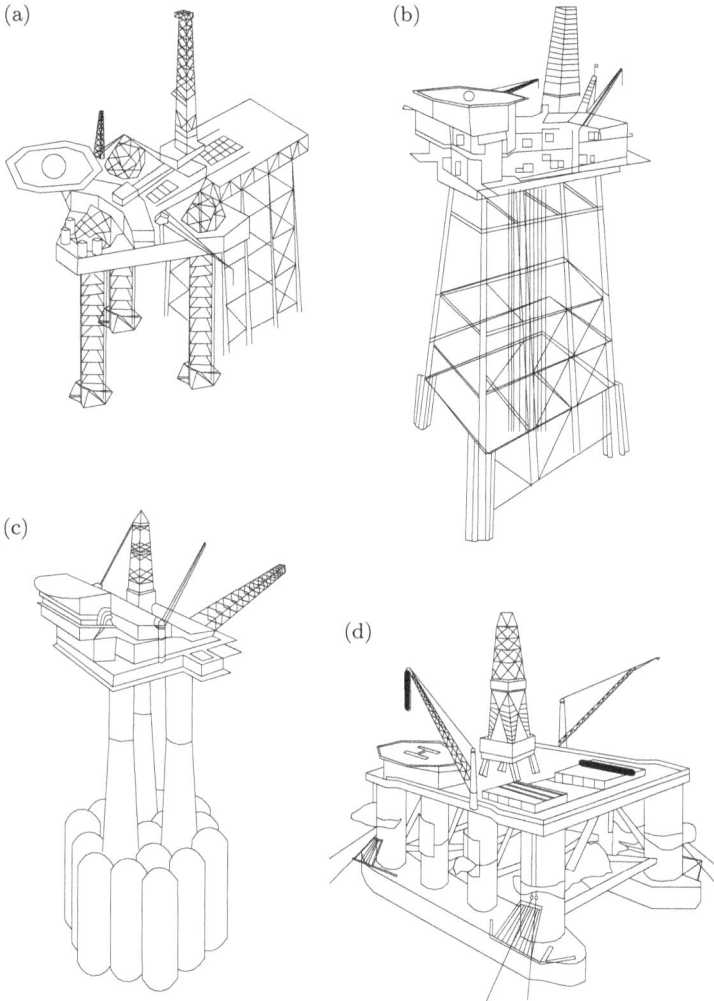

Fig. 1.6 Views of (a) jack-up rig, (b) jacket platform, (c) gravity platform, and (d) semi-submersible.

performed above the water level. It is usually anchored to the seafloor with multiple taut mooring lines.

An overview of the development of different kinds of offshore platforms with increasing water depths is shown in Fig. 1.7. For further details on offshore structures, refer to Chakrabarti (1987) and Reddy (2013).

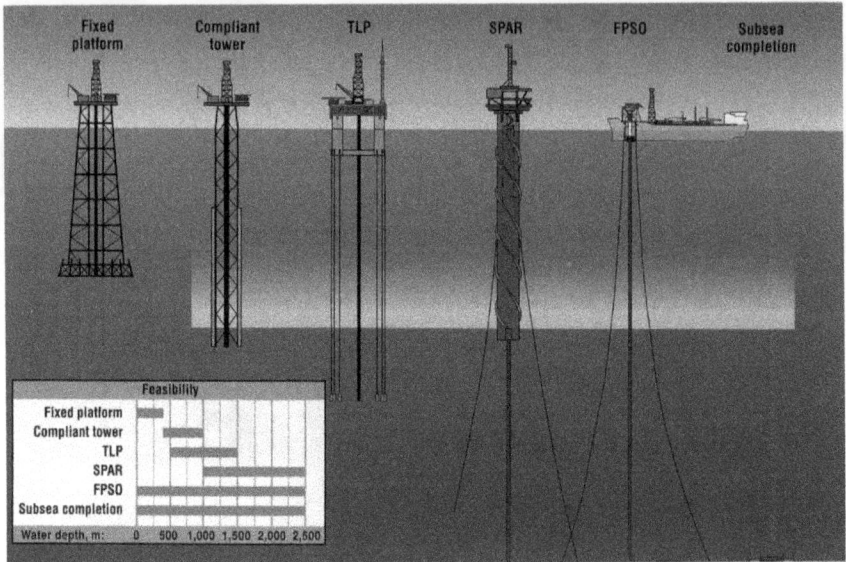

Fig. 1.7   Developments of platforms from fixed to compliant and floating systems with increasing water depths.

1.3   Environmental Parameters

1.3.1   *General*

The most important parameters such as wind, waves, tides and currents are discussed in this section. A detailed knowledge of the aforesaid parameters are essential for the design of marine structures as well as to conserve and manage coastal developmental activities. Depending on the site conditions one process augments another or may act against another. Nevertheless, it is convenient to separate the three components — astronomically-generated tidal movements, winds, wind-generated wave action and various forms of current flow, for the purpose of discussion.

1.3.2   *Wind*

One of the important factors in the design of marine structures is the characteristics of the wind blowing the ocean surface. Extreme water levels, wave climate, may simultaneously occur with wind loading which need to be considered for the survival of the structures. The probability of marine structures being exposed to extreme winds compared to land-based structures is quite high, as in the former there is no sheltering effect of terrain or vegetation. The important parameters that dictate the wind characteristics are its direction, frequency, and intensity. The wind velocity usually referred by Beaufort scale that ranges between 0 and 12 can be expressed as

$$V = 2.99\sqrt{B^3}, \tag{1.1}$$

where "V" is wind velocity in kmph and "B" is the Beaufort number.

When wind speed is measured $10\,\mathrm{m}$ above the earth, its speed V_s at any elevation "s" is given by

$$V_s = V_{10} \times (s/10)^{1/7}. \tag{1.2}$$

The pressure intensity in $\mathrm{kg/m^2}$ due to wind, "p" with "C" defined as a coefficient $= 0.00481$ is given by

$$p = CV^2. \tag{1.3}$$

The total force due to wind F_w can be evaluated as

$$F_w = K_{sf}pA. \tag{1.4}$$

Here, A is area of the exposed surface and K_{sf} is a shape factor varying between 1.3 and 1.6. The wind data is usually presented in terms of percentage of occurrence of wind intensity in a particular direction either monthly, season-wise, or annual. Typical wind rose diagram is shown in Fig. 1.8.

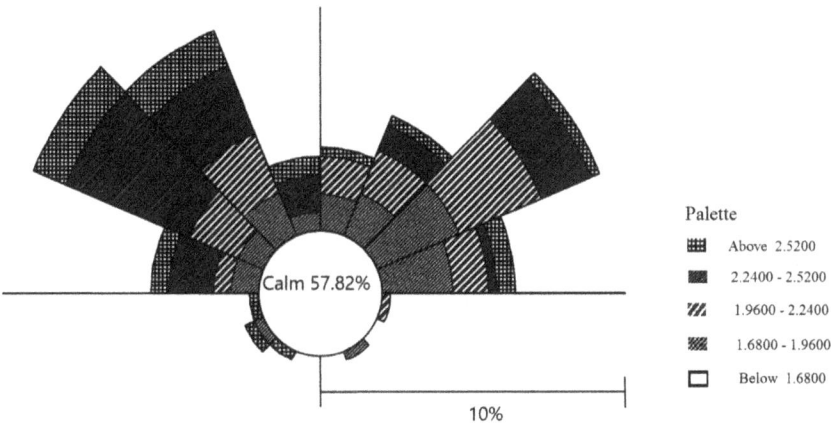

Fig. 1.8   Typical wind rose diagram.

1.3.3   *Waves*

1.3.3.1   *General*

Waves are undulations of the ocean water surface about a reference level, which may be the mean sea level, high tide line, or low tide line, the details of which will be discussed later. The undulations are due to the wind blowing over the ocean's surface. The characteristics of waves are defined by its height which is the distance between its crest and the trough, the distance between successive crests termed as its length, while the time taken for the wave to travel one wavelength is its period. The ultimate state of wave growth depends primarily on three parameters of wind fetch (F) or the distance over which it blows on the water surface, its velocity (v) and the duration (t) of time for which the wind blows. Although the theoretical and simplistic wave form is a sinusoidal curve, that form is not common in nature. Wave shape depends on the conditions of the wind, water depth, and the progression of the wave itself. Waves are basically surface phenomena characterized by its oscillatory motion, the magnitude of which reduces rapidly towards the seabed. This would mean that a body floating on the surface of the ocean will undergo an oscillatory motion and when structures are exposed to such motions, they will experience cyclic loading.

1.3.3.2   *Wave generation*

Waves propagate in an area under the influence of the winds, i.e., within fetch region are called wind waves, or a sea. These have relatively peaked crests and broad troughs. After the waves leave the generating area of the fetch, principally they are smoother losing their rough appearance. These waves are called swell. Swell waves commonly have a long wavelength and small wave height, thus having a very low steepness value, much lower than sea waves. For a given steady wind speed, the development of waves may be limited by the fetch or the duration. If the wind blows over a sufficient distance for a given period of time, a more or less steady state condition is attained. The steady state is achieved when any additional transfer of energy from the wind to the ocean surface does not result in the growth of waves. This condition is called a fully developed sea (FDS). Figure 1.9 shows development of ocean waves within generating area. Hence, the three parameters that dictate the growth of waves are fetch, F, wind velocity, V and its duration, t_d.

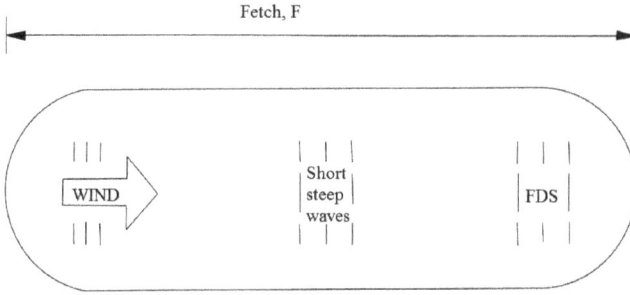

Fig. 1.9   Development of waves within generating area.

A few empirical relationships for estimating the maximum wave height and period are as follows:

$$H_{max} = 0.026V^2, \tag{1.5}$$

$$H_{max} = 0.0555\sqrt{V^2 F}, \tag{1.6}$$

$$T_{max} = 0.5(V^2 F)^{1/4}, \tag{1.7}$$

where H_{max} is the maximum wave height in feet, F the fetch in nautical miles, V the wind speed in knots, and T_{max} the maximum wave period in seconds. These relations are particularly useful for short fetches and high wind speeds such as found in hurricanes and other intense storms.

Wind force and direction are of importance in the study of establishing weather window, design, construction, and maintenance of structures in the marine environment. The relationship between wind and wave characteristics has been brought as a scale that was devised in 1805 by Rear admiral, Francis Beaufort, an Irish Navy officer which is called as Beaufort scale and is given in Table 1.1. The average wave height along the Indian coast was reported by Narasimha Rao and Sundar (1982) to range between 1 m and 2.8 m, whereas the wave period was found to vary between 5 s and 8 s. The wave data are usually presented in terms of percentage of occurrence of its height and period in a particular direction either monthly, season-wise, or annual. A typical wave rose diagram is shown in Fig. 1.10.

1.3.3.3   *Importance of wave characteristics*

The information on the wave characteristics is essential and has a wide application in the studies related to oceanography and ocean engineering. A few of them are listed below.

Table 1.1   Beaufort wind and wave scale.

Beaufort number	Description	Wind speed (kmph)	Wave height (m)	Sea conditions	Land conditions
0	Calm	<1	0	Flat	Calm. Smoke rises vertically
1	Light air	1.1–5.5	0–0.2	Ripples without crests	Smoke drift indicates wind direction, still wind vanes
2	Light breeze	5.6–11	0.2–0.5	Small wavelets. Crests of glassy appearance, not breaking	Wind felt on exposed skin. Leaves rustle, vanes begin to move
3	Gentle breeze	12–19	0.5–1	Large wavelets. Crests begin to break; scattered whitecaps	Leaves and small twigs constantly moving, light flags extended
4	Moderate breeze	20–28	1–2	Small waves with breaking crests. Fairly frequent white horses	Dust and loose paper will rise. Small branches begin to move
5	Fresh breeze	29–38	2–3	Moderate waves of some length. Many white horses. Small amounts of spray	Branches of a moderate size move. Small trees in leaf begin to sway
6	Strong breeze	39–49	3–4	Long waves begin to form. White foam crests are very frequent. Some airborne spray is present	Large branches in motion. Whistling heard in overhead wires. Umbrella use becomes difficult. Empty plastic garbage can strip over
7	High wind, Moderate Gale Near gale	50–61	4–5.5	Sea heaps up. Some foam from breaking waves is blown into streaks along wind direction. Moderate amounts of airborne spray	Whole trees in motion. Effort needed to walk against the wind

Table 1.1 (*Continued*)

Beaufort number	Description	Wind speed (kmph)	Wave height (m)	Sea conditions	Land conditions
8	Gale Fresh gale	62–74	5.5–7.5	Moderately high waves with breaking crests forming spindrift. Well-marked streaks of foam are blown along wind direction. Considerable airborne spray	Some twigs broken from trees. Cars veer on road. Progress on foot is seriously impeded
9	Strong gale	75–88	7–10	High waves whose crests sometimes roll over. Dense foam is blown along wind direction. Large amounts of airborne spray may begin to reduce visibility	Some branches break off trees, and some small trees blow over. Construction/ temporary signs and barricades blow over
10	Storm whole gale	89–102	9–12.5	Very high waves with overhanging crests. Large patches of foam from wave crests give the sea a white appearance. Considerable tumbling of waves with heavy impact. Large amounts of airborne spray reduce visibility	Trees are broken off or uprooted, saplings bent and deformed. Poorly attached asphalt shingles and shingles in poor condition peel off roofs

(*Continued*)

Table 1.1   (*Continued*)

Beaufort number	Description	Wind speed (kmph)	Wave height (m)	Sea conditions	Land conditions
11	Violent storm	103–117	11.5–16	Exceptionally high waves. Very large patches of foam, driven before the wind, cover much of the sea surface. Very large amounts of airborne spray severely reduce visibility	Widespread damage to vegetation. Many roofing surfaces are damaged; asphalt tiles that have curried up and/or fractured due to age may break away completely
12	Hurricane-force	≥118	≥14	Huge waves. Sea is completely white with foam and spray. Air is filled with driving spray, greatly reducing visibility	Very widespread damage to vegetation. Some windows may break; mobile homes and poorly constructed sheds and barns are damaged. Debris may be hurled about

- establishment of weather window needed for the planning of marine operations,
- design of structures and ships,
- design and stability of floating structures,
- coastal process studies,
- coastal protection measures,
- development of fishing and commercial harbors,
- to verify wave forecasting and hind-casting models,
- wave energy potential and performance evaluation of wave energy convertors.

Major applications of wave data include offshore platform design, breakwater jetty design, site selection criteria, operation planning, military operations, transportation studies, environmental impact assessment, real time input to wave forecasting, search and rescue pollution, and cleanup.

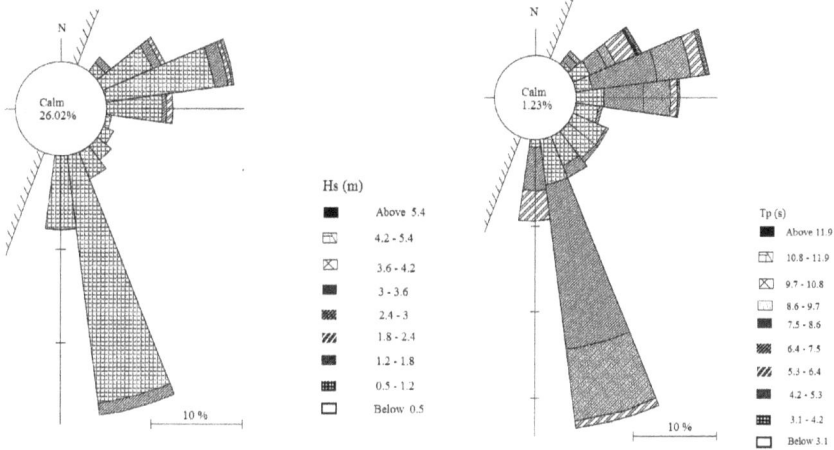

Fig. 1.10   Typical wave rose diagram.

Wave data also find important applications in the specification of ship responses, ship design, and ship operation and planning. In coastal processes studies, measured wave data are needed for studies and applications involving sediment structure interaction, shoreline erosion, design of navigation channels, beach nourishment, coastal hydrodynamics, coastal zone planning, coastal operations, harbor surging, dredging, recreation and marine facilities designs. Wave data are also important in testing newly developed wave forecasting models and in verifying existing hind-casting and forecasting models. Wave data analysis forms the most important component in studies relating to estimation of wave energy potential for different offshore and coastal regions.

1.3.4   *Tides*

1.3.4.1   *General*

A tide is gradual oscillation of the water surface in the ocean that is caused by the gravitational attraction between sun, moon, and the earth, which modified by the rotation of the earth, friction forces, and the ocean boundaries. The moon being closer to the earth exerts more force and is believed that the sun exerts only 46% of the force of the moon. Hence, the tides generally follow more than that of the sun. The net effect produced is that the acceleration due to gravity at the earth's surface directly below the moon is effectively reduced because of the moon's attractive force, and water is then heaped up towards the moon. They are essentially long waves with its dynamic component having almost negligible effect on the loading of structures. The tidal levels at a site dictates the top levels of fixed and floating structures. In order to avoid overtopping and to prevent coastal flooding during extreme events, the highest high tide level (HHTL) is considered over which an additional buffer accommodating the wave run-up that will be over the HHTL and a free board will be considered. Further, the scope of mooring lines of floating structures and the minimum water level may affect the stability of bulk heads. The tidal range affects the vertical extent to which the structure is affected by degrees of corrosion and deterioration. For the construction of all structures, the establishment of some vertical reference plane is necessary and it is very important for the engineer to understand the chart and survey datum in use in the site of interest and the relation of changes in water level to that datum. The various tidal levels as per NOAA are shown in Table 1.2.

1.3.4.2 *Types of tides*

The time taken for the moon to make a complete revolution around the earth is about 24 h 50 min. This period is called a lunar tidal day. The highest tide for a location does not occur during the new moon and the full moon but occurs after a constant interval. This interval may be as much as $2\frac{1}{2}$ days and is known as the age of the tide.

The tides which occur twice during each lunar day are called semi-diurnal tides in which case there will be two high tides and two low tides during a tidal day. If there is only one high tide during a tidal day, it is called a diurnal tide, e.g., at Pensacola, Florida. If one of the two daily high tides do not reach the height of the previous tide, it is called a mixed diurnal tide, e.g., at San Francisco, USA The representation of these different types of tides is shown in Fig. 1.11. Mean sea level (MSL) can be defined as that average sea level about which the tides oscillate. Even though MSL is important in establishing a fundamental benchmark for land surveys, it is not generally used as a reference plane for marine structures. Different datum levels are used in different parts of the world. However, it has been determined by international agreement that the chart datum shall be some level below which the tides will not fall frequently. The chart datum used in India is the Indian Spring Low water. The mean low water is the average of the low water over a 19-year period. The mean high water is the average of the high water over a 19-year period. The highest high water and lowest

Table 1.2   Definitions of tidal datum.

HAT — highest astronomical tide	The elevation of the highest predicted astronomical tide expected to occur at a specific tide station over the national tidal datum epoch
MHHW — mean higher high water	The average of the higher high-water height of each tidal day observed over the national tidal datum epoch. For stations with shorter series, comparison of simultaneous observations with a control tide station is made in order to derive the equivalent datum of the national tidal datum epoch
MHW — mean high water	The average of all the high-water heights observed over the national tidal datum epoch. For stations with shorter series, comparison of simultaneous observations with a control tide station is made in order to derive the equivalent datum of the national tidal datum epoch
MTL — mean tide level	The arithmetic mean of mean high water and mean low water
MSL — mean sea level	The arithmetic mean of hourly heights observed over the national tidal datum epoch. Shorter series are specified in the name; e.g., monthly mean sea level and yearly mean sea level
MLW — mean low water	The average of all the low water heights observed over the national tidal datum epoch. For stations with shorter series, comparison of simultaneous observations with a control tide station is made in order to derive the equivalent datum of the national tidal datum epoch
MLLW — mean lower low water	The average of the lower low water height of each tidal day observed over the national tidal datum epoch. For stations with shorter series, comparison of simultaneous observations with a control tide station is made in order to derive the equivalent datum of the national tidal datum epoch
LAT — lowest astronomical tide	The elevation of the lowest astronomical predicted tide expected to occur at a specific tide station over the national tidal datum epoch Co-ops.nos.noaa.gov
Storm surge	Is a temporary water level increase (surge) due to persistent action of wind over water, as during cyclones. They may last for up to a couple of hours

low water are the highest and lowest, respectively, of the spring tides of record.

1.3.4.3   *Tide tables*

Present-day tide predictions are done by performing harmonic analysis, in which, the movements of the sun and the moon are broken into various

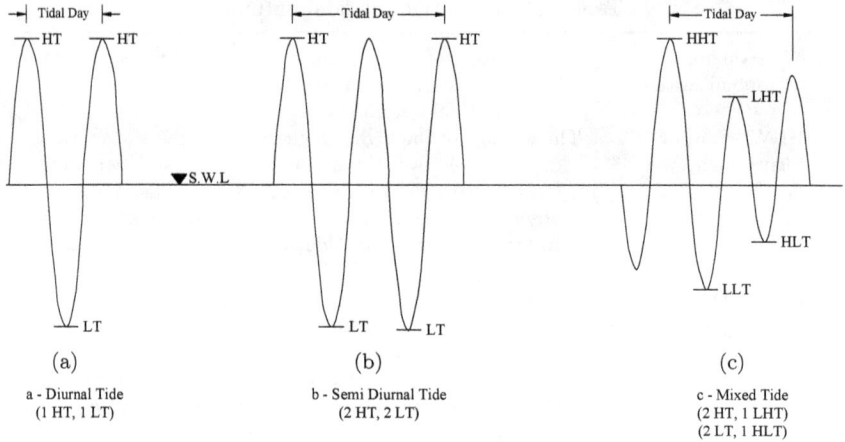

Fig. 1.11   Classification of tides.

periods during which some cycles of motion are repeated. Tidal tables are prepared which give the times and levels of high and low water at the standard ports in the world. Tidal computations for a comprehensive harmonic analysis based on hourly tide reading used to be a most laborious occupation when carried out manually. The tidal level at various locations along the Indian coastline is found using national hydrographic charts.

1.3.4.4   Tides along Indian coast

The vertical difference in height between the high water and low water at a place is called the tidal range. If it is the mean high water and mean low water, then the difference between them will be the mean tidal range. The highest tidal range in India occurs at Bhavanagar-exceeding 11 m. At the Bay of Fundy in Canada, the highest tides in the world occur; a rise of 30 m has been recorded. A high tide of 18 m has been observed in the Bristol Channel.

Along the Indian coast, the maximum tidal range is observed in the Gulf of Kutch and Gulf of Khambhat region along the west coast with range of 10–11 m. The next higher tidal range observed along the Sundarbans area along the east coast is about 5.5 m, a magnitude of which is experienced in several other locations. Apart from this, the regions south of Gujarat and West Bengal also experience moderate tidal range of 3–5 m. The tidal range is less than 1.5 m along the coast of the southern maritime states of India.

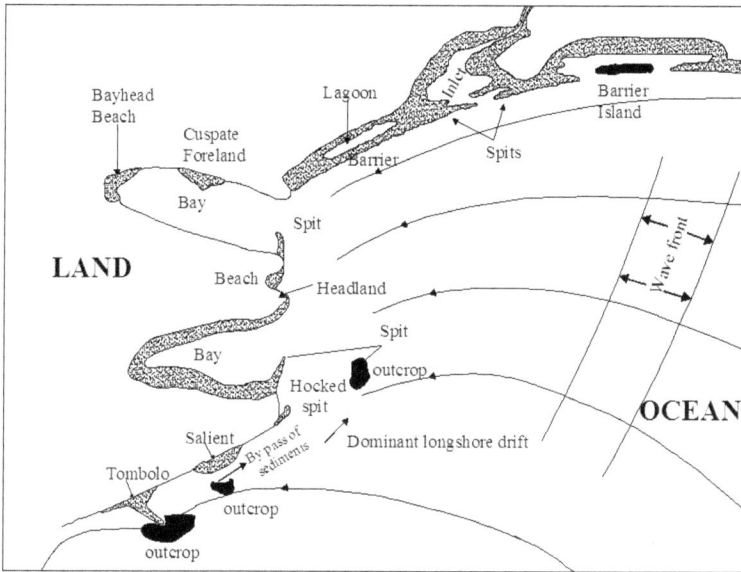

Fig. 1.12 General coastal features.

1.3.5 *Currents*

1.3.5.1 *General*

Current is defined as movement of mass of water due to the existence of a gradient. The gradient could be due to changes in salinity, density, temperature, pressure, wave height, and tidal levels. Coastal currents often accelerate deposition of sediments or leading to erosion depending on the coastal features and wave and sediment characteristics. Currents are also responsible for the development of a variety of coastal features like formation of inter-tidal spits and sand bars, tombolo, salient, various kinds of ripple patterns on foreshores and tidal flats, on the floors or lagoons and seabed as schematically presented in Fig. 1.12. Currents inside harbor basins can lead to shoaling of approach channels, seafloor scour, increased corrosion rates, and lead to modification of wave characteristics. Hence, it is important to consider the flow characteristics particularly in coastal zone. They are also sources of significant loads on marine structures, especially upon moored vessels and when additive to wave loads on offshore structures.

Ocean currents can broadly be classified according to (i) the forces by which they are created, (ii) the time of appearance, (iii) the locality, and

As per the Forces by which they are Created

Wind force	Tides	Waves	Density Difference
Permanent	Rotating	Shoreward	Surface
Periodical	Reversing	Longshore	Sub Surface
Accidental	Hydraulic	Seaward	Deep

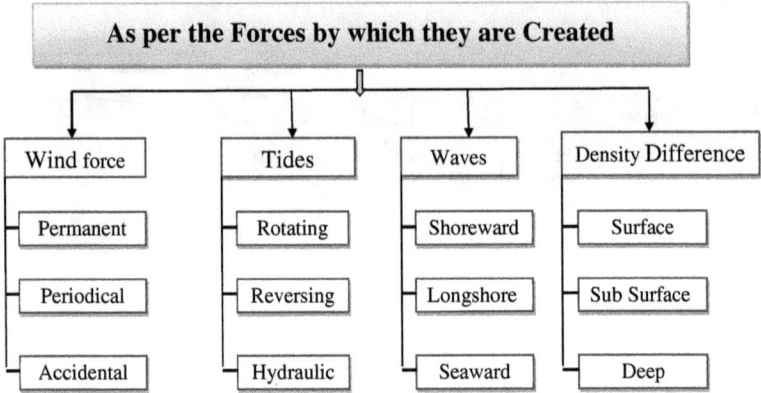

Fig. 1.13   Classification of currents as per forces by which they are created.

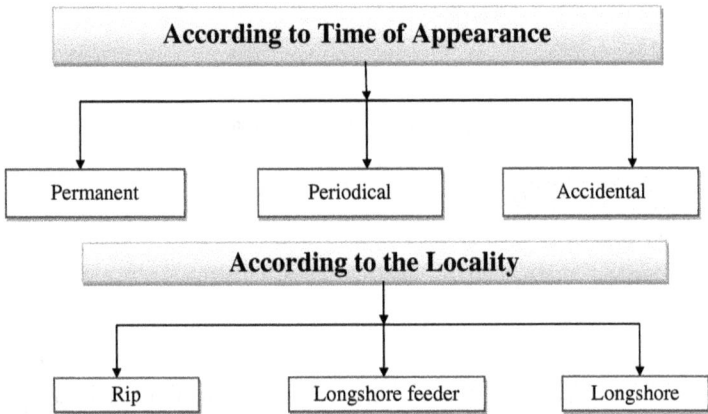

According to Time of Appearance

Permanent	Periodical	Accidental

According to the Locality

Rip	Longshore feeder	Longshore

Fig. 1.14   Classification of currents as per time of appearance and locality.

(iv) the region in which they are situated. The details of these classifications are projected in Figs. 1.13–1.15.

1.3.6 *Nearshore currents*

Consider the action of waves on any beach in plan, the wave crests reaching the shore are seldom parallel to the shoreline or the underwater depth contours. The effect of this oblique attack of the waves in the break zone is to generate two components of the fluid velocity, of which, one along the direction parallel to the shore is called as "longshore currents" responsible

According to the region in which situated

- Tropical current
- Subtropical current
- Arctic current
- Antarctic current
- Equatorial current

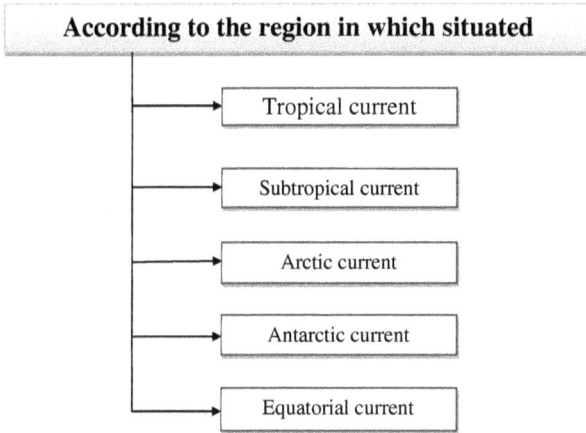

Fig. 1.15   Classification of currents as per regions.

for the transport of sediment along the shore and another termed as "cross-shore currents" that set the sediments in the coastal in offshore onshore direction, i.e., normal to the shoreline. These currents may be even of the order of 0.5 m/s. The magnitude and direction of the longshore currents depends mainly on the angle between the wave crest and the shoreline.

The longshore sediment transport or "littoral drift", however, is more dominant and mainly responsible for the shoreline instabilities. As the sea waves approach the shoaling water (i.e., lesser depth of water), they become waves of translation and pile up water against the coast. A hydraulic grade line is thus established and hence, somewhere along the beach, there must be a return flow to the sea. This water is met by the next oncoming wave. The oncoming wave flows over the returning water, building up a hydraulic head at the beach line. This hydraulic head forces the returning water down and outward, thus setting up what is called as the "undertow". The undertow in some instances becomes a longshore current of considerable importance. As waves propagate towards the coast, there is considerable mass transfer.

Along certain stretches of the coast, in particular, along a barred coast or along some of the steeper coasts, there could be large gradient in the height along the wave crest and this in addition to the setup produces seaward flowing currents that are rather narrow and that create circulation cells within the surf zone as shown in Fig. 1.16. These narrow currents with a magnitude of even up to few tens of centimeters per second which can

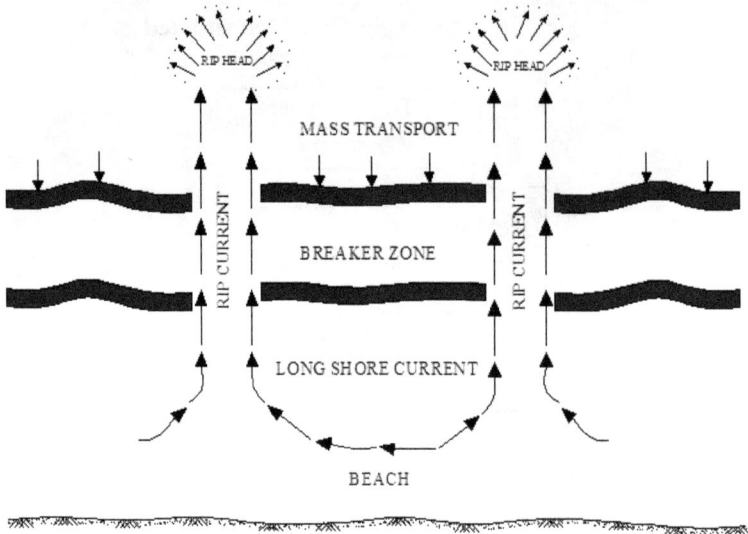

Fig. 1.16   Schematic representation of nearshore currents.

transport sediment are called "rip currents". Rip currents can be a danger to swimmers. In actuality they should not be a problem to swimmers because they are narrow and the swimmer simply needs to move a short distance along shore to escape their seaward path.

1.4   Storm Tides, Tidal Bores, Tsunamis, and Seiches

1.4.1   *Storm surge*

Coastlines may be subject to higher (and lower) than normal water levels associated with meteorological conditions such as strong winds and low atmospheric pressures occurring during periods of intense storm activity and these are referred to as storm tides or surges. A storm surge is an abnormal rise of seawater due to tropical cyclone and is greatly amplified where the coastal water is shallow, in the estuarine region and where the shape of the coast is like a funnel. One of the peculiar characteristics and having a very high damage potential is storm surge.

The major factors include: (i) a fall in the atmospheric pressure over the sea surface, (ii) the effect of the wind, (iii) the influence of the seabed, (iv) a funneling effect, (v) the angle and speed at which the storm approaches the cost, and (vi) the tides.

The low-pressure area or "eye" of the cyclone allows the sea level to rise. The high-speed winds surrounding the "eye" drive more water over this rise. The sloping bed of the sea and contours off the shoreline add further to the height. A further contribution to the height of the storm surge is added if the cyclone arrives at high tide. Storm surges are not waves though they may look like them. A storm surge is a mass of water, which will submerge everything in its path, till it recedes back into the sea. It moves at the same speed as that of the cyclone. It travels up to the point where the ground height is equal to the height of the surge.

The coastal areas of North Bay of Bengal satisfy most of the above-mentioned criteria and the storm surge gets enormously amplified there. Due to several favorable factors in these areas, the world's highest storm surge of over 13 m was reported from the area in 1876 near Bakerganj. Cyclonic storms are sometimes accompanied by tidal waves with heights of 5 m and sometimes hit 20 km inland with wind speed of 150 kmph.

Indian Nation Centre for Ocean Information Services (INCOIS) has categorized the entire Indian coast into four zones

(1) very high risk zones, VHRZ (surge height >5 m),
(2) high risk zone, HRZ (surge height between 3–5 m), and
(3) moderate risk zone, MRZ (surge height between 1.5 and 3 m),
(4) minimal risk zone (surge height <1.5 m).

Accordingly, the coastal areas and offshore islands of Bengal and adjoining Bangladesh are the most storm-surge prone ($\sim$10–13 m), the east coast of India between Paradip and Balasore in Orissa ($\sim$5–7 m) — and the Andhra coast between Bapatla and Kakinada holding estuaries of two major rivers Krishna and Godavari ($\sim$5–7 m) are classified under VHRZ. Tamil Nadu coast between Pamban and Nagapattinam ($\sim$3–5 m) is considered as HRZ, whereas Gujarat along the west coast of India ($\sim$2–3 m) falls under MRZ.

1.4.2   *Tidal bores*

Tidal bores occur when waves propagate in channels particularly with converging sides. Under such conditions, the tidal energy of the waves act on a decreasing width and their height is set to keep on increasing. In exceptional cases, a tidal wave may form a steep wall of water and rush up in the channel or river estuary. This phenomenon is usually referred to as a tidal bore.

1.4.3   *Tsunamis*

A tsunami is basically a long wave in deep offshore waters that has the characteristics of wave in shallow waters. Ordinary wind generated waves possess a period varying between 5 s and 30 s and a wavelength of few hundred meters, whereas a tsunami possess a prolonged time period in the range of 10 min–2 h and a wavelength exceeding 500 km. Tsunami waves relative behavior to shallow water waves is due to their longer wavelengths. *Tsunami* is a Japanese word which translates as *harbor wave*.

One major reason for tsunami outbreak is resulted due to volcanoes, earthquakes, and landslides occurring above or below sea surface, although these activities produce tsunamis with comparatively much lesser impact than those produced by submarine faulting (earthquake). During such an instance, huge water mass in the ocean in the vicinity of the disturbance is displaced from its equilibrium position that travels in concentric circles. It can propagate over several km uninterrupted. The speed of the tsunami is a function of only the water depth and in the event it occurs in a water depth of about 4 km, its speed will be 713 kmph. Moving with such orders of tremendous speeds, they pile up along shores causing great destruction.

Among about 100 tsunamis around the globe that has occurred in the past, the great Indian Ocean tsunami of 2004 caused by an underwater earthquake of strength 9.3 on the Richter scale on 26 December off the coast of Banda Acèh, Indonesia lead the worst disaster in history. Its effect was felt in Bangladesh, India, Malaysia, Myanmar, Singapore, Thailand, and the Maldives. The two maritime states that were severely affected were Tamil Nadu and Kerala. The height of water that propagates along the beach or a structure from mean sea level termed as *run-up* is referred to as inundation height. Similarly, the distance up to which the seawater rushes into the land is the inundation distance. For Tamil Nadu, the tsunami run-up was up to a maximum of about 5 m and inundation distance into the land was up to about 15 km (Sundar *et al.*, 2007), whereas, for Kerala, the maximum run-up and inundation distance as per Kurian and Praveen (2010) were reported to be about 4.5 m and 2.5 km, respectively. To understand the clear difference between the hydrodynamic behavior of a tsunami and a wave close to the seabed, Problem 3.6 in Chapter 3 may be referred.

1.4.4   *Seiche*

A seiche is a surge generated within an enclosed or partially enclosed body of water which has a resonance similar to that of the disturbing force.

The wave in seiche is non-progressive in nature and remains stationary in the horizontal plane. Unlike the wind waves forward motion in the sea, the waves in a seiche moves up and down. Hence they are called as *standing waves* instead of/as opposed to progressive waves.

The characteristics of progressive waves and standing waves are discussed in Chapters 3 and 4. Seiches tend to be triggered in a still body of water by strong winds, variation in atmospheric pressure, earthquakes, etc., this is most frequently encountered in an enclosed or semi-enclosed basin such as lakes, bays, harbors, and bathtubs.

During the tsunami of 2004, a number of boats within the harbors along the affected coast experienced this phenomenon which resulted in their damages.

Chapter 2

Basic Fluid Mechanics

2.1 General

Matter exists in three states namely solid, liquid, and gas. Liquid and gas are commonly called fluids. Fluids can be defined as "a substance, which undergoes continuous deformation under the action of shear forces regardless of their magnitude". The main distinction between a liquid and a gas lies in their rate of change in density. The density of gas changes more readily than that of liquid. However, gases can be treated in the same way without taking into account the change of density, provided that the speed of flow is low as compared with the speed of sound propagation in the fluid. The fluid is called incompressible if the change of the density is negligible.

2.1.1 *Ideal fluids*

Fluids with no viscosity, no surface tension and that are incompressible are called as ideal fluids. They do not offer any resistance when they flow. Ideal fluids do not exist in nature. They are imaginary fluids. Fluids with low viscosity such as air, water may be considered as ideal fluid.

2.1.2 *Real or practical fluids*

Fluids with viscosity, surface tension and that are compressible are called as real fluids. They exist in nature.

2.2 Types of Flow

  (I) Steady and unsteady flow;
 (II) Uniform and Non-uniform;
(III) Rotational and irrotational;
(IV) Laminar and turbulent.

Steady flow: Fluid characteristics such as velocity "u", pressure "p", density "ρ", temperature "T", etc., do not change with time at any point

(x, y, z), i.e.,

$$\text{at } (x, y, z), \quad \frac{\partial u}{\partial t} = 0, \quad \frac{\partial p}{\partial t} = 0, \quad \frac{\partial \rho}{\partial t} = 0.$$

Unsteady flow: Fluid characteristics do change with time at any point (x, y, z):

$$\frac{\partial u}{\partial t} \neq 0, \quad \frac{\partial p}{\partial t} \neq 0, \quad \frac{\partial \rho}{\partial t} \neq 0, \quad \frac{\partial T}{\partial t} \neq 0.$$

Most of the practical problems of engineering involve only steady-flow conditions and it is simpler to solve compared to problems of unsteady flow, e.g., behavior of ocean waves.

Uniform flow: When fluid properties does not change both in magnitude and direction, from point to point in the fluid at any given instant of time, the flow is said to be uniform, i.e., velocity with respect to distance

$$\left(\frac{\partial u}{\partial s} \right)_{t=t_1} = 0,$$

e.g., liquid flow through pipes of fixed cross-section/diameter.

Non-uniform flow: If velocity of fluid changes from point to point at any instant, the flow is non-uniform:

$$\left(\frac{\partial u}{\partial s} \right)_{t=t_1} \neq 0,$$

e.g., liquid flow through pipes of changing cross-section.

Any combinations of these types of flow can exist independent of each other. Thus for the liquid flowing at a constant rate in a long horizontal pipeline of constant cross-section can be termed as steady-uniform flow, whereas liquid flowing at a constant rate in a pipeline of varying cross-section is steady-non-uniform flow. For the above two types of flows, if the rate of flow varies, we would have unsteady uniform and unsteady non-uniform, respectively.

Rotational flow: In a rotational flow, the fluid particles rotate about their mass centers while moving along the direction of flow, e.g., liquid in a rotating tank.

Irrotational flow: In an irrotational flow, the fluid particles do not rotate about their mass centers. This type of flow exists only in the case of ideal fluid for which no tangential or shear stresses occur. But the flow of real fluids may also be assumed to be irrotational if the viscosity of the fluid has little significance. For a fluid flow to be irrotational, the following conditions are to be satisfied.

It can be proved that the rotation components about the axes parallel to x- and y-axes can be obtained as

$$W_x = \frac{1}{2}\left(\frac{\partial w}{\partial y} - \frac{\partial v}{\partial z}\right),$$

$$W_y = \frac{1}{2}\left(\frac{\partial u}{\partial z} - \frac{\partial w}{\partial x}\right),$$

$$W_z = \frac{1}{2}\left(\frac{\partial v}{\partial x} - \frac{\partial u}{\partial y}\right), \tag{2.1}$$

and if at every point in the flowing fluid the rotation components W_x, W_y, and W_z are equal to zero, then

$$W_x = 0; \quad \frac{\partial w}{\partial y} = \frac{\partial v}{\partial z},$$

$$W_y = 0; \quad \frac{\partial u}{\partial z} = \frac{\partial w}{\partial x},$$

$$W_z = 0; \quad \frac{\partial v}{\partial x} = \frac{\partial u}{\partial y}, \tag{2.2}$$

where u, v and w are velocities in the x, y, and z directions, respectively.

Laminar flow: A flow is said to be laminar if the fluid particles move along straight parallel paths in layers, such that the path of the individual fluid particles do not cross those of the neighboring particles. In other words, the fluids appear to move in layers sliding over adjoining layers. This type of flow occurs when the viscous forces dominate the inertia forces at low velocities. Laminar flow can occur in flow through pipes, open channels, and through porous media.

Turbulent flow: A fluid motion is said to be turbulent when the fluid particles move in an entirely random or disorderly manner, which results in a rapid and continuous mixing of the fluid leading to momentum transfer as flow take place. Eddies or Vortices of different sizes and shapes are present moving over large distances in such a fluid flow. Flow in natural streams, artificial channels, sewers, etc., are a few examples of turbulent flow.

2.3 Continuity Equation Conservation of Mass

In a real fluid, mass can neither be created nor be destroyed.

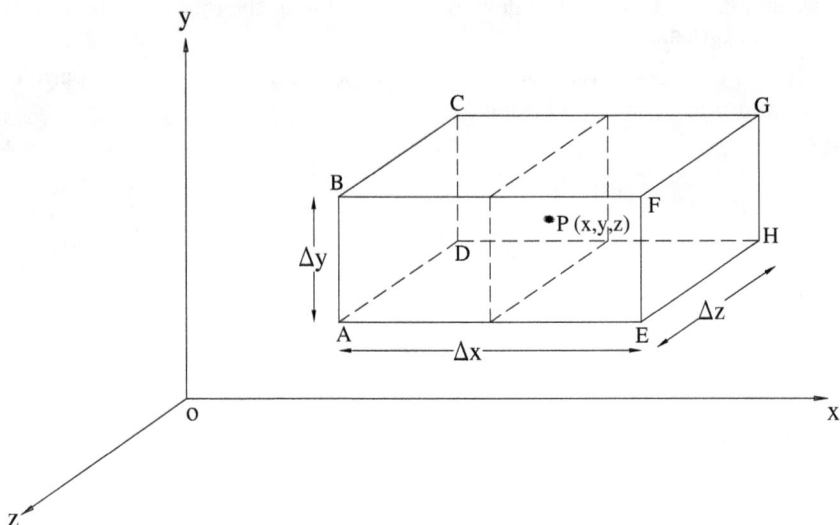

Fig. 2.1   Definition sketch of coordinate system.

Consider an elementary rectangular parallelepiped with sides of length $\Delta x, \Delta y, \Delta z$ as shown in Fig. 2.1

Let the center of the fluid medium be at a point $P(x, y, z)$ where the velocity components in the x, y, z directions are u, v, w, respectively.

Let "ρ" be the mass density of the fluid.

The mass of fluid passing *per unit time* through the face of area $\Delta y \Delta z$ normal to the x-axis through point P is $(\rho u \Delta y \Delta z)$.

Then mass of fluid flowing *per unit time* into the parallelepiped through the face ABCD is

$$\left[(\rho u \Delta y \Delta z) + \frac{\partial}{\partial x} (\rho u \Delta y \Delta z) \left(-\frac{\Delta x}{2} \right) \right]. \tag{2.3}$$

In the above expression, the negative sign has been used since face ABCD is on the left of point P.

Similarly, the mass of fluid flowing per unit time out of the fluid medium through the face EFGH is

$$\left[(\rho u \Delta y \Delta z) + \frac{\partial}{\partial x} (\rho u \Delta y \Delta z) \left(\frac{\Delta x}{2} \right) \right]. \tag{2.4}$$

Therefore, the net mass of fluid that has remained in the fluid medium per unit time through the pair of faces ABCD and EFGH is obtained as

$$\frac{-\partial}{\partial x}(\rho u \Delta y \Delta z)\,\Delta x = \frac{-\partial}{\partial x}(\rho u)\,\Delta x \Delta y \Delta z. \tag{2.5}$$

The area $(\Delta y \Delta z)$ has been taken out of the parenthesis as it is not a function of x. Similarly, the net mass of fluid that remains in the cube per unit time through the other two pairs of faces of the cube may be obtained as

$$= \frac{-\partial}{\partial y}(\rho v)\,(\Delta x \Delta y \Delta z)\ \text{through pair of faces AEHD and BFGC,}$$

$$= \frac{-\partial}{\partial z}(\rho w)\,(\Delta x \Delta y \Delta z)\ \text{through pair of faces DHGC and AEFB.}$$

The net total mass of fluid that has remained in the cube per unit time can be determined by adding the above expressions:

$$-\left[\frac{\partial(\rho u)}{\partial x} + \frac{\partial(\rho v)}{\partial y} + \frac{\partial(\rho w)}{\partial z}\right]\Delta x \Delta y \Delta z. \tag{2.6}$$

As we know, the mass of the fluid can neither be created nor destroyed in the cube, any increase in the mass of the fluid contained in this space per unit time will be equal to the net total mass of fluid, which has remained in the cube per unit time.

The rate of increase in the mass of the fluid $(\rho \Delta x \Delta y \Delta z)$ with time is

$$\frac{\partial}{\partial t}(\rho \Delta x \Delta y \Delta z) = \frac{\partial \rho}{\partial t}(\Delta x \Delta y \Delta z). \tag{2.7}$$

Equating Eqs. (2.6) and (2.7)

$$-\left[\frac{\partial(\rho u)}{\partial x} + \frac{\partial(\rho v)}{\partial y} + \frac{\partial(\rho w)}{\partial z}\right]\Delta x \Delta y \Delta z = \frac{\partial \rho}{\partial t}(\Delta x \Delta y \Delta z).$$

On simplification, the continuity equation is obtained as

$$\frac{\partial \rho}{\partial t} + \frac{\partial(\rho u)}{\partial x} + \frac{\partial(\rho v)}{\partial y} + \frac{\partial(\rho w)}{\partial z} = 0. \tag{2.8}$$

This is the *generalized form of continuity equation*. This equation is applicable for steady and unsteady flow, uniform and non-uniform flow, and compressible and incompressible fluids.

In the case of a steady flow, $\frac{\partial \rho}{\partial t} = 0$.

Equation (2.8) becomes

$$\frac{\partial(\rho u)}{\partial x} + \frac{\partial(\rho v)}{\partial y} + \frac{\partial(\rho w)}{\partial z} = 0.$$

For incompressible fluids, the "ρ" does not change with x, y, z and "t". This leads to

$$\frac{\partial u}{\partial x} + \frac{\partial v}{\partial y} + \frac{\partial w}{\partial z} = 0. \tag{2.9}$$

2.4   Forces Acting on Fluids in Motion

The different forces influencing the fluid motion are due to gravity, pressure, viscosity, turbulence, surface tension, and compressibility and are listed below:

F_g (Gravity force)　　　　　　Due to weight of fluid

　　　　　　　　　　　　　　　$=$ (mass $*$ gravitational constant)

F_p (Pressure force)　　　　　　Due to pressure gradient

F_v (Viscous force)　　　　　　Due to viscosity

F_t (Turbulent force)　　　　　Due to turbulence

F_s (Surface tension force)　　Due to surface tension

F_e (Compressibility force)　　Due to elastic property of the fluid

If all the above-mentioned forces are responsible for a given mass of fluid in motion, according to Newton's second law of motion, the equation of motion can be written as

$$Ma = F_g + F_p + F_v + F_t + F_s + F_e. \tag{2.10}$$

Further resolving these forces in x, y, z direction

$$Ma_x = F_{gx} + F_{px} + F_{vx} + F_{tx} + F_{sx} + F_{ex},$$

$$Ma_y = F_{gy} + F_{py} + F_{vy} + F_{ty} + F_{sy} + F_{ey},$$

$$Ma_z = F_{gz} + F_{pz} + F_{vz} + F_{tz} + F_{sz} + F_{ez}. \tag{2.11}$$

Herein, "M" is the mass of fluid and a_x, a_y, a_z are fluid acceleration in the x, y and z directions, respectively.

In most fluid problems, F_e and F_s may be neglected.

Hence,

$$Ma = F_g + F_p + F_v + F_t. \tag{2.12}$$

This is known as *Reynold's equation of motion*.

For laminar flows F_t is negligible.

Hence,

$$Ma = F_g + F_p + F_v. \tag{2.13}$$

This is known as *Navier–Stokes equation*.

In case of ideal fluids $F_v = 0$.

Hence,

$$Ma = F_g + F_p. \tag{2.14}$$

This is known as the *Euler's equation of motion*.

2.5 Euler's Equation of Motion

It is assumed that only the fluid weight (body force) and pressure forces act on a certain mass of fluid to set it in motion. Now, consider a point $P(x, y, z)$ in a fluid flow in which, u, v and w be the velocity components in directions x, y and z, respectively. The body force per unit mass at the same point be X, Y and Z in the x, y and z, directions, respectively.

If the mass of fluid in the medium as shown in Fig. 2.2 is $(\rho \Delta x \Delta y \Delta z)$, the total component of the body force acting on the cube in "x" direction

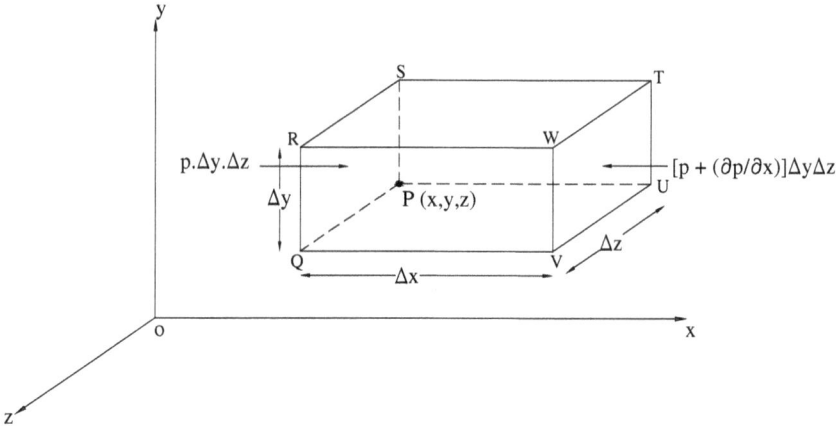

Fig. 2.2   Definition sketch of coordinate system.

is equal to

$$(X\rho\Delta x\Delta y\Delta z)(p + \frac{\partial p}{\partial x}\Delta x)\Delta y\Delta z.$$

Similarly, in "y" and "z" direction are $(Y\rho\Delta x\Delta y\Delta z)$ and $(Z\rho\Delta x\Delta y\Delta z)$ "p" is pressure intensity at point "P". Since the lengths of the edges of the fluid medium are extremely small, it may be assumed that the "p" on the face PQRS is uniform and equal to p.

Therefore, the total pressure force acting on face PQRS in x direction = $p\Delta y\Delta z$.

Since the "p" varies with x, y and z, the pressure intensity on the face USWT will be is equal to

$$\left(p + \frac{\partial p}{\partial x}\Delta x\right).$$

Therefore, the total pressure force acting on the face USWT in the x direction is equal to

$$\left(p + \frac{\partial p}{\partial x}\Delta x\right)\Delta y\Delta z.$$

The net pressure force F_{px} acting on the fluid mass in the x direction, the magnitude of which is obtained as

$$F_{px} = p\Delta y\Delta z - \left(p + \frac{\partial p}{\partial x}\partial x\right)\Delta y\Delta z. \tag{2.15}$$

Hence,

$$F_{px} = -\frac{\partial p}{\partial x}\Delta x\Delta y\Delta z, \quad F_{py} = -\frac{\partial p}{\partial y}\Delta x\Delta y\Delta z, \quad F_{pz} = -\frac{\partial p}{\partial z}\Delta x\Delta y\Delta z. \tag{2.16}$$

Further, pressure force per unit volume are

$$F_{px} = -\frac{\partial p}{\partial x}, \quad F_{py} = -\frac{\partial p}{\partial y}, \quad F_{pz} = -\frac{\partial p}{\partial z}.$$

Adding the pressure forces and the body forces and then equating to mass * acceleration as per the Newton's second law of motion, we get

$$X\rho(\Delta x\Delta y\Delta z) - \frac{\partial p}{\partial x}\Delta x\Delta y\Delta z = (\rho\Delta x\Delta y\Delta z)a_x \tag{2.17}$$

i.e.,

$$X - \frac{1}{\rho}\frac{\partial p}{\partial x} = a_x$$

Likewise,

$$Y - \frac{1}{\rho}\frac{\partial p}{\partial y} = a_y,$$

$$Z - \frac{1}{\rho}\frac{\partial p}{\partial z} = a_z. \tag{2.18}$$

The above equations are known as *Euler's equation of motion*.

Herein $a_x = \frac{du}{dt}$, $a_y = \frac{dv}{dt}$ and $a_z = \frac{dw}{dt}$ may be expressed in terms of the velocity components u, v and w as

$$a_x = \frac{\partial u}{\partial t} + u\frac{\partial u}{\partial x} + v\frac{\partial u}{\partial y} + w\frac{\partial u}{\partial z},$$

$$a_y = \frac{\partial v}{\partial t} + u\frac{\partial v}{\partial x} + v\frac{\partial v}{\partial y} + w\frac{\partial v}{\partial z},$$

$$a_z = \frac{\partial w}{\partial t} + u\frac{\partial w}{\partial x} + v\frac{\partial w}{\partial y} + w\frac{\partial w}{\partial z},$$

where $\frac{\partial u}{\partial t}$, $\frac{\partial v}{\partial t}$, $\frac{\partial w}{\partial t}$ are known as local or temporal acceleration and $\frac{\partial u}{\partial x}$, $\frac{\partial u}{\partial y}$, $\frac{\partial u}{\partial z}$, $\frac{\partial v}{\partial x}$, $\frac{\partial v}{\partial y}$, $\frac{\partial v}{\partial z}$, $\frac{\partial w}{\partial x}$, $\frac{\partial w}{\partial y}$, $\frac{\partial w}{\partial z}$ are convective acceleration.

As "ρ" is considered in these equations, these are applicable to compressible or incompressible, non-viscous fluid in steady or unsteady state of flow.

2.6 Pathlines and Streamlines

The path of a single fluid particle in a fluid domain at any given instant of time is termed as a "pathline" which also represents its velocity direction. Streamlines are a group of curves which are acting tangential to the velocity vectors in a fluid flow.

In a steady flow, pathlines and streamlines are identical. The equation of a streamline can be represented as

$$\frac{u}{dx} = \frac{v}{dy} = \frac{w}{dz}. \tag{2.19}$$

2.7 Velocity Potential

A scalar function of space and time, defined as velocity potential, "φ" whose derivative with respect to a direction provides the velocity in that direction.

Hence, for any direction α, in which the velocity is V_α

$$\frac{\partial \phi}{\partial \alpha} = V_\alpha, \tag{2.20}$$

$$u = \frac{\partial \phi}{\partial x}, \quad v = \frac{\partial \phi}{\partial y}. \tag{2.21}$$

When substituted in continuity Eq. (2.8) we obtain the Laplace equation as follows:

$$\nabla^2 \phi = 0. \tag{2.22}$$

2.8   Stream Function

A scalar function of space and time defined as "ψ" such that its partial derivative with respect to any direction gives the velocity component at right angles (in the counter clockwise direction) to this direction.

$$u = -\frac{\partial \psi}{\partial y}, \quad v = \frac{\partial \psi}{\partial x}. \tag{2.23}$$

Thus for an irrotational flow,

$$\nabla^2 \psi = 0. \tag{2.24}$$

2.9   Bernoulli Equation

The Bernoulli's equation is simply based on the principle of conservation of energy. For a steady flow, as per this principle, the summation of the different forms of energy (kinetic energy, potential energy and static head) along a streamline is the same at all points on that streamline which is a constant. The Bernoulli's principle is applicable to various types of fluid flow.

Let "Ω" be force potential such that

$$X = \frac{-\partial \Omega}{\partial x}, \quad Y = \frac{-\partial \Omega}{\partial y}, \quad Z = \frac{-\partial \Omega}{\partial z},$$

$$u = \frac{\partial \varphi}{\partial x}, \quad v = \frac{\partial \varphi}{\partial y}, \quad w = \frac{\partial \varphi}{\partial z}$$

Substituting these in Euler's Eq. (2.14) and applying the irrotational flow conditions, the following set of equations can be derived.

$$\frac{\partial^2 \phi}{\partial x \partial t} + u\frac{\partial u}{\partial x} + v\frac{\partial v}{\partial x} + w\frac{\partial w}{\partial x} = \frac{-\partial \Omega}{\partial x} - \frac{1}{\rho}\frac{\partial p}{\partial x},$$

$$\frac{\partial^2 \phi}{\partial y \partial t} + u\frac{\partial u}{\partial y} + v\frac{\partial v}{\partial y} + w\frac{\partial w}{\partial y} = \frac{-\partial \Omega}{\partial y} - \frac{1}{\rho}\frac{\partial p}{\partial y},$$

$$\frac{\partial^2 \phi}{\partial z \partial t} + u\frac{\partial u}{\partial z} + v\frac{\partial v}{\partial z} + w\frac{\partial w}{\partial z} = \frac{-\partial \Omega}{\partial z} - \frac{1}{\rho}\frac{\partial p}{\partial z}. \qquad (2.25)$$

If "ρ" is constant, integration with respect to x, y, and z of the above set of equations yields

$$\frac{1}{2}(u^2 + v^2 + w^2) + \frac{\partial \phi}{\partial t} + \Omega + \frac{p}{\rho} = F_1(y, z, t),$$

$$\frac{1}{2}(u^2 + v^2 + w^2) + \frac{\partial \phi}{\partial t} + \Omega + \frac{p}{\rho} = F_2(z, x, t),$$

$$\frac{1}{2}(u^2 + v^2 + w^2) + \frac{\partial \phi}{\partial t} + \Omega + \frac{p}{\rho} = F_3(x, y, t). \qquad (2.26)$$

If u, v, ω are the resolvents of V,

$$\frac{1}{2}V^2 + \frac{\partial \varphi}{\partial t} + \Omega + \frac{p}{\rho} = F(t). \qquad (2.27)$$

This equation is applicable for both compressible and incompressible flows.
For steady flow, "t" will disappear
We also have $-g = -\frac{\partial \Omega}{\partial h}$ if "h" is positive upwards.
Hence $\Omega = gh + $ constant.

$$\text{Total head} = \frac{V^2}{2} + gh + \frac{p}{\rho} = \text{constant.}$$

(Kinetic head) + (Potential head) + (Pressure head)

Worked Out Examples

Problem 2.1

The rate at which water flows through a horizontal 20 cm pipe is increased linearly from 40 to 170 liters/s in 5 s. What pressure gradient must exist to produce this acceleration? What difference in pressure intensity will prevail between sections 10 m apart? Take $\rho = 1000 \text{ kg/m}^3$ in SI unit.

Solution

Euler's equation along the pipe axis may be written as

$$X - \frac{1}{\rho}\frac{\partial p}{\partial x} = \frac{\partial u}{\partial t} + u\frac{\partial u}{\partial x}$$

Since the pipe has a constant diameter $\frac{\partial u}{\partial x} = 0$ and since it is horizontal, the body force per unit volume, X along the flow direction is also zero. The above equation of motion, therefore, reduces to

$$\frac{\partial u}{\partial t} = -\frac{1}{\rho}\cdot\frac{\partial \rho}{\partial x}.$$

The changes in velocity as the flow changes from 40 to 170 liters/s are given by

$$\Delta u = \frac{170}{1000 * \frac{\pi}{4} * (0.20)^2} - \frac{40}{1000 * \frac{\pi}{4} * (0.20)^2} = 4.14\,\text{m/s},$$

therefore

$$\frac{\partial u}{\partial t} = \frac{4.14}{5} = 0.828\,\text{m/s}^2,$$

and the pressure gradient

$$\frac{\partial p}{\partial x} = -\rho\frac{\partial u}{\partial t} = -1000 * 0.828 = -828\,\text{N/m}^2/\text{m}.$$

Difference in pressure between sections 10 m apart is

$$\frac{\partial \rho}{\partial x} * 10 = -828 * 10\,\text{N/m}^2 = -8.28\,\text{kN/m}^2.$$

Problem 2.2

The wind velocity in a cyclone may be assumed to vary according to free vortex law. If the velocity is 15 km/h, 45 km from the center of the cyclone, what pressure gradient should obtain at this point? What reduction in barometric pressure should occur over a radial distance of 15 km from this point towards the centre of the storm? Take mass density of air as 1.208 kg (mass) per m^3.

Solution

The velocity distribution in a free vortex is given by

$$Vr = C. \tag{1}$$

At a radial location defined by

$$r_1 = 45 \, \text{km}, \quad V_1 = 15 \, \text{kmph}.$$

Substituting this in Eq. (1)

$$C = 15 * 45 = 675 \, \text{km}^2/\text{h}.$$

Velocity at a radial distance $(45 - 15) = 30 \, \text{km}$.
Velocity from the centre of cyclone, V_2

$$V_2 = C/r_2 = 675/30 = 22.5 \, \text{kmph}.$$

From Bernoullis equation

$$p + \rho \frac{V^2}{2} = \text{constant}.$$

By differentiating with respect to r

$$\frac{dp}{dr} + \rho V \frac{dV}{dr} = 0 \quad \text{or} \quad \frac{dp}{dr} = -\rho V \frac{dV}{dr}.$$

Differentiating Eq. (2.27)

$$r \frac{dV}{dr} + V = 0 \quad \text{or} \quad \frac{dV}{dr} = -\frac{V}{r},$$

$$\frac{dp}{dr} = -\rho V \frac{dV}{dr} = \rho \frac{V^2}{r}.$$

The pressure gradient at a radial distance of 45 km where the velocity is 15 kmph

$$\left(\frac{dp}{dr}\right)_{r=45\,\text{km}} = 1.208 * \frac{\left[\frac{15*1000}{3600}\right]^2}{45 * 1000} = 4.66 * 10^{-4} \, \text{kg/m}^2/\text{m}$$

$$= 0.466 \, \text{kg/m}^2/\text{km}$$

In SI units:

$$\left(\frac{dp}{dr}\right)_{r=45\,\text{km}} = 0.466 * 9.81 \, \text{N/m}^2/\text{km}$$

$$= 4.57 \, \text{N/m}^2/\text{km}.$$

Reduction in barometric pressure over a radial distance of 15 km between $r_1 = 45 \, \text{km}$, $V_1 = 15 \, \text{kmph}$, and $r_2 = 30 \, \text{km}$, $V_2 = 22.5 \, \text{kmph}$.

Using Bernoulli's equation,

$$p_1 + \rho\frac{V_1^2}{2} = p_2 + \rho\frac{V_2^2}{2}$$

or

$$(p_1 - p_2) = (1/2)\rho[V_2^2 - V_1^2]$$

$$= \frac{1.208}{2}\left[\left(\frac{22.5 * 10^3}{3600}\right)^2 - \left(\frac{15 * 10^3}{3600}\right)^2\right]$$

$$= 13.11\,\text{N/m}^2 \text{ (Reduction in barometric pressure).}$$

Problem 2.3

A 15 cm diameter pipe carries oil of specific gravity 0.85 at the rate of 110 liters/s and the pressure at a point "A" is 0.15 kg/cm^2 (gage). If the point is "A" 4 m above the datum line, calculate the total energy at point A in meters of oil.

Solution

Total energy in terms of meters of oil is given by $\frac{p}{\gamma} + \frac{V^2}{2g} + z$. Then

$$\frac{p}{\gamma} = \frac{0.15 * 10^4}{0.85 * 10^3} = 1.76\,\text{m of oil.}$$

By continuity equation, $Q = AV$, where

$$Q = \frac{110 * 10^3}{10^6} = 0.11\,\text{m}^3/\text{s}$$

$$A = \frac{\pi}{4}(0.15)^2 = 0.018\,\text{m}^2,$$

$$\therefore V = \frac{Q}{A} = \frac{0.11}{0.018} = 6.11\,\text{m/s},$$

$$\text{and } \frac{V^2}{2g} = \frac{(6.11)^2}{2 * 9.81} = 1.9\,\text{m of oil,}$$

$$z = 4\,\text{m.}$$

Total energy $= (1.76 + 1.9 + 4) = 7.66\,\text{m of oil.}$

Problem 2.4

A 22 cm pipe carries water at a velocity of 22.5 m/sec. At point A measurement of pressure and elevation were 2.98 kg/cm^2 and 27.15 m, respectively.

The pressure and elevation at point B were $2.97\,\text{kg/cm}^2$ and $31.2\,\text{m}$, respectively. For steady flow, find the loss of head between A and B.

Solution

Total energy in terms of meters of water is given by $\frac{p}{\gamma} + \frac{V^2}{2g} + z$.

At point A,

$$(p/\gamma) = (2.98 * 10^4)/1000 = 29.80\,\text{m of water},$$

$$(V^2/2g) = (22.5)^2/(2 * 9.81) = 25.80\,\text{m of water},$$

$$z = 27.15\,\text{m},$$

$\therefore$ Total energy at A $= (29.80 + 25.80 + 27.15) = 82.75\,\text{m}$.

At point B,

$$(p/\gamma) = (2.97 * 10^4)/1000 = 29.7\,\text{m of water},$$

$$(V^2/2g) = (22.5)^2/(2 * 9.81) = 25.80\,\text{m of water},$$

$$z = 31.2\,\text{m},$$

$\therefore$ Total energy at B $= (29.70 + 25.80 + 31.2) = 86.7\,\text{m}$,

$\therefore$ Loss of head $= (82.75 - 86.7) = -3.95\,\text{m}$.

Problem 2.5

A pipe $320\,\text{m}$ long has a slope of 1 in 112 and tapers from $1.4\,\text{m}$ diameter at the high end to $0.40\,\text{m}$ diameter at the low end. Quantity of water flowing is $5225\,\text{liters/min}$. If the pressure at higher end is $0.68\,\text{kg/cm}^2$, find the pressure at the lower end. Neglect losses.

Solution

Discharge $Q = (5225 * 10^3)/(60 * 10^6) = 0.087\,\text{m}^3/\text{s}$,
Area of flow section at higher end $= (\pi/4)(1.4)^2 = 1.54\,\text{m/s}$,
Velocity at higher end $= Q/A = (0.087/1.54) = 0.056\,\text{m/s}$,
Area of flow at lower end $= (\pi/4)(0.40)^2 = 0.126\,\text{m/s}$,
Velocity at lower end $= (0.087/0.126) = 0.690\,\text{m/s}$,
Applying Bernoulli's equation between the higher and the lower ends of the pipe

$$\frac{p_1}{\gamma} + \frac{v_1^2}{2g} + z_1 = \frac{p_2}{\gamma} + \frac{v_2^2}{2g} + z_2.$$

Assuming datum to be passing through the lower end,

$$z_2 = 0, \quad z_1 = (320/112) = 2.857\,\text{m}.$$

On substitution,

$$\frac{0.68 * 10^4}{1000} + \frac{(0.056)^2}{2 * 9.81} + 2.857 = \frac{p_2}{\gamma} + \frac{(0.69)^2}{2 * 9.81} + 0,$$

$$6.8 + 0.16 * 10^{-3} + 2.857 = \frac{p_2}{\gamma} + 0.024,$$

$$\frac{p_2}{\gamma} = 9.63 \text{ m of water},$$

$$\therefore \ p_2 = 9630 \text{ kg/m}^2.$$

Problem 2.6

A conical tube is fixed vertically with its smaller end upwards. The velocity of flow down the tube is $3.88\,\text{m/s}$ at the upper end and $1.66\,\text{m/s}$ at the lower end. The tube is $2.7\,\text{m}$ long and the pressure head at the upper end is $2.75\,\text{m}$ of the liquid. The loss in the tube expressed as a head is $0.35\frac{(V_1-V_2)^2}{2g}$, where V_1 and V_2 are the velocities at the upper and lower ends, respectively. What is the pressure head at the lower end?

Solution

Applying Bernoulli's equation between the upper and the lower ends,

$$\frac{p_1}{\gamma} + \frac{V_1^2}{2g} + z_1 = \frac{p_2}{\gamma} + \frac{V_2^2}{2g} + z_2 + \frac{0.35(V_1 - V_2)^2}{2g},$$

$$\frac{p_2}{\gamma} = \frac{p_1}{\gamma} + (z_1 - z_2) + \frac{V_1^2}{2g} - \frac{V_2^2}{2g} - \frac{0.35(V_1 - V_2)^2}{2g}.$$

Here, $\frac{p_1}{\gamma} = 2.75\,\text{m}$, $(z_1 - z_2) = 2.7\,\text{m}$, $V_1 = 3.88\,\text{m/s}$, and $V_2 = 1.66\,\text{m/s}$. Substituting the values in the Bernoulli's equation gives

$$\frac{p_2}{\gamma} = 2.75 + 2.7 + \frac{(3.88)^2}{2 * 9.81} - \frac{(1.66)^2}{2 * 9.81} - \frac{0.35(3.88 - 1.66)^2}{2 * 9.81},$$

$$\frac{p_2}{\gamma} = (2.75 + 2.7 + 0.767 - 0.140 - 0.088) = 5.989.$$

Pressure head at lower end, $\frac{p_2}{\gamma} = 5.989\,\text{m}$ of liquid.

Problem 2.7

The velocity components in a two-dimensional flow field for an incompressible fluid are

$$u = \frac{y^3}{3} - x^2 + 2x - \log t \quad \text{and} \quad v = e^t - \frac{x^3}{3} + xy^2 - 2y.$$

(i) Show that these functions represent an irrotational flow.

(ii) Obtain an expression for stream function ψ.

Solution

(i)

$$\frac{\partial u}{\partial x} = 2 - 2xy, \quad \frac{\partial v}{\partial y} = 2xy - 2.$$

For a two-dimensional flow of incompressible fluid, the continuity Eq. (2.8) is expressed as

$$\frac{\partial u}{\partial x} + \frac{\partial v}{\partial y} = 0.$$

By substituting for the two terms

$$2 - 2xy + 2xy - 2 = 0.$$

Thus, continuity equation is satisfied. Hence fluid flow is possible. Further,

$$\frac{\partial v}{\partial x} = y^2 - x^2, \quad \frac{\partial u}{\partial y} = y^2 - x^2,$$

$$\frac{\partial v}{\partial x} - \frac{\partial u}{\partial y} = y^2 - x^2 - y^2 + x^2 = 0.$$

On substituting the above in Eq. (2.1) for the condition of irrotational flow, we get $W_z = 0$.

Hence, the flow is irrotational.

(ii)

$$\frac{\partial \psi}{\partial x} = v = xy^2 - 2y - \frac{x^3}{3}, \tag{1}$$

$$\frac{\partial \psi}{\partial y} = -u = -\left(\frac{y^3}{3} + 2x - x^2 y\right). \tag{2}$$

Integrating Eq. (1) we get

$$\psi = \frac{x^2 y^2}{2} - 2xy - \frac{x^4}{12} + f(y). \tag{3}$$

Differentiating Eq. (3) with respect to y, we get

$$\frac{\partial \psi}{\partial y} = x^2 y - 2x + f'(y). \tag{4}$$

Equating the values of $(\frac{\partial \psi}{\partial y})$ from Eqs. (2) and (4), we get

$$\left(\frac{y^3}{3} - 2x - x^2 y\right) = x^2 y - 2x + f'(y),$$

$$f'(y) = \frac{-y^4}{12} + C,$$

where "C" is a numerical constant of integration.
Therefore,

$$\psi = \frac{x^2 y^2}{2} - 2xy - \frac{x^4}{12} - \frac{y^4}{12} + C.$$

Since "C" is a numerical constant, it may also be considered as zero.

Problem 2.8

Find the stream function ψ for the velocity field $u = 4x^2$ and $v = 8xy + 3$?

Solution

Before finding the ψ, first check whether the fluid flow is incompressible.
Since $\frac{\partial u}{\partial x} = 8x$ and $\frac{\partial v}{\partial y} = 8x$, the flow fails to satisfy the continuity equation $\frac{\partial u}{\partial x} + \frac{\partial v}{\partial y} = 0$ and hence the stream function does not exist.

Problem 2.9

Let the velocity field $\mathbf{u}(\mathbf{x}, t)$ satisfy the continuity equation $\nabla \cdot \mathbf{u} = 0$.

(i) What does the coordinate transformation $\mathbf{X} = \mathbf{x} + \mathbf{V} \cdot t$, where $\mathbf{V}$ is a constant vector, correspond physically?
(ii) Check whether the continuity equation holds under the transformed coordinate system.

Solution

(i) In physical terms, the transformation moves the frame of reference at the constant speed $\mathbf{V}$.
(ii) Differentiating the transformation $\mathbf{X} = \mathbf{x} + \mathbf{V} \cdot t$, with respect to "$t$" gives

$$\mathbf{U}(\mathbf{X}, t) = \mathbf{u}(\mathbf{x}, t) + \mathbf{V}.$$

Now, $\nabla \cdot \mathbf{U} = \nabla \cdot (\mathbf{u} + \mathbf{V}) = \nabla \cdot \mathbf{u} + \nabla \cdot \mathbf{V} = 0$.

Since the first term is zero by continuity and the second term is due to V is constant, thus, the continuity equation holds under the transformed coordinate system and is invariant to an observer moving in a constant speed.

Chapter 3

Ocean Wave Mechanics

3.1 Introduction

Among all the environmental parameters as discussed in the previous chapter that are needed for the design of structures in the marine environment, the effect of ocean waves is the most important, and hence this chapter is devoted to understand its kinematics. A few worked examples are also illustrated at the end of the chapter. The ocean waves can broadly be categorized based on their generation and propagation in the presence or absence of natural or man-made obstructions. A finer classification is according to the apparent shape, relative water depth and origin. According to the apparent shape, waves can be classified as progressive and standing waves. Progressive waves may be oscillatory or solitary. According to the relative water depth, two types of waves exist, namely small amplitude waves and finite amplitude waves. Finite amplitude waves may be further classified as intermediate depth waves (Stokes' wave) and shallow water waves (Cnoidal waves). Figure 3.1 shows the classification of ocean waves and Fig. 3.2 shows the classification of waves as per its frequency.

The broad classification of ocean waves based on the causes for its generation as well as period are as follows:

(a) capillary wave (due to surface tension) T up to 0.1 s;
(b) ultra-gravity waves (surface tension and gravity) T up to 0.1 s;
(c) gravity wave (wind) T up to 30 s;
(d) infra gravity waves (storms) $T = 30$ s to 5 min;
(e) long period waves (storms) $T = 5$ min to 12 h;
(f) ordinary tidal waves (attraction of astronomical bodies) $T = 12$ h to 24 h;
(g) trans tidal waves (storms, tsunamis — due to undersea explosives).

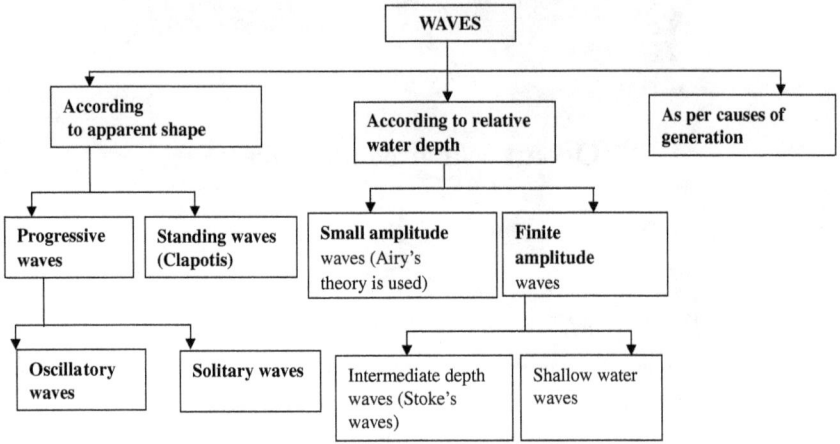

Fig. 3.1   Classification of ocean waves.

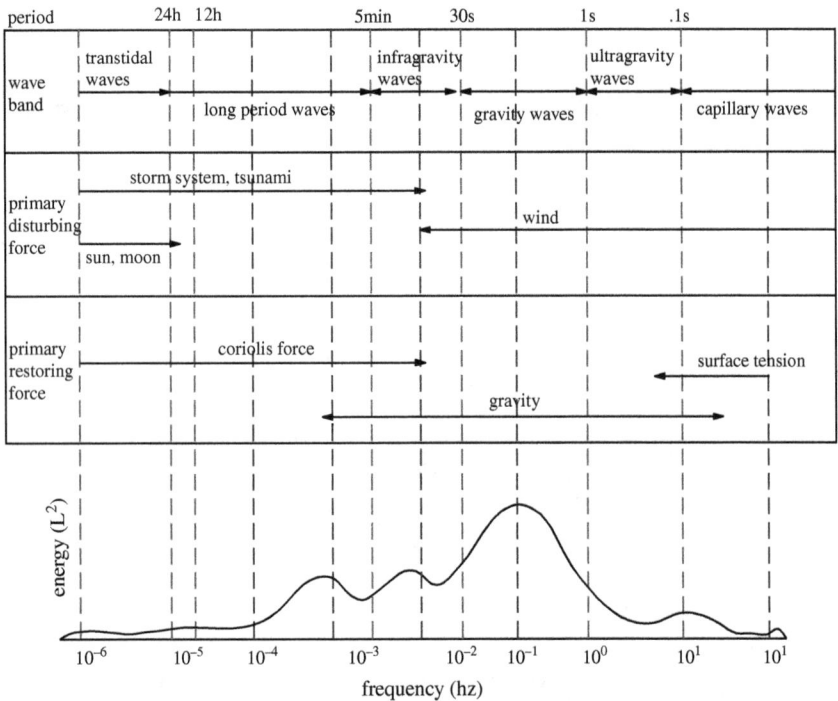

Fig. 3.2   Classification of waves as per its frequency.

3.2 Small Amplitude Wave Theory

3.2.1 *General*

Any primary representation of simple sinusoidal oscillatory wave consists of
its length, L or period, T and height, H. The velocity potential, Laplace's
equation and Bernoulli's dynamic equation — jointly with the suitable
boundary conditions postulate the required information in establishing the
small amplitude wave formulas.

   The assumptions in deriving the expression for the velocity potential
due to propagating ocean waves are as follows:

(a) Flow is said to be irrotational.
(b) Fluid is ideal.
(c) Surface tension is neglected.
(d) Pressure at the free surface is uniform and constant.
(e) The seabed is rigid, horizontal, and impermeable.
(f) Wave height is small compared to its length.
(g) Potential flow theory is applicable. A velocity potential Φ exists and the
    velocity components u and w in the x and z directions can be obtained
    as $\frac{\partial \phi}{\partial x}$ and $\frac{\partial \phi}{\partial z}$.

3.2.2 *Derivation for velocity potential*

The governing equation is the Laplace equation and is given by

$$\nabla^2 \phi = 0. \tag{3.1}$$

The continuity equation and Bernoulli's equation given by Eqs. (3.2) and
(3.3) are applied in the solution procedure:

$$\frac{\partial u}{\partial x} + \frac{\partial v}{\partial y} + \frac{\partial w}{\partial z} = 0, \tag{3.2}$$

$$\frac{-\partial \phi}{\partial t} + \frac{1}{2}(u^2 + v^2 + w^2) + \frac{p}{\rho} + gz = 0. \tag{3.3}$$

The sketch defining the boundaries, seabed and wave surface elevation is
given in Fig. 3.3.

3.2.3 *Boundary conditions*

(I) Equation (3.1) is to be satisfied in the region $-d \le z \le \eta - \infty < x < \infty$
    where "η" is the water surface elevation measured from the still water
    level (SWL).

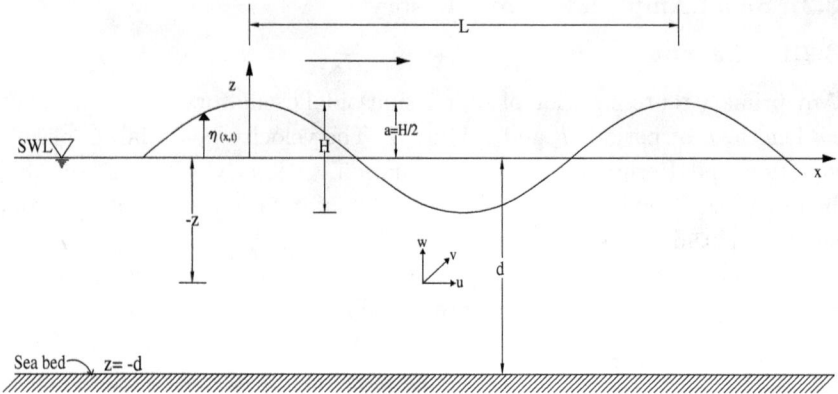

Fig. 3.3   Definition sketch for wave motion.

(II) The *kinematic bottom boundary condition* states that the vertical velocity component at the sea bottom is zero:

$$w = \frac{\partial \phi}{\partial z} = 0 \quad \text{at } z = -d,$$

since "z" is negative downwards from SWL.

(III) The pressure at the free surface is zero at $z = \eta$, and linearizing the Bernoulli's equation results in

$$\frac{-\partial \phi}{\partial t} + \frac{p}{\rho} + gz = 0, \tag{3.4}$$

when $z = \eta$ and taking $p = 0$ using Eq. (3.4) we get

$$\eta = \frac{1}{g} \left[\frac{\partial \phi}{\partial t} \right]_{z=\eta}.$$

This is the *dynamic free surface boundary condition*. Since we assume that amplitude of the waves is small, the above equation can be rewritten as follows:

$$\eta = \frac{1}{g} \left[\frac{\partial \phi}{\partial t} \right]_{z=0}. \tag{3.5}$$

This is applicable only when "η" is small and is valid for $\frac{H}{d}$ and $\frac{H}{L} < 1$.

3.2.4   Solution to the Laplace equation

$$\frac{\partial^2 \phi}{\partial x^2} + \frac{\partial^2 \phi}{\partial z^2} = 0. \tag{3.6}$$

Method of separable is used to obtain the solution to the above equation.

Let us assume

$$\phi(x, z, t) = \overline{X}(x)\overline{Z}(z)\overline{T}(t). \tag{3.7}$$

Substituting Eq. (3.7) in Eq. (3.6) we get

$$\overline{X}''\overline{Z}\overline{T} + \overline{X}\overline{Z}''\overline{T} = 0.$$

Herein, each prime denotes differentiation once with respect to the independent variable.

Dividing both sides of the above $\overline{X}\overline{Z}\overline{T}$ gives $\frac{\overline{X}''}{\overline{X}} = \frac{-\overline{Z}''}{\overline{Z}}$ and let this be a constant $= -k^2$.

Then

$$\overline{X}'' + k^2\overline{X} = 0, \tag{3.8}$$
$$\overline{Z}'' - k^2\overline{Z} = 0, \tag{3.9}$$
$$\overline{X} = A\cos kx + B\sin kx,$$
$$\overline{Z} = Ce^{kz} + De^{-kz}.$$

Hence,

$$\phi(x, z, t) = (A\cos kx + B\sin kx)(Ce^{kz} + De^{-kz})\overline{T}(t).$$

The solutions to ϕ are simple harmonic in time requiring $\overline{T}(t)$ be expressed as $\cos(\sigma t)$ or $\sin(\sigma t)$, thus leading to four forms of solutions to ϕ, such that

$$\phi_1 = C_1(Ce^{kz} + De^{-kz})\cos kx \, \cos \sigma t,$$
$$\phi_2 = C_2(Ce^{kz} + De^{-kz})\sin kx \, \sin \sigma t,$$
$$\phi_3 = C_3(Ce^{kz} + De^{-kz})\sin kx \, \cos \sigma t,$$
$$\phi_4 = C_4(Ce^{kz} + De^{-kz})\cos kx \, \sin \sigma t.$$

3.2.5 Determination of the constants

The constants are determined by using the dynamic free surface boundary condition and the kinematic bottom boundary condition.

Considering ϕ_2

$$\phi_2 = C_2(Ce^{kz} + De^{-kz}) \sin kx \sin \sigma t. \tag{3.10}$$

Applying the kinematic bottom boundary condition i.e.,

$$\frac{\partial \phi}{\partial z} = 0 \quad \text{at } z = -d$$

$$\frac{\partial \phi_2}{\partial z}\bigg|_{z=-d} = C_2(Cke^{kz} - Dke^{-kz}) \sin kx \sin \sigma t = 0,$$

$C_2 \neq 0, \sin kx \sin \sigma t \neq 0$ [since velocity potential exists],

$$\therefore \ C = De^{2kd}.$$

Substituting for C in Eq. (3.10) and simplifying, we get

$$\phi_2 = 2C_2De^{kd}\left[\frac{e^{k(d+z)} + e^{-k(d+z)}}{2}\right] \sin kx \sin \sigma t$$

$$\phi_2 = 2C_2De^{kd} \cosh k(d+z) \sin kx \sin \sigma t \tag{3.11}$$

and $\quad \dfrac{\partial \phi_2}{\partial t}\bigg|_{z=0} = (2C_2D\sigma e^{kd} \cosh kd \sin kx \cos \sigma t).$

On assuming,

$$\eta = a \sin kx \cos \sigma t.$$

Now applying the free surface boundary condition $\{\eta = \frac{1}{g}\frac{\partial \phi_2}{\partial t}|_{z=0}\}$, we get

$$a \sin kx \cos \sigma t = \frac{2C_2D\sigma e^{kd}}{g} \cosh kd \sin kx \cos \sigma t,$$

$$2C_2De^{kd} = \frac{ag}{\sigma}\frac{1}{\cosh kd}.$$

Substituting in Eq. (3.11), we get

$$\phi_2 = \frac{ag}{\sigma}\frac{\cosh k(d+z)}{\cosh kd} \sin kx \sin \sigma t. \tag{3.12}$$

Similarly, ϕ_1, ϕ_3 and ϕ_4 can be evaluated as follows:

$$\phi_1 = \frac{-ag}{\sigma}\frac{\cosh k(d+z)}{\cosh kd} \cos kx \cos \sigma t, \tag{3.13}$$

$$\phi_3 = \frac{-ag}{\sigma} \frac{\cosh k(d+z)}{\cosh kd} \sin kx \cos \sigma t, \tag{3.14}$$

$$\phi_4 = \frac{ag}{\sigma} \frac{\cosh k(d+z)}{\cosh kd} \cos kx \sin \sigma t. \tag{3.15}$$

Considering the wave moving in positive direction, we have the following condition.

If $\phi^+ = \phi_2 - \phi_1$

$$= \frac{ag}{\sigma} \frac{\cosh k(d+z)}{\cosh kd} [\cos kx \cos \sigma t + \sin kx \sin \sigma t],$$

then

$$\phi = \frac{ag}{\sigma} \cdot \frac{\cosh k(d+z)}{\cosh kd} \cos(kx - \sigma t). \tag{3.16}$$

This is the expression for the velocity potential for a propagating wave in a constant water depth.

Since

$$\eta = \frac{1}{g} \cdot \frac{\partial \phi}{\partial t} \Big|_{z=0},$$

$$\eta = \frac{1}{g} \frac{ag}{\sigma} \frac{\cosh k(d+z)}{\cosh kd} \sigma \sin(kx - \sigma t).$$

Hence

$$\eta = a \sin(kx - \sigma t). \tag{3.17}$$

Here "η" is periodic in x and t. If we locate a point on the wave and traverse along with it, in such a way, at all time "t", our position relative to the wave form remains fixed. The phase difference then is zero or

$$(kx - \sigma t) = \text{constant}.$$

The speed with which we must move to accomplish this is given by $kx = \sigma t + \text{Constant}$. That is,

$$k \frac{dx}{dt} = \sigma \quad \text{or} \quad \frac{dx}{dt} = \frac{\sigma}{k} = \frac{2\pi}{T} \cdot \frac{L}{2\pi} = \frac{L}{T} = C,$$

$$C = \frac{L}{T} = \textit{celerity} \text{ or speed of the wave.} \tag{3.18}$$

3.2.6   *Wave moving in negative direction*

If

$$\phi^- = \phi_2 + \phi_1$$

$$= \frac{-ag}{\sigma} \frac{\cosh k(kx + \sigma t)}{\cosh kd} \cos(kx + \sigma t),$$

then

$$\eta = \frac{1}{g} \cdot \frac{\partial \phi}{\partial t}\Big|_{z=0}$$

$$\eta = \frac{1}{g} \left[-\frac{ag}{\sigma} \frac{\cosh k(d + z)}{\cosh kd}(-\sigma \sin(kx + \sigma t)) \right]$$

$$= a \sin(kx + \sigma t).$$

For celerity of the wave, $kx + \sigma t = $ Constant. That is,

$$\frac{dx}{dt} = \frac{-\sigma}{k} = \frac{-L}{T} = -C. \tag{3.19}$$

The water surface elevation, $\eta = a \sin(kx - \sigma t)$ is a function of x and t. If one considers a wave propagating in a tank, at a given instant of time, "t", a snapshot would show the locations of crests along the flume, x. Similarly, at any location in the flume, x, the variation of the wave profile could be obtained from recording.

3.3   Dispersion Relationship

The relationship between wavelength, period and water depth is the dispersion relationship which is obtained as discussed below. As we deal with small amplitude waves, the slope of the wave profile is small so that $(\frac{d\eta}{dt})$ can be approximately said as equal to the vertical velocity component, w.

     Hence,

$$w = \frac{d\eta}{dt} = \frac{\partial \eta}{\partial t} + \frac{\partial \eta}{\partial x} \cdot \frac{\partial x}{\partial t}.$$

Since, the wave slope is small ($\frac{\partial \eta}{\partial x} = 0$) and since $w = \frac{\partial \eta}{\partial t}$ and $w = \frac{-\partial \phi}{\partial z}$

     Hence,

$$\frac{\partial \eta}{\partial t} = \frac{-\partial \phi}{\partial z}. \tag{3.20}$$

Differentiating Eq. (3.16), we get

$$\frac{\partial \eta}{\partial t} = \frac{1}{g} \frac{\partial^2 \phi}{\partial t^2}\bigg|_{z=0}.$$

Hence,

$$\frac{\partial \eta}{\partial t} = \frac{-R\sigma^2}{g} \cosh kd \cos(kx - \sigma t), \tag{3.21}$$

where

$$R = \frac{H}{2} \frac{g}{\sigma} \frac{1}{\cosh kd},$$

$$w = \frac{-\partial \phi}{\partial z} = -Rk \sinh kd \cos(kx - \sigma t). \tag{3.22}$$

Using Eq. (3.20), and equating Eqs. (3.21) and (3.22), we get

$$\frac{R\sigma^2}{g} \cosh kd \cos(kx - \sigma t) = Rk \sinh kd \cos(kx - \sigma t),$$

$$\frac{\sigma^2}{g} = \frac{k \sinh kd}{\cosh kd}.$$

The dispersion equation can then be written as

$$\sigma^2 = gk \tanh kd, \tag{3.23}$$

where $\sigma = \frac{2\pi}{T}$ is the wave angular frequency and $k = \frac{2\pi}{L}$ is the wave number. Here σ is also referred to as "ω".

The above equation can be simplified as

$$\left(\frac{2\pi}{T}\right)^2 = g\left(\frac{2\pi}{L}\right) \tanh kd,$$

$$C^2 = \frac{g}{k} \tanh kd. \tag{3.24}$$

The speed at which a wave moves in its direction of propagation as a function of water depth is given by the above equation. Since $C = \frac{L}{T}$, from the above equation we get

$$C = \sqrt{\frac{gL}{2\pi} \tanh kd} \tag{3.25}$$

or

$$L = \frac{gT^2}{2\pi} \tanh kd. \tag{3.26}$$

Since the unknown "L" occurs on both sides, Eq. (3.26) is an implicit equation which has to be solved by trial and error.

3.4   Celerity in Different Water Depth Conditions

3.4.1   *General*

Classification of waves according to relative water depth, d/L which is controlled by the hyperbolic function $\tanh kd$ is given in Table 3.1.

The variation of the hyperbolic functions as a function of d/L_0 shown in Fig. 3.4 permits us the determination of 'L' and the needed hyperbolic functions.

3.4.2   *Deep water conditions*

In the case of deep water Eq. (3.24) becomes

$$C_0 = \sqrt{\frac{gL_0}{2\pi}}.$$

Table 3.1.   Classification of ocean waves according to water depth.

Classification	d/L	$\frac{2\pi d}{L}$	$\tanh \frac{2\pi d}{L}$
Deep waters	$> 1/2$	$> \pi$	~ 1
Intermediate waters	$1/2$ to $1/20$	$\pi/10$ to π	$\tanh(2\pi d/L)$
Shallow waters	$\leq 1/20$	0 to $\pi/10$	$\sim 2\pi d/L$

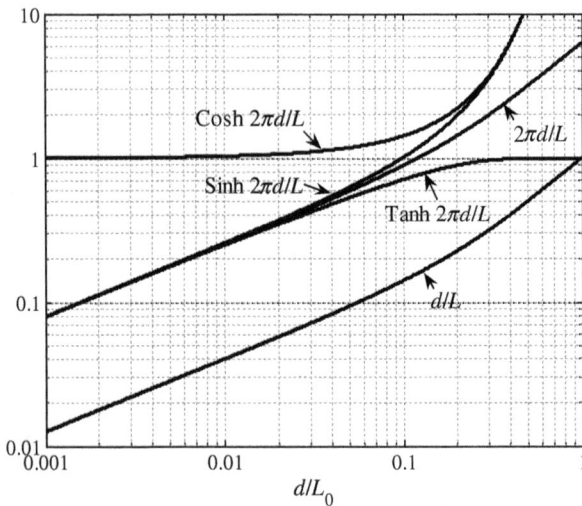

Fig. 3.4   Relationship between d/L_0 and functions of d/L.

Since $\tanh kd = 1$, Eq. (3.25) becomes

$$C_0 = \frac{gT}{2\pi} \qquad (3.27)$$

and

$$L_0 = \frac{gT^2}{2\pi}, \qquad (3.28)$$

i.e., when $\frac{d}{L} \geq \frac{1}{2}$, $\tanh kd$ approaches unity and the wave characteristics are independent of the water depth, "d", while the wave period remains constant. Hence,

$$L_0 = \frac{gT^2}{2\pi} = 1.56T^2 \text{m}.$$

If Eq. (3.28) is used to compute the wave celerity for shallow water conditions ($d/L < 1/20$), an error of about 20–50% would exist.

3.4.3 *Shallow water conditions*

When $kd = \frac{\pi}{10}$, $\frac{d}{L} \leq \frac{1}{20}$,

$$C^2 = \frac{gL}{2\pi} \tanh kd.$$

Herein, $\tanh kd \sim kd = \frac{2\pi d}{L}$, and hence

$$C = \sqrt{gd}. \qquad (3.29)$$

The above relation indicates that the wave celerity depends on the water depth for shallow waters.

3.4.4 *Relationship between d/L and d/L_0*

It can be shown by dividing Eq. (3.25) by Eq. (3.27) and dividing Eq. (3.26) by Eq. (3.28) that

$$\frac{C}{C_0} = \frac{L}{L_0} = \tanh kd.$$

Multiplying both sides by $\frac{d}{L}$, we get

$$\frac{d}{L_0} = \frac{d}{L} \tanh kd. \qquad (3.30)$$

The relation between $\frac{d}{L}$ and $\frac{d}{L_0}$ is given in the wave tables (Fig. 3.4).

3.4.5   *Approximate solutions to the dispersion equation*

An approximate solution for wave number "k" in the dispersion relationship given by Eq. (3.29) for a given "σ" and "d" proposed by Hunt (1979) can be solved directly for kd:

$$(kd)^2 = y^2 + \frac{y}{1 + \sum_{n=1}^{6} d_n Y^n}, \tag{3.31}$$

where $y = \frac{\sigma^2 d}{g} = k_0 d$ and

$$d_1 = 0.6666666666, \quad d_2 = 0.3555555555,$$

$$d_3 = 0.1608465608, \quad d_4 = 0.0632098765,$$

$$d_5 = 0.0217540484, \quad d_6 = 0.0065407983.$$

The wave celerity can be obtained as

$$\frac{C^2}{gd} = [y + (1 + 0.6522y + 0.4622y^2 + 0.0864y^4 + 0.0675y^5)^{-1}]^{-1}, \tag{3.32}$$

which is reported to be accurate to 0.1% for $0 < y < \infty$.

The other equations proposed for the calculation of wavelength are given below:

- Fenton and Mckee (1990)

$$kd = \frac{\sigma^2 d}{g}(\coth((\sigma\sqrt{d/g})^{3/2}))^{2/3}; \tag{3.33}$$

- Guo (2002)

$$kd = \frac{\sigma^2 d}{g}(1 - e^{-(\sigma\sqrt{d/g})^{5/2}})^{-2/5}; \tag{3.34}$$

- You (2008)

$$kd = k_0 d/\tanh(\xi_0), \tag{3.35}$$

where

$$k_0 d = \frac{\sigma^2 d}{g}, \quad \xi_0 = (k_0 d)^{0.5}\left[1 + \frac{k_0 d}{6} + \frac{(k_0 d)^2}{30}\right].$$

The difference in the values of the wavelength obtained from the different approaches is illustrated with a worked-out example at the end of this chapter.

3.5 Local Fluid Particle Velocities and Accelerations Under Progressive Waves

In the evaluation of wave forces on offshore structures, it is essential to know the fluid particle kinematics, i.e., orbital velocities and accelerations in the horizontal and vertical directions.

We know

$$\phi = \frac{ag}{\sigma} \frac{\cosh k(d+z)}{\cosh kd} \cos(kx - \sigma t). \tag{3.36}$$

The horizontal water particle velocity or orbital velocity u is given by

$$u = \frac{-\partial \phi}{\partial x} = \frac{ag}{\sigma} k \frac{\cosh k(d+z)}{\sinh kd} \sin(kx - \sigma t)$$

$$= \frac{ag}{\sigma} k \frac{\cosh k(d+z)}{\sinh kd} \tanh kd \sin(kx - \sigma t). \tag{3.37}$$

Substituting the relationship $C^2 = \frac{g}{k} \tanh kd$ (Eq. (3.24)) and $a = \frac{H}{2}$ in the above expression, we get

$$u = \frac{H}{2\sigma} C^2 k^2 \frac{\cosh k(d+z)}{\sinh kd} \sin(kx - \sigma t).$$

Simplifying, we get

$$u = \frac{\pi H}{T} \frac{\cosh k(d+z)}{\sinh kd} \sin(kx - \sigma t). \tag{3.38}$$

The vertical fluid particle velocity w is given by

$$w = \frac{-\partial \phi}{\partial z} = \frac{-agk}{\sigma} \frac{\sinh k(d+z)}{\cosh kd} \cos(kx - \sigma t)$$

$$= \frac{-agk}{\sigma} \frac{\sinh k(d+z)}{\sinh kd} \tanh kd \cos(kx - \sigma t). \tag{3.39}$$

Using Eq. (3.24), we get

$$w = \frac{-\pi H}{T} \frac{\sinh k(d+z)}{\sinh kd} \cos(kx - \sigma t). \tag{3.40}$$

The above equation expresses the velocity component with in the wave at any elevation, "z". At a given z, the velocities are harmonic in "x" and "t" as shown in Fig. 3.5.

The variations of "u" and "w" with respect to phase are shown in Fig. 3.6. At a given phase angle θ, ($\theta = kx - \sigma t$), the hyperbolic functions cause an exponential decay of "u" and "w" from the free surface towards

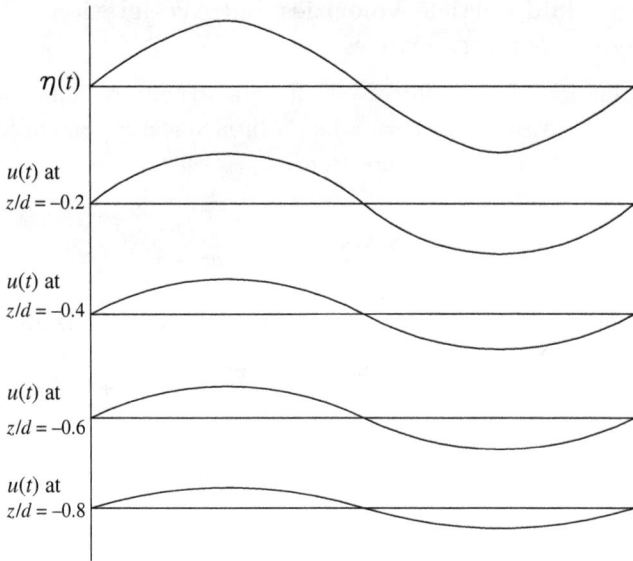

Fig. 3.5   Schematic representation of the variation of $u(t)$ along the depth.

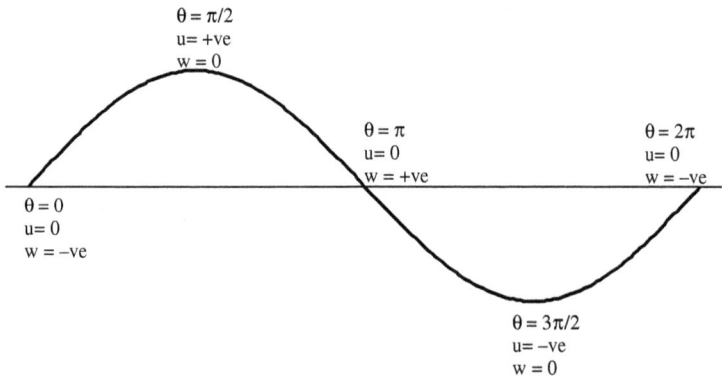

Fig. 3.6   Phase variation of $u(t)$ and $w(t)$.

the seabed. This is indicated schematically in Fig. 3.7 for the phase angles at which the components are largest.

The local acceleration in x and z directions can be obtained as follows:

$$\dot{u} = \frac{\partial u}{\partial t} = \frac{-2\pi^2 H}{T^2} \frac{\cosh k(d+z)}{\sinh kd} \cos(kx - \sigma t), \qquad (3.41)$$

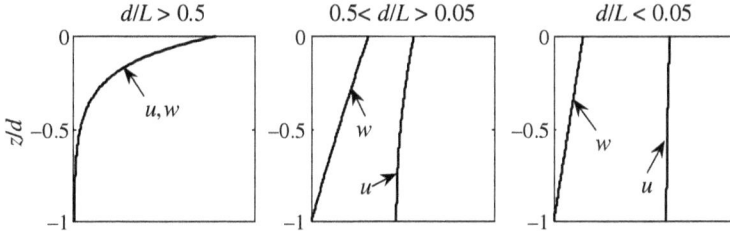

Fig. 3.7   Variation of maximum u and w.

$$\dot{w} = \frac{\partial w}{\partial t} = \frac{-2\pi^2 H}{T^2} \frac{\sinh k(d+z)}{\sinh kd} \sin(kx - \sigma t). \qquad (3.42)$$

3.6   Water Particle Displacement Under Progressive Waves

The expressions for individual horizontal and vertical water particle displacements are obtained as follows:

$$\delta_x = \int u\, dt = \frac{H}{2} \frac{\cosh k(d+z)}{\sinh kd} \cos(kx - \sigma t), \qquad (3.43)$$

$$\delta_z = \int w\, dt = \frac{H}{2} \frac{\sinh k(d+z)}{\sinh kd} \sin(kx - \sigma t), \qquad (3.44)$$

$$\delta_x = D \cos(kx - \sigma t) \quad \text{where } D = \frac{H}{2} \frac{\cosh k(d+z)}{\sinh kd}.$$

Let

$$\delta_z = B \sin(kx - \sigma t) \text{ where } B = \frac{H}{2} \frac{\sinh k(d+z)}{\sinh kd},$$

$$\cos^2(kx - \sigma t) = \left(\frac{\delta_x}{B}\right)^2, \quad \sin^2(kx - \sigma t) = \left(\frac{\delta_z}{B}\right)^2.$$

Since $[\sin^2(kx - \sigma t) + \cos^2(kx - \sigma t) = 1]$, we have

$$\left(\frac{\delta_x}{D}\right)^2 + \left(\frac{\delta_z}{B}\right)^2 = 1. \qquad (3.45)$$

This is the equation of an ellipse showing that the water particles move in an elliptical orbit, where D is the semi-major axis (horizontal measure of particle displacement) and B is semi-minor axis (vertical measure of particle displacement).

3.6.1   *Shallow water condition*

If $\frac{d}{L} < \frac{1}{20}$, then $\cosh k(d+z)$ and $\sinh k(d+z)$ reduce to $k(d+z)$ and $\sinh kd$ reduces to kd. Hence,

$$D = \frac{H}{2}\frac{1}{kd} \quad \text{and} \quad B = \frac{H}{2}\frac{k(d+z)}{kd} = \frac{H}{2}\frac{(d+z)}{d}.$$

Thus, the water particles move in elliptical orbits (paths) in shallow and intermediate waters with the equation of the form:

$$\left(\frac{\delta x}{\frac{H}{2}\frac{1}{kd}}\right)^2 + \left(\frac{\delta z}{\frac{H}{2}\frac{(d+z)}{d}}\right)^2 = 1. \tag{3.46}$$

3.6.2   *Deep water condition*

If $\frac{d}{L} > \frac{1}{2}$, then $D = \frac{H}{2}\frac{\cosh k(d+z)}{\sinh kd} = \frac{H}{2}\left(\frac{e^{k(d+z)}+e^{-k(d+z)}}{e^{kd}-e^{-kd}}\right)$.

As "d" (depth of water or d/L) is very large, $e^{-k(d+z)}$ and e^{-kd} will be very small compared to $e^{k(d+z)}$.

Hence,

$$D = \frac{H}{2}\frac{e^{k(d+z)}}{e^{kd}} = \frac{H}{2}e^{kz}.$$

Similarly,

$$B = \frac{H}{2}e^{kz}.$$

Thus, the water particles move in circular orbits in deep waters (since $D = B$) with equation of the form:

$$\left(\frac{\delta x}{\frac{H}{2}e^{kz}}\right)^2 + \left(\frac{\delta z}{\frac{H}{2}e^{kz}}\right)^2 = 1. \tag{3.47}$$

This shows that for deep water conditions, the water particle paths are circular. In deep water regions, the amplitude variation of the water particle displacement decreases exponentially along with the depth. The water particle displacement at $z = -L_0/2$ is very less compared to the incident wave height. The variation of the water particle displacements under different water depth conditions is illustrated in Fig. 3.8.

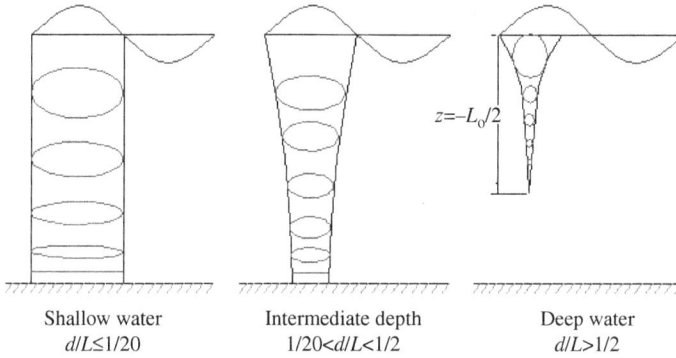

$z=-L_o/2$

Shallow water	Intermediate depth	Deep water
$d/L \leq 1/20$	$1/20 < d/L < 1/2$	$d/L > 1/2$

Fig. 3.8   Schematic representation of fluid particle trajectories.

3.7   Pressure Distribution Under Progressive Waves

The linearized Bernoulli's equation is given by

$$\frac{-\partial \phi}{\partial t} + \frac{p}{\rho} + gz = 0.$$

Multiplying throughout by "ρ", the total pressure is given by

$$p = \rho \frac{\partial \phi}{\partial t} + (-\gamma z).$$

$$(\text{dynamic}) + (\text{static})$$

Substituting for "ϕ" from Eq. (3.16), we get

$$p = \frac{\gamma H}{2} \cdot \frac{\cosh k(d+z)}{\cosh kd} \cdot \sin(kx - \sigma t) - \gamma z,$$

$$\eta = \frac{H}{2} \sin(kx - \sigma t) \text{ and let } \frac{\cosh k(d+z)}{\cosh kd} = K_p,$$

where K_p is the pressure response factor. Then,

$$p = \gamma \eta K_p - \gamma z \quad \text{or} \quad \frac{p}{\gamma} = (K_p \eta - z). \tag{3.48}$$

It is to be mentioned that p was set to zero to define the free surface boundary condition in the Bernoulli equation. However, ϕ was determined by setting $p = 0$ at $z = 0$ instead of $z = \eta$ [see Eq. (3.5)]. Hence, Eq. (3.48) is valid only for negative z.

Applying the above equation, we get

$$\text{Pressure at } z = 0, \quad \frac{p}{\gamma} = \eta,$$

$$\text{Pressure at } z = -d, \quad \frac{p}{\gamma} = \frac{\eta}{\cosh kd} + d, \tag{3.49}$$

that is,

$$\left[d + \frac{\eta}{\cosh kd} \right] < d + \eta,$$

since $\cosh kd$ is always greater than 1.

3.7.1   *Under the trough at seabed*

Substituting $z = -d$, $\eta = -\eta$ leads to

$$K_p = \frac{\cosh k(d - d)}{\cosh kd} = \frac{1}{\cosh kd},$$

and hence from Eq. (3.54)

$$\frac{p}{\gamma} = \frac{-\eta}{\cosh kd} + d, \tag{3.50}$$

$$\frac{p}{\gamma} = \left[d - \frac{\eta}{\cosh kd} \right] > (d - \eta).$$

In order to determine the surface wave height based on subsurface measurement of pressure, Eq. (3.48) is represented as

$$\eta = \frac{N(p + \rho g z)}{\rho g K},$$

where "K" is pressure response factor at the seabed given by $\frac{1}{\cosh kd}$, "N" is the correction factor depending on the period, depth, wave amplitude, etc, i.e.,

$$N > 1 \text{ for long period waves,}$$

$$N < 1 \text{ for short period waves,}$$

$$N = 1 \text{ for linear waves.}$$

The pressure distribution under a progressive wave is shown in Fig. 3.9.

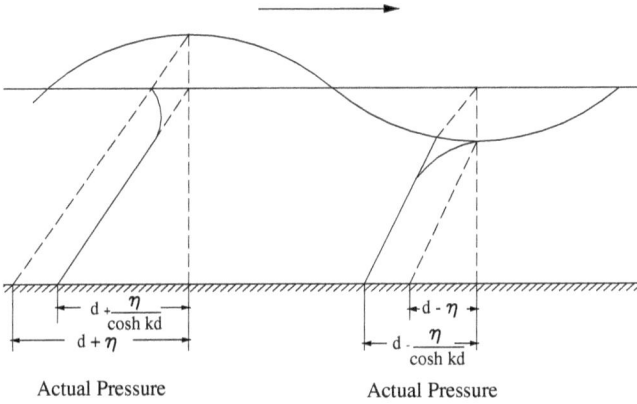

Fig. 3.9   Pressure distribution under a progressive wave.

3.8   Group Celerity

When a group of waves propagates, its speed will in general not be similar to that of the speed with which individual waves within the group travel. If any two wave trains of the same amplitude, but, slightly different wavelengths or periods progress in the same direction, the resultant surface disturbance can be represented as the sum of the individual disturbances. For waves propagating in deep or transitional waters, the group velocity is determined as follows:

$$\eta_T = \eta_1 + \eta_2 = a\sin(k_1 x - \sigma_1 t) + a\sin(k_2 x - \sigma_2 t),$$

$$\eta_T = 2a\cos\left[\left(\frac{k_1 - k_2}{2}\right)x - \left(\frac{\sigma_1 - \sigma_2}{2}\right)t\right]$$

$$\times \sin\left[\left(\frac{k_1 + k_2}{2}\right)x - \left(\frac{\sigma_1 + \sigma_2}{2}\right)t\right].$$
(3.51)

This is a form of a series of sine waves the amplitude of which varies slowly from 0 to 2a according to the cosine factor.

The points of zero amplitude (nodes) of the wave envelope η_T are located by finding the zeros of the cosine factor, i.e.,

$$\eta_{T\max} = 0 \text{ occurs when } \left(\frac{k_1 - k_2}{2}\right)x - \left(\frac{\sigma_1 - \sigma_2}{2}\right)t = (2m+1)\frac{\pi}{2}.$$

In other words, the nodes will occur on "x" axis at distances as follows:

$$x_{node} = \frac{(2m+1)\pi}{k_1 - k_2} + \left(\frac{\sigma_1 - \sigma_2}{k_1 - k_2}\right)t.$$

Since the position of all the nodes is a function of time, they are not stationary. At $t = 0$, there will be nodes at

$$\frac{\pi}{k_1 - k_2}, \quad \frac{3\pi}{k_1 - k_2}, \quad \frac{5\pi}{k_1 - k_2}, \quad \text{etc. i.e., at } m = 0, 1, 2, 3, \ldots.$$

The distances between the nodes are given by

$$x = \frac{2\pi}{k_1 - k_2} = \frac{L_1 L_2}{L_2 - L_1}. \tag{3.52}$$

The speed of propagation of the nodes and hence the speed of propagation of the wave group is called the "*Group Velocity*" and is given by

$$\frac{dx_{\text{node}}}{dt} = \text{Wave Group Velocity } C_G$$

$$= \frac{\sigma_1 - \sigma_2}{k_1 - k} = \frac{d\sigma}{dk}, \quad \text{since } \sigma = kC = \frac{2\pi}{L} \cdot \frac{L}{T} = \frac{2\pi}{T},$$

$$C_G = \frac{d(kC)}{dk} = C + k\frac{dC}{dk} = C + \frac{k.dC}{dL}\left(\frac{1}{\frac{dk}{dL}}\right).$$

Since $k = \frac{2\pi}{L}$ and $C^2 = \frac{g}{k}\tanh(kd)$, substituting and on simplification we get

$$\frac{C_G}{C} = n = \frac{1}{2}\left[1 + \frac{2kd}{\sinh 2kd}\right]. \tag{3.53}$$

Note: Observe the waves generated in a wave flume as shown in Fig. 3.10. Consider the wave between locations (1) and (2), which is a single with length L and time period T, and hence the speed will be $C = L/T$. With a snapshot, we can measure L and note down the time T. Now consider a wave train in between locations (3) and (4). Several waves will be present over this stretch. If through a snapshot, the distance between these points is measured as L_g and the time taken for the waves to travel from locations (3) to (4), say, T_g, is also noted, then the velocity of the wave train will be L_g/T_g, which is Group celerity, C_g. Then C and C_g will be found to be same if the waves generated in the tank is basically shallow water waves. However, for other water depth conditions, they will be different.

For Deep waters, as $\frac{2kd}{\sinh 2kd}$ is zero,

$$C_G = \frac{1}{2}\frac{L_0}{T} = \frac{1}{2}C_0. \tag{3.54}$$

The group celerity is one-half of the phase velocity in deep waters. Further, it should be noted that variables if associated with a suffix '0' refer to deep water conditions. For example, C_0 is deep water celerity.

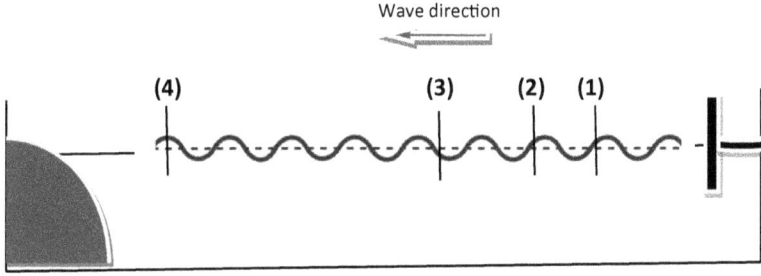

Fig. 3.10   Understanding group celerity through tests.

Table 3.2.   Variation of asymptotic functions.

	Asymptotes	
Function	Shallow waters	Deep waters
$\sinh kd$	kd	$(e^{kd})/2$
$\cosh kd$	1	$(e^{kd})/2$
$\tanh kd$	kd	1

For shallow waters, since $\sinh 2kd = 2kd$,

$$C_G = C = \sqrt{gd}. \tag{3.55}$$

Hence, in shallow waters, the group and phase velocities are same and is a function of only depth of water and in deep waters, the C_G is a function of wavelength. The variation of asymptotic functions in deep and shallow waters are given in Table 3.2.

3.9   Wave Energy

The total energy under a progressive wave is a sum of potential and kinetic energies, the details of which are explained below.

3.9.1   *Potential energy*

In order to determine the total energy under progressive waves, the potential energy of the wave above $z = -d$ with a wave form present is determined from which the potential energy of the water in the absence of a wave form is subtracted. See Fig. 3.11 for definitions.

The potential energy (with respect to $z = -d$) of a small column of water $(d + \eta)$ high, dx long and 1 m wide is

$$dPE_1 = \gamma A\bar{x} = \gamma dx \, (d + \eta) \left(\frac{d + \eta}{2} \right) = \gamma \frac{(d + \eta)^2}{2} \, dx. \tag{3.56}$$

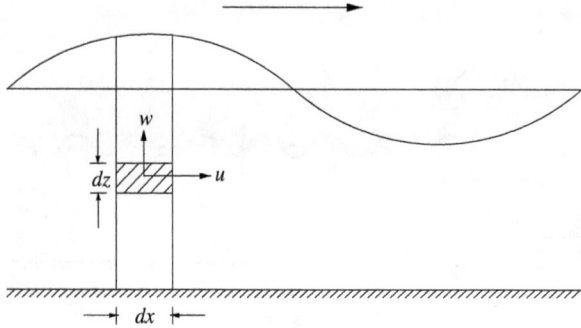

Fig. 3.11   Definition sketch for kinetic energy under a progressive wave.

The *average potential energy* per unit surface area (sometimes called the average potential energy density) is

$$\overline{PE_1} = \frac{\gamma}{2} \frac{1}{L} \frac{1}{T} \int_t^{t+T} \int_x^{x+L} (d + \eta)^2 \, dx \, dt.$$

The integration is from a time "t" over a wave period, T and from a certain distance x over a distance $x + L$:

$$\overline{PE_1} = \frac{\gamma}{2LT} \int_t^{t+T} \int_x^{x+L} (d + \eta)^2 \, dx \, dt. \tag{3.57}$$

Using $\eta = a \sin(kx - \sigma t)$, the above equation becomes

$$\overline{PE_1} = \frac{\gamma}{2LT} \int_t^{t+T} \int_x^{x+L} [(d^2 + 2ad \sin(kx - \sigma t)$$
$$+ a^2 \sin^2(kx - \sigma t)] dx \, dt.$$

On simplification,

$$\overline{PE_1} = \frac{\gamma d^2}{2} + \frac{\gamma a^2}{4}, \tag{3.58}$$

which is the average potential energy per unit surface area of all the water above $z = -d$.

The potential energy in the absence of a wave would be

$$\overline{PE_2} = \frac{\gamma}{2LT} \int_t^{t+T} \int_x^{x+L} d^2 dx \, dt = \gamma d^2 / 2. \tag{3.59}$$

The average potential energy density, $\overline{PE}$ which is attributable to the presence of the progressive wave on the free surface, is

$$\overline{PE} = \overline{PE}_1 - \overline{PE}_2 = \text{Average Potential Energy}$$

$$= \frac{\gamma d^2}{2} + \frac{\gamma a^2}{4} - \frac{\gamma d^2}{2} = \frac{\gamma a^2}{4}. \tag{3.60}$$

3.9.2 Kinetic energy

The kinetic energy, $KE = \frac{1}{2}mv^2$, where "m" is the mass of the fluid and "v" is the resultant velocity. For a two-dimensional wave flow,

$$d(KE) = \frac{1}{2}(\sqrt{u^2 + w^2})^2 dM = \frac{1}{2}(u^2 + w^2)\rho dz\, dx.$$

The average KE per unit of surface area is then given by

$$\overline{KE} = \frac{\rho}{2LT} \int_t^{t+T} \int_x^{x+L} \int_{-d}^{\eta} (u^2 + w^2) dz\, dx\, dt$$

with $\eta = a\sin(kx - \sigma t)$ and substituting for u and w, it can be shown that

$$\overline{KE} = \frac{\gamma a^2}{4}. \tag{3.61}$$

Hence, the total energy is given by

$$E = \overline{PE} + \overline{KE} = \frac{\gamma a^2}{2}. \tag{3.62}$$

The average total energy per unit surface area is the sum of the average potential and kinetic energy densities, often called as *specific energy* or *energy density*.

3.10 Wave Power

The wave energy flux is the rate at which energy is transmitted in the direction of wave propagation across a vertical plane perpendicular to the direction of the wave advance and extending down the entire depth. The average energy flux per unit wave crest width transmitted across a plane perpendicular to wave advance is given by

$$\overline{P} = \text{Wave Power} = \text{Average energy flux per unit wave crest width,}$$

$$\overline{P} = \overline{E}nC = \overline{E}C_g, \tag{3.63}$$

where $n = \frac{1}{2}[1 + \frac{2kd}{\sinh 2kd}]$.

In *deep waters* $\frac{2kd}{\sinh 2kd} = 0$ and $C_G = \frac{1}{2}C_0$ and $n = \frac{1}{2}$

$$\bar{P}_0 = \frac{1}{2}\bar{E}_0 C_0. \tag{3.64}$$

In *shallow waters*

$$\bar{P} = \bar{E}C = \bar{E}C_g (\text{since } \sinh 2kd = 2kd). \tag{3.65}$$

Assume the wave propagates from deepwater towards the shore. The ocean bottom slope is gradual and there are no undulations and has parallel bottom slope contours. According to the conservation of energy, equating the power in the deep waters (Eq. (3.64)) to that in shallow waters (Eq. (3.65)), we get

$$\frac{\gamma H^2}{8} \cdot C_G = \frac{\gamma H_0^2}{8} \cdot \frac{C_0}{2}.$$

On substituting for C_G and simplifying, we obtain

$$\frac{H}{H_0} = \sqrt{\frac{C_0}{C}} \cdot \frac{1}{2n} = K_s. \tag{3.66}$$

The above equation is the ratio between wave height at any water depth in shallower waters and the deep water height. This relationship obtained without considering the irregular variation in the sea bottom contours is called as *shoaling coefficient*. The variations of K_s, n, C/C_0 and d/L as a function of d/L_0 for small amplitude waves are shown in Fig. 3.12.

The variations of all the parameters that are shown in Figs. 3.4 and 3.12 along with other parameters that are needed for solving problems are available in the form of wave tables of Shore Protection Manual, (SPM) (1984), which is provided in Appendix A. A MATLAB code to arrive at the above parameters is given below.

```
An example calculation of all parameters
clc;clear;
t=input('enter wave period (s)...');
d=input('enter water depth (m)...');
g=9.81;
L0= 1.56*t^2;
L0_old=L0;
  for i = 1:10000
  L_new =((g*t^2)/(2*pi))*(tanh((2*pi)*d/L0_old));
  err = abs(L_new-L0_old);
    if err < 1e-6
```

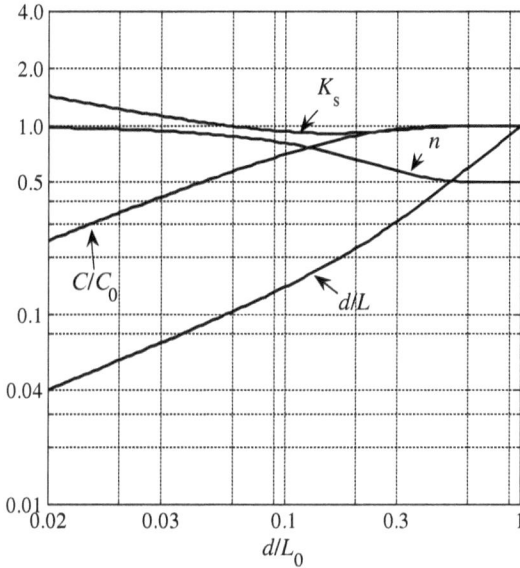

Fig. 3.12 Properties of small amplitude waves.

```
      L=L_new;
   break;
   else
   L0_old = L_new;
   end
  end
k=(2*pi)/L;
n=0.5*(1+((2*k*d)/(sinh(2*k*d))));
C0=L0/t;  C=L/t;Cg=n*C;
ks=sqrt(0.5*1/n*1/(C/C0));
K=1/(cosh(k*d));
fprintf('d/L0              = %f\n',d/L0);
fprintf('d/L               = %f\n',d/L);
fprintf('2(pi)d/L          = %f\n',k*d);
fprintf('tanh(2(pi)d/L)    = %f\n',tanh(k*d));
fprintf('sinh(2(pi)d/L)    = %f\n',sinh(k*d));
fprintf('cosh(2(pi)d/L)    = %f\n',cosh(k*d));
fprintf('H/H0_dash         = %f\n',ks);
fprintf('K                 = %f\n',K);
```

```
fprintf('4(pi)d/L)          = %f\n',2*(k*d));
fprintf('sinh(4(pi)d/L)     = %f\n',sinh(2*k*d));
fprintf('cosh(4(pi)d/L)     = %f\n',cosh(2*k*d));
fprintf('n                  = %f\n',n);
fprintf('Cg/C0              = %f\n',Cg/C0);
```

3.11   Mass Transport Velocity

It is a usual sight to see a ball thrown into the ocean, particularly within the surf zone it gets back. If it is thrown offshore of the breaker zone, it will be undergoing oscillatory motion and being gradually pushed towards the shore, which in fact would be at a faster rate when the sea is rough characterized by steep waves. This is in fact contrary to waves being stated as undergoing oscillatory motion. It is to be mentioned that when the waves are in motion, the particles upon completion of nearly an elliptical motion would have advanced a short distance in the direction of propagation (Fig. 3.13), which is true also in deep waters. That is, the predominant direction of wave propagation dictates the direction of mass transport. The mass transport velocity at any depth z below SWL is given by

$$\bar{u}(z) = \left(\frac{\pi H}{L}\right)^2 \frac{C}{2} \cdot \frac{\cosh 2k(d+z)}{\sinh^2 kd}. \tag{3.67}$$

The mass transport speed is appreciable for high steep waves and is very small for waves of long period.

3.12   Constancy of Wave Period

When propagate from deep to shallow waters, its period will remain constant while its length decreases with reduction in the water depth (Fig. 3.14). This could be explained as follows.

Let a wave with period, T_1 enter a domain in a different water depth with its period as T_2. After an arbitrary time, t, $n_1 = dt/T_1$, $n_2 = dt/T_2$

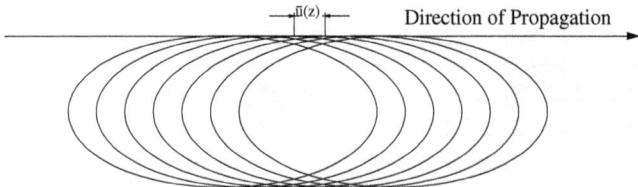

Fig. 3.13   Definition of mass transport velocity.

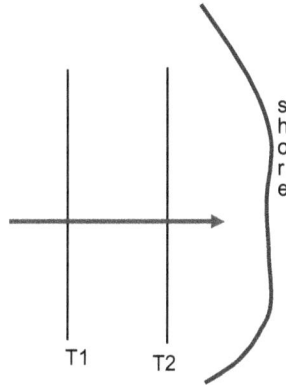

Fig. 3.14 Definition for constancy of wave period.

and the number of waves accumulated is $n_2 - n_1 = dt(1/T_2 - 1/T_1)$. At dt tends to infinity, $T_1 > T_2$ or $T_2 > T_1$ is unrealistic

Hence, $T_1 = T_2 = T$.

Worked Out Examples

Note: Wave table needed for the worked-out examples is provided in Appendix A

Problem 3.1

A wave flume is filled with fresh water to a depth of 1.5 m. A wave of height 0.2 m and period 3.0 s is generated. Calculate the wave celerity, group celerity, energy, and power.

Solution

$$\frac{d}{L_0} = \frac{d}{L} \tanh kd,$$

$$L_0 = 1.56T^2 = 1.56 * 3.0^2 = 14.04 \text{ m},$$

$$\frac{d}{L_0} = \frac{1.5}{14.04} = 0.107.$$

From wave tables, $\frac{d}{L} = 0.147$ corresponding to $\frac{d}{L_0} = 0.107$, $L = 10.2$ m.

$$\text{Celerity} = \frac{L}{T} = \frac{10.2}{3.0} = 3.4 \text{ m/s}.$$

$$\text{Group Celerity} = \frac{C}{2} \left[1 + \frac{2kd}{\sinh 2kd} \right], \text{ where } kd = \frac{2\pi d}{L}$$

$$= \frac{3.4}{2} \left[1 + \frac{2*0.924}{\sinh(2*0.924)} \right]$$

$$= 2.72 \text{ m/s.}$$

$$\text{Energy} = \frac{\gamma H^2}{8} = \frac{9810 \times 0.2^2}{8} = 49.05 \text{ N/m.}$$

$$\text{Power} = \frac{\gamma H^2}{8} C_G = 49.05 \times 2.72 = 133.42 \text{ N/s.}$$

Problem 3.2

Oscillatory surface waves were observed in deep water and the wave period was found to be 12 s.

(i) At what bottom depth would the phase velocity begin to change with decrease in water depth?
(ii) What is the celerity at a bottom depth of 20 m and 5 m?
(iii) Compute the ratio of celerity at the above water depth to deep water celerity?

Solution

(i) $L_0 = 1.56T^2 = 1.56 \times 12^2 = 224.64$ m.
     The celerity will change when water depth is less than

$$\frac{L_0}{2} = \frac{224.64}{2} = 112.32 \text{ m.}$$

(ii) At $d = 20$ m

$$\frac{d}{L_0} = \frac{20}{224.64} = 0.089.$$

From wave tables, $\frac{d}{L} = 0.1313$ corresponding to $\frac{d}{L_0} = 0.089$,

$$L = 152.32 \text{ m.}$$

Celerity $= \frac{L}{T} = \frac{152.32}{12} = 12.69$ m/s.
     At $d = 5$ m

$$\frac{d}{L_0} = \frac{5}{224.64} = 0.022.$$

From wave tables, $\frac{d}{L} = 0.06057$ corresponding to $\frac{d}{L_0} = 0.022$

$$L = 82.55 \text{ m.}$$

Celerity $= \frac{L}{T} = \frac{82.55}{12} = 6.88$ m/s.

(iii) Ratio of Celerity

$$C_0 = \frac{L_0}{T} = 1.56 \times T = 1.56 \times 12 = 18.72 \text{ m/s}.$$

$$\text{At } d = 20 \text{ m}, \quad \frac{C}{C_0} = \frac{12.69}{18.72} = 0.678.$$

$$\text{At } d = 5 \text{ m}, \quad \frac{C}{C_0} = \frac{6.88}{18.72} = 0.368.$$

Problem 3.3

Calculate the wavelength using the different approaches for wave with period 10 s propagating from 100 m to 10 m. Calculations may be carried out with 10 m interval.

Solution

Water depth (m)	d/L	d/L_0	Linear wave theory	Wavelength, L (m)			
				Fenton & McKee (1990)	Bob You (2008)	Hunt (1979)	Guo (2002)
10	0.1675288	0.0641025	92.37387	91.1839	92.2845	92.297	91.7905
20	0.1920156	0.1282051	121.2369	120.7184	121.1239	121.1388	121.0339
30	0.2297933	0.1923076	137.2949	137.4261	137.1757	137.1965	137.3736
40	0.2774721	0.2564102	146.3735	146.6361	146.2436	146.2819	146.4665
50	0.3318614	0.3205128	151.2983	151.4424	151.16	151.2025	151.3031
60	0.3904602	0.3846153	153.8264	153.8376	153.6874	153.718	153.7566
70	0.4515427	0.4487179	155.0612	154.9921	154.9253	154.9416	154.953
80	0.5140218	0.5128205	155.6427	155.536	155.5096	155.5166	155.5192
90	0.5772583	0.5769230	155.9103	155.7882	155.7786	155.7812	155.7815
100	0.6408945	0.6410256	156.0318	155.9039	155.9006	155.9015	155.9014

It is claimed that the error between linear wave theory and Fenton & McKee method is with in 1.5% accuracy, whereas with that of Bob You is within 0.1% particularly in shallow waters.

Problem 3.4

Consider a particle initially 10 m below the SWL and 24 m above the seabed. After the wave motion is established, what is the size and character of the orbit of the particle? Repeat the calculation for a particle at the surface and at the seabed. The wave period $T = 12$ s and deep-water wave height $H_0 = 4$ m.

Solution

Water depth, $d = 10 + 24 = 34$ m.

$$L_0 = 1.56T^2 = 1.56 \times 12^2 = 224.64 \text{ m},$$

$$\frac{d}{L_0} = \frac{34}{224.64} = 0.1514.$$

The wave is in intermediate water depth condition $0.05 \leq \frac{d}{L_0} \leq 0.5$. Therefore, the particles will move in elliptical orbit. From wave tables,

$$\frac{d}{L} = 0.1841 \text{ and } \frac{H}{H_0} = 0.9133 \text{ for } \frac{d}{L_0} = 0.1514,$$

$$\therefore L = \frac{34}{0.1841} = 184.68 \text{ m, and}$$

$$H = 0.9133 * 4 = 3.65 \text{ m}.$$

For the size and character of the orbit of the particle, compute

$$k = \frac{2\pi}{L} = \frac{2\pi}{184.68} = 0.034, \text{ and } kd = 1.156.$$

At $z = -10$ m

$$D = \frac{H}{2} \frac{\cosh k(d+z)}{\sinh kd} = 1.724 \text{ m}, \ B = \frac{H}{2} \frac{\sinh k(d+z)}{\sinh kd} = 1.16 \text{ m}.$$

Similarly, at surface $z = 0$ m,

$$D = 2.23 \text{ m and } B = 1.825 \text{ m},$$

and at seabed $z = -34$ m,

$$D = 1.275 \text{ m and } B = 0.$$

Problem 3.5

A wave of height $H = 3$ m and wave period $T = 10$ s, propagates in a water depth of 12 m, the corresponding deep-water wave height $H_0 = 3.5$ m. Estimate,

(a) The horizontal and vertical displacements from its mean position at $z = 0$ and $z = -d$.

(b) The maximum water particle displacements at a depth of 7.50 m below SWL, where the wave is in deep water.

(c) For the deep-water conditions of above show that the water particle displacements are small, relative to the wave height at $z = -L_0/2$.

(d) Also, compare the water particle displacements in deep water conditions for the corresponding deep-water wave height, $H_0 = 3.5$ m and wave period $T = 15$ s at $z = L_0/2$ and $z = -7.5$ m.

Solution

Given Data: $H = 3$ m, $T = 10$ s and $H_0 = 3.5$ m

(a) The maximum horizontal (D) and vertical (B) displacements of water particle.

$$L_0 = 1.56\,T^2 = 1.56(10)^2 = 156 \text{ m}, \ d/L_0 = 12/156 = 0.07692.$$

From wave tables, $d/L = 0.1205$ and hence $L = 99.58$ m, $k = 2\pi/L = 0.0631$
at $z = 0$

$$D = \frac{H}{2}\frac{\cosh k(d+z)}{\sinh kd}, B = \frac{H}{2}\frac{\sinh k(d+z)}{\sinh kd}.$$

substituting for the parameters, $D = 2.34$ m and $B = 1.5$ m
at $z = -12$ m

$$D = \frac{H}{2}\frac{1}{\sinh kd} = \frac{3}{2}\left(\frac{1}{0.8316}\right) = 1.80 \text{ m and } B = 0.$$

(b) In deep water conditions, at $z = -7.5$ m

$$k_0 = \frac{2\pi}{L_0} = \frac{2\pi}{156} = 0.0402 \quad D = B = \frac{H_0}{2}e^{k_0 z}$$

$$= \frac{3.5}{2}e^{(0.0402\times(-7.5))} = 1.2938 \text{ m}$$

(c) In deep water conditions at $z = -L_0/2$

$z = -156/2 = -78$ m

As $H_0 = 3.5$ m, $D = B = \frac{H_0}{2}e^{k_0 z} = \frac{3.5}{2}e^{[0.0402\times(-78)]} = 0.07562$ m.

It is inferred that for deep-water conditions, the particle displacements are very small at $z = -7.5$ m compared to at $z = L_0/2$ by a factor of $1.2938/0.07562 = 17.1$. Thus the water particle displacements at $z = -7.5$ m is 17 times greater than the water particle displacements at $z = -78$ m.

(d) $T = 15$ s, $H_0 = 3.5$ m.

$$L_0 = 1.56\, T^2 = 1.56(15)^2 = 351 \text{ m}, \quad k_0 = 2\pi/L_0 = 0.0179$$

At $z = -L_0/2 = -351/2 = -175.5$ m

$$D = B = \frac{H_0}{2} e^{k_0 z} = \frac{3.5}{2} e^{(0.0179 \times (-175.5))} = 0.07563 \text{ m.}$$

at $z = -7.5$ m

$$D = B = \frac{H_0}{2} e^{k_0 z} = \frac{3.5}{2} e^{(0.0179 \times (-7.5))} = 1.53 \text{ m.}$$

displacement at $z = -7.5$ m / displacement at $z = 175.5$ m $= 1.53/0.07563 = 20.2$.

The problem proves that the displacements are larger for long period waves.

Problem 3.6

In order to understand the effect of tsunami on the seabed, determine the water particle displacement of a wave of height 1 m propagating in a water depth of 4500 m. Carry out the evaluation of the displacements at $z = -L_0/2$ for the wave with periods (a) 12 s, (b) 30 s, (c) $T = 7300$ s (tsunami) at $z = -L_0/2$ for the wave, $T = 30$ s, and (d) for the tsunami with the period $= 7300$ s.

Solution

(a) For $H_0 = 1$ m, $T = 12$ s and $d = 4500$ m

$$L_0 = 1.56\, T^2 = 1.56(12)^2 = 224.64 \text{ m.}$$

For deep water conditions at $z = -L_0/2 = -112.32$ m, and $k_0 = 2\pi/L_0 = 0.028$,

$$D = B = \frac{H_0}{2} e^{k_0 z} = \frac{1}{2} e^{[0.028 \times (-112.32)]} = 0.0215 \text{ m.}$$

(b) For $H_0 = 1$ m, $T = 30$ s and $d = 4500$ m,

$$L_0 = 1.56\, T^2 = 1.56(30)^2 = 1404 \text{ m.}$$

For deep water conditions at $z = -L_0/2 = -702$ m, $k_0 = 2\pi/L_0 = 0.00447$,

$$D = B = \frac{H_0}{2} e^{k_0 z} = \frac{1}{2} e^{[0.0044702 \times (-702)]} = 0.02168 \text{ m.}$$

(c) For $H_0 = 1$ m, $T = 7300$ s (2 h, period of a tsunami) and $d = 4500$ m, $z = -702$ m ($-L_0/2$ for the wave with the probable longest period of 30 s),

$$L_0 = 1.56\, T^2 = 1.56(7300)^2 = 83132400 \text{ m},$$

$d/L_0 = 0.00005413$, for which the corresponding $d/L = 0.00294$ (from wave tables), $L = 1533051.45$ m,

$$k = 2\pi/L = 2\pi/1533051.45 = 0.000004098,$$

$$D = \frac{H}{2}\frac{\cosh k(d+z)}{\sinh kd} = \frac{1}{2}\left(\frac{1}{0.0184}\right) = 27.17 \text{ m},$$

$$B = \frac{H}{2}\frac{\sinh k(d+z)}{\sinh kd} = \frac{1}{2}\left(\frac{0.0156}{0.0184}\right) = 0.424 \text{ m}.$$

From the above problem, it can be inferred that when a wave of $T = 7300$ s, i.e., a tsunami propagates in deep water of 4500 m, the horizontal and vertical displacement determined at $z = -702$ m would be 27.17 m and 0.424 m, respectively. The above equations were found to be negligible even for the longer wave of $T = 30$ s as seen in (b).

(d) For $H = 1$ m, $T = 7300$ s and $d = 4500$ mm, $L_0 = 1.56\, T^2 = 1.56(7300)^2 = 83132400$ m. In deep water conditions at $z = \frac{-L_0}{2}$, $k_0 = 2\pi/L_0 = 0.0000000756$; $z = -L_0/2 = -41566200$ m. It is seen that $z = -L_0/2$ is greater than the available water depth of 4500 m. Hence, at seabed $z = -4500$ m, the displacements are evaluated.

$d/L_0 = 0.00005413$, $d/L = 0.00294$ and $L = 1533051.45$ m, from these parameters,

$$D = \frac{H}{2}\frac{1}{\sinh kd} = \frac{1}{2}\left(\frac{1}{0.0184}\right) = 27.17 \text{ m}; \quad B = 0.$$

It is seen from this section of the problem that at an elevation of 702 m below SWL as well as at the seabed, the horizontal water displacement is of the same order of 27.17 m, which reveal that a tsunami will behave as a shallow water wave in deep waters.

Problem 3.7

For a wave of height 1.0 m and period 8.5 s. Estimate the maximum velocities and accelerations of a fluid particle at a position 2.5 m below SWL and 11 m above the seabed, at SWL and at the seabed.

Solution

$d = 2.5 + 11 = 13.5$ m, $\frac{d}{L_0} = \frac{13.5}{1.56*8.5^2} = 0.1197$. From wave table, $\frac{d}{L} = 0.1579$.

From which $L = 85.49$ m.

At $z = 0$ m,

$$\cosh k(d+z) = 1.5341, \quad \sinh k(d+z) = 1.1635, \quad \text{and} \quad \sinh kd = 1.1635.$$

Substituting the values, we get

$$u_{max} = \frac{\pi H}{T} \frac{\cosh k(d+z)}{\sinh kd} = 0.487 \text{ m/s},$$

$$w_{max} = -\frac{\pi H}{T} \frac{\sinh k(d+z)}{\sinh kd} = -0.37 \text{ m/s},$$

$$\dot{u}_{max} = -\frac{2\pi^2 H}{T^2} \frac{\cosh k(d+z)}{\sinh kd} = -0.36 \text{ m/s}^2,$$

$$\dot{w}_{max} = -\frac{2\pi^2 H}{T^2} \frac{\sinh k(d+z)}{\sinh kd} = -0.273 \text{ m/s}^2.$$

At $z = -2.5$ m

$$\cosh k(d+z) = 1.345, \quad \sinh k(d+z) = 0.899, \quad \text{and} \quad \sinh kd = 1.1635.$$

Substituting the values, we get

$$u_{max} = \frac{\pi H}{T} \frac{\cosh k(d+z)}{\sinh kd} = 0.4273 \text{ m/s},$$

$$w_{max} = -\frac{\pi H}{T} \frac{\sinh k(d+z)}{\sinh kd} = -0.2858 \text{ m/s},$$

$$\dot{u}_{max} = -\frac{2\pi^2 H}{T^2} \frac{\cosh k(d+z)}{\sinh kd} = -0.3159 \text{ m/s}^2,$$

$$\dot{w}_{max} = -\frac{2\pi^2 H}{T^2} \frac{\sinh k(d+z)}{\sinh kd} = -0.2112 \text{ m/s}^2.$$

At $z = -13.5$ m

$$\cosh k(d+z) = 1, \quad \sinh k(d+z) = 0, \quad \text{and} \quad \sinh kd = 1.1635.$$

Substituting the values, we get

$$u_{max} = \frac{\pi H}{T} \frac{\cosh k(d+z)}{\sinh kd} = 0.3177 \text{ m/s},$$

$$w_{max} = -\frac{\pi H}{T} \frac{\sinh k(d+z)}{\sinh kd} = 0 \text{ m/s},$$

$$\dot{u}_{max} = -\frac{2\pi^2 H}{T^2} \frac{\cosh k(d+z)}{\sinh kd} = -0.2348 \text{ m/s}^2,$$

$$\dot{w}_{max} = -\frac{2\pi^2 H}{T^2} \frac{\sinh k(d+z)}{\sinh kd} = 0 \text{ m/s}^2.$$

Problem 3.8

For a wave of height 2 m and period 7 s, plot the variation of orbital velocity and acceleration in the vertical and horizontal directions of a particle at a position 4 m below SWL and 20 m above the seabed.

Solution

$d = 20 + 4 = 24$ m; $\dfrac{d}{L_0} = 0.314$; corresponding $\dfrac{d}{L} = 0.325$; $L = 73.96$ m.

At $z = -4.0$ m,

$$kd = 2.039, \ \cosh k(d+z) = 2.826, \ \cosh(kd) = 3.907,$$

$$\sinh k(d+z) = 2.643 \text{ and } \sinh(kd) = 3.776.$$

Substituting the values in the expressions for u and w (Eqs. (3.38) and (3.40)), the phase variations are projected in Fig. 3.15.

Problem 3.9

A wave with a height 4.5 m and wavelength 75 m propagates in a water depth of 20 m. Determine the local horizontal and vertical velocities at a

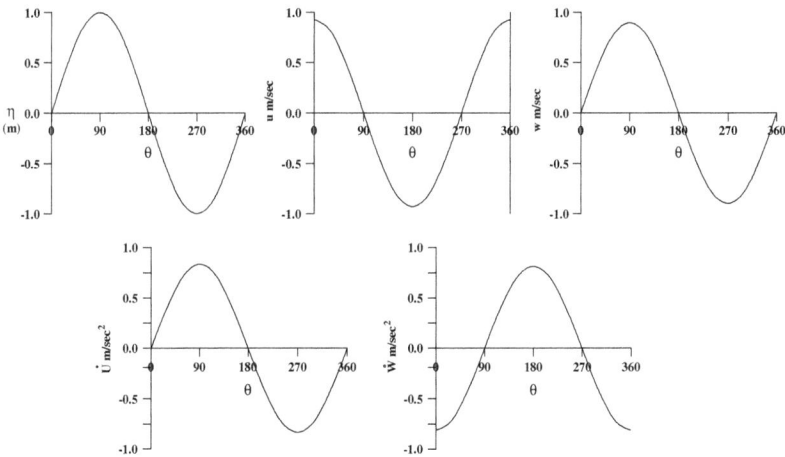

Fig. 3.15   The phase variation of u, w, $\overset{\circ}{u}$ and $\overset{\circ}{w}$.

depth of 6 m below the still water at a position one-sixth ahead of a wave crest.

Solution

$$L = 75 \text{ m}, \quad \frac{1}{6}L = 12.5 \text{ m},$$

$$\theta = \frac{12.5}{75} \times 360° + 90° = 150°.$$

At $z = -6$ m,

$$u = \frac{\pi H}{T} \frac{\cosh k(d+z) \sin \theta}{\sinh kd} = 0.782 \text{ m/s},$$

$$w = \frac{\pi H}{T} \frac{\sinh k(d+z) \cos \theta}{\sinh kd} = -0.907 \text{ m/s}.$$

Problem 3.10

Aerial photographs of a coastal line displayed the presence of two wave systems. One with crest 150 m apart and another with crest at 30 m spacing. Timing of major breaking on the beach in the same period indicated the wave period to be 10 s for the longer wave. What was the depth of water in the zone of wave observation and what was the period of the minor wave system?

Solution

$$L_{\text{major}} = 150 \text{ m}, \quad L_{\text{minor}} = 30 \text{ m},$$

$$T_{\text{major}} = 10 \text{ s},$$

$$L = \frac{g}{2\pi} T^2 \tanh kd,$$

$$150 = \frac{9.81}{2\pi} \times 10^2 \tanh kd.$$

We get, $\tanh kd = 0.96$.

Corresponding to this value, from the wave tables, $d/L = 0.3103$. Since $L_{\text{major}} = 150$ m,

$$d = 0.3103 * 150 = 46.545 \text{ m},$$

$$L_{\text{minor}} = \frac{g}{2\pi} T^2_{\text{minor}} \tanh \left(\frac{2\pi}{L} d \right),$$

$$30 = \frac{9.81}{2\pi} T^2_{\text{minor}} \tanh \left(\frac{2\pi}{30} * 46.545 \right).$$

We get $T^2_{minor} = 19.21$ s,

$T_{minor} = 4.38$ s.

Problem 3.11

If a pressure sensing instrument is set up at 3 m below SWL in a water depth of 15 m, determine the phase distribution of pressure head this instrument would record and the maximum dynamic pressure. The wave height is 2 m and period is 12 s. and $\gamma = 10$ kN/m^3. Also determine the above for a wave with period 4 s.

Solution

The pressure head under a progressive wave is

$$\frac{p}{\gamma} = \left(\frac{H}{2}\sin\theta\right)k_p - z.$$

For $T = 12$ s, $d = 15$ m,

L is calculated as 135.01 m, $k = 0.0465$, $kd = \frac{2\pi}{L}d = 0.6975$.

$$K_p = \frac{\cosh k(d+z)}{\cosh kd} = 0.925,$$

Dynamic pressure head, $\frac{p}{\gamma} = \frac{2}{2} \times 0.925 \times \sin\theta = 0.925$,

Maximum Dynamic Pressure, $p = 0.925 \times 10006.2 = 9255.435$ N/m^2,

Total Pressure, $p = \left[\left(\frac{2}{2}\sin\theta\right)0.925 + 3\right]10006.2 = 39274.34$ N/m^2.

For $T = 4$ s, $d = 15$ m, $d/L = 0.6$,

$L = L_0 = 25$ m, $k = \frac{2\pi}{L} = 0.2512$, $kd = 3.768$,

$$K_p = \frac{\cosh k(d+z)}{\cosh kd} = 0.472,$$

Maximum dynamic pressure, $p = \left(\frac{2}{2} \times 0.472 \times \sin 90°\right)10006.2 = 4722.9$ N/m^2.

Note: For a wave period of 4 s, the total pressure is less when compared to the wave period of 12 s for the other conditions not being changed. So, for longer waves the pressure is more.

Problem 3.12

A subsurface pressure type recorder is installed at a depth of 4 m at the point where water depth is 12 m. The average pressure and the period registered by the recorder are 3 bar and 9 s respectively. Compute η, $\gamma = 10006.2$.

Solution

$L_0 = 1.56\, T^2 = 1.56 \times 9^2 = 126.36,$

$z = 4$ m, $d = 12$ m, $P = 3$ bar,

$d/L_0 = 12/126.36 = 0.095,$

$\dfrac{d}{L_0} = \dfrac{12}{126.36} = 0.095,\ d/L = 0.1366,\ L = 87.85,$

$k = \dfrac{2\pi}{L} = 0.0715,\ k = 0.0715,\ K_p = \dfrac{\cosh k(d+z)}{\cosh kd} = 1.2077,$

$\dfrac{p}{\gamma} = \eta K_p - z,$

$\dfrac{3 \times 10^5}{10006.2} = \eta 1.177 - (-4),\ 1\ \text{bar} = 10^5 \text{N/m}^2,$

$\eta = 21.51$ m.

Note: Problem incorrect due to wave breaking condition. The pressure recorded indicated is too high and unrealistic.

Problem 3.13

A maximum pressure of 1 bar is measured by a subsurface pressure recorder located at 1.0 m above, the seabed in a water depth of 12 m. The average wave frequency is 0.10 cycles/s and $\gamma = 10006.2$ N/m^3. Determine the wave height.

Solution

$$\frac{p}{\gamma} = \eta k_p - z.$$

The maximum pressure would occur at $\eta = H/2$ (Crest of wave).

$$p_{\max} = 100000 \ \text{N/m}^2,\ \gamma = 10006.2 \ \text{N/m}^3,$$

$$z = -(12 - 1.0) = -11.0$$

$$T = 1/0.10 = 10 \text{ s}, \ L = 99.59 \text{ m},$$

$$K_p = \frac{\cosh k(d+z)}{\cosh kd} = 0.77.$$

Substituting for the variables in the above formula

$$\frac{100000}{10006.2} = 0.77 \times \frac{H}{2} - (-11.0).$$

We get $H = 2.61$ m.

Problem 3.14

Water depth, $d = 12$ m, wave period, $T = 10$ s and wave height, $H = 1.0$ m.

(a) Calculate mass transport velocity, $\bar{u}(z)$ for $z = 0$ to $-d$ and find the variation $\bar{u}(z)$ with z/d.
(b) $d = 12$ m, $T = 10$ s and $H = 0.5$ to 9 m at 0.5 m interval. Find the variation $\bar{u}(z)$ with H/L at $z = 0.0$ m.
(c) $d = 12$ m, $H = 2.0$ m and $T = 5$ to 15 s at every 1 s interval. Find the variation $\bar{u}(z)$ with d/L at $z = 0.0$ m. Also plot the variation of $u_{max}/\bar{u}_{max}$.

Solution

Mass transport velocity,

$$\overline{u(z)} = \left(\frac{\pi H}{L}\right)^2 \frac{C}{2} \frac{\cosh 2k(d+z)}{\sinh^2 kd}.$$

Orbital velocity,

$$u_{max} = \frac{\pi H}{T} \cdot \frac{\cosh k(d+z)}{\sinh kd}.$$

(a) Effect of z/d on $\bar{u}(z)$:

z	0	−1	−2	−3	−4	−5	−6	−7	−8	−9	−10	−11	−12
z/d	0	0.083	0.167	0.25	0.333	0.417	0.5	0.583	0.667	0.75	0.833	0.917	1
$\bar{u}(z)$	0.017	0.015	0.014	0.012	0.011	0.01	0.009	0.009	0.008	0.008	0.007	0.007	0.007

It is observed that the mass transport velocity decreases from the free surface to the sea floor. The variation is shown in Fig. 3.16.

$d = 12$ m; $t = 10$ s; $H = 1.0$ m; $z = 0$ to d at 1 m interval

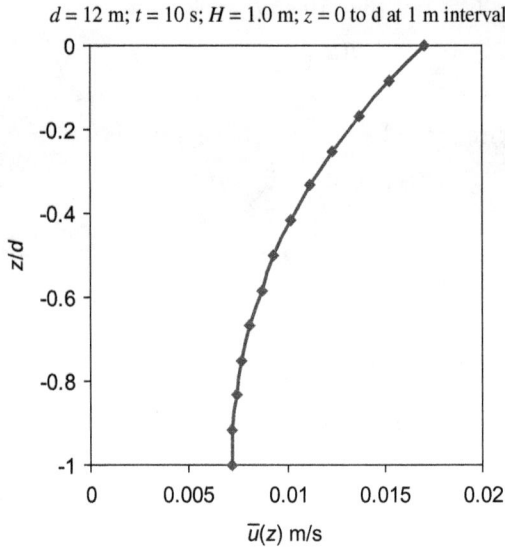

Fig. 3.16   Variation of $\bar{u}(z)$ along the depth.

d=12m; t=10sec;z=0; H =0.5 to 9m @0.5m interval

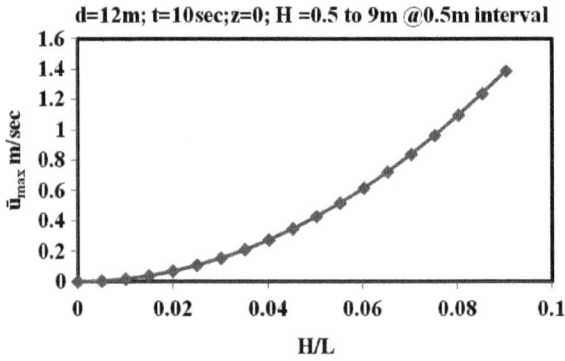

Fig. 3.17   Variation of $\bar{u}_{max}$ with H/L.

(b) Effect of H/L on $\bar{u}_{max}$ at $z = 0.0$ (see Fig. 3.17):

H	0.5	1	1.5	2	2.5	3	3.5	4	4.5	5	5.5	6	6.5	7	7.5	8	8.5	9
H/L	0.01	0.01	0.02	0.02	0.03	0.03	0.04	0.04	0.05	0.05	0.06	0.06	0.07	0.07	0.08	0.08	0.09	0.09
$\bar{u}(z)$	0	0.02	0.04	0.07	0.11	0.15	0.21	0.27	0.35	0.43	0.52	0.62	0.72	0.84	0.96	1.09	1.24	1.39

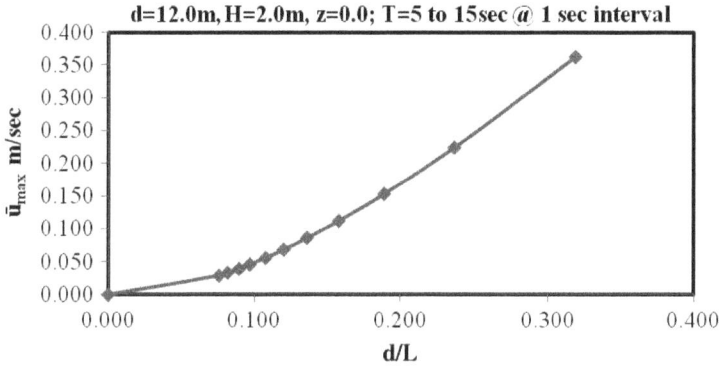

Fig. 3.18   Variation of $\bar{u}_{\max}$ for various values of d/L.

From the plot, it is seen $\bar{u}(z)$ increases with an increase in H/L values. This means that, as the wave steepness increases, the mass transport velocity also increases.

(c) Effect of d/L on $\bar{u}(z)$ (Fig. 3.18):

T	L_0	d/L_0	d/L	L	C	$\bar{u}_{\max}$	$u_{\max}$	$u_{\max}/\bar{u}_{\max}$
5	39	0.308	0.319	37.6412	7.528	0.362	1.30267	3.59804
6	56.16	0.214	0.237	50.6971	8.450	0.224	1.15952	5.17618
7	76.44	0.157	0.189	63.4585	9.065	0.153	1.08113	7.04799
8	99.84	0.120	0.158	75.9013	9.488	0.112	1.03482	9.22155
9	126.36	0.095	0.137	87.8477	9.761	0.086	1.00378	11.64692
10	156	0.077	0.121	99.5851	9.959	0.068	0.98255	14.35976
11	188.76	0.064	0.108	110.9057	10.082	0.056	0.96569	17.28953
12	224.64	0.053	0.097	123.3806	10.282	0.046	0.96080	20.87657
13	263.64	0.046	0.090	133.4668	10.267	0.039	0.94460	24.05266
14	305.76	0.039	0.082	146.0743	10.434	0.033	0.94529	28.37048
15	351	0.034	0.076	157.2739	10.485	0.029	0.93961	32.53078

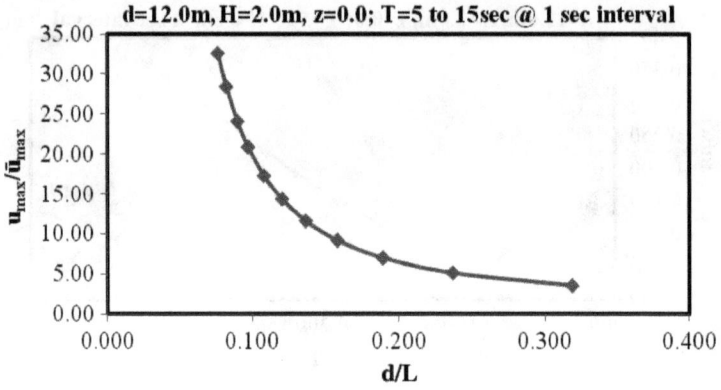

Fig. 3.19   Variation of $u_{max}/\bar{u}_{max}$ for various values of d/L.

The plot shows that with an increase in d/L values $\bar{u}_{max}$ also increases. If wave period increases, then the mass transport decrease (Fig. 3.19).

Chapter 4

Deformation of Waves

4.1　General

Ocean waves traveling in the offshore with no net gain or loss of energy will travel in straight lines with same height. However, while they propagate from offshore towards the coast will undergo several deformations due to the variations in the bathymetry, i.e., water depth, nature of the seabed and presence of structures. The wave deformation may occur due to

  (i) lateral diffraction of wave energy,
 (ii) by the process of attenuation,
(iii) air resistance encountered by the waves or by directly opposing winds
(iv) by the tendency of the waves to overrun the currents.

These deformations are broadly classified as (1) shoaling, (2) refraction, (3) diffraction, (4) reflection and (5) breaking. This chapter deals with the above aspects together with the changes taking place in propagation of waves when it encounters currents.

4.2　Different Zones based on Behavior of Ocean Waves

The different zones as per the classification on the behavior of waves provided in Fig. 4.1 are discussed below.

Deep water zone: In this zone, the water depth is large compared to the wavelength and the waves do not feel the presence of the seabed. Wave crests are straight and the velocity of wave as well as the angle of incidence relative to the shore is constant.

Refraction zone: Waves feel the presence of the seabed and the wave fronts are parallel to each other and the speed of the wave along its crest varies resulting in bending and aligning itself parallel to the bottom depth contours.

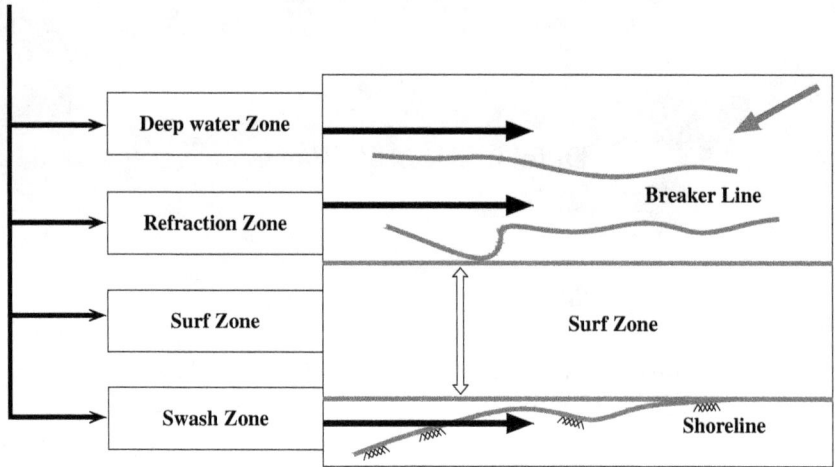

Fig. 4.1   Different zones based on behavior of ocean waves.

Surf zone: The zone between the breaker point and the shore, where the longshore sediment transport is active. When a wave propagates closer to the coast, it loses its stability and breaks during which the sediments from the seabed that are brought to the surface undergo the process of mixing and later set in motion.

Swash zone: It is defined by the highest point on the beach that the breaking waves run up on shore and the lowest point to which the water recedes between waves.

4.3   Shoaling

When waves propagate over a sloping seabed its length gets decreases. The power transmission in a wave is proportional to $C_g H^2$. In case of pure shoaling, there is no transfer of power in the lateral direction, i.e., normal to the wave direction. Now, let us examine Eq. (3.65). If the energy flux "$\overline{P}$" is a constant and C_g decreases, the energy density of the wave is bound to increase. This results in gradual increase in the wave height, as the waves propagate towards the shallow waters. On assuming no energy losses as stated above, the relationship of wave height in shallow water to that in deep waters, as derived in Eq. (3.66) is reproduced here. The variation K_s

with d/L_0 is shown in Fig. 3.12.

$$\frac{H}{H_0} = \sqrt{\frac{C_0}{C}\frac{1}{2n}} = K_s.$$ (4.1)

4.4 Wave Refraction

The waves in the offshore usually has a well-defined direction and their crests will be parallel to each other as their celerity is a function of the wave period only and not a function of depth as stated earlier. However, when they enter the transitional or the intermediate water depths at an angle, they feel the presence of the seabed and it usual that a part of the crest is in deep water while the remaining part of the same crest is in transitional water depth, wherein the speed of the wave is as function of both period and water depth. This creates a situation that the part of the wave in deeper water move faster compared to that in shallower water leading to bending of the wave crests and tending to become parallel to the water depth contours. This phenomena of bending of the wave crests is termed as refraction of waves. Wave celerity and wavelength are related to wave period (which is the only parameter which remains constant). This can be understood by postulating a change in wave period (from T_1 to T_2) over an area of sea. The number of waves entering the area in a fixed time t would be t/T_1, and the number leaving would be t/T_2. Unless T_1 equals T_2, the number of waves within the region could increase or decrease indefinitely. Thus,

$$C/C_0 = \tanh(kd) \quad \text{and} \quad C/C_0 = L/L_0.$$ (4.2)

To find the wave celerity and wavelength at any depth, d the two expressions in Eq. (4.2) must be solved simultaneously. The solution is always such that $C < C_0$ and $L < L_0$ for $d < d_0$ (where the subscript "0" refers to deep water conditions).

Consider a deep water wave approaching the transitional depth limit $(d/L_0 = 0.5)$, as shown in Fig. 4.2(a). A wave traveling from A to B (in deep water) traverses a distance L_0 in one wave period T. However, the wave traveling from C to D travels a smaller distance, L, over the same time, as it is in the transitional depth region. Hence, the new wave crest is now BD, which has rotated with respect to AC. If the angle α represent the angle of the wave crest to the depth contour, then

$$\sin \alpha = L/BC \quad \text{and} \quad \sin \alpha_0 = L_0/BC$$

and combining these two equations, we have

$$\frac{\sin \alpha}{\sin \alpha_0} = \frac{L}{L_0} \tag{4.3}$$

$$\frac{\sin \alpha}{\sin \alpha_0} = \frac{L}{L_0} = C/C_0 = \tanh kd. \tag{4.4}$$

The bending effect of the wave called refraction depends on the relation of water depth to wavelength. This is the *Snell's law*.

Waves approaching the shore at right angles for which $\alpha = 0$ are slowed down but not refracted.

Waves approaching at an oblique angle are refracted in such a way that the angle α is decreased, corresponding to the wave fronts tend to become more nearly parallel to the shore. If two-wave rays, defined as orthogonal or ray of the wave fronts, a distance "b", apart is considered (Fig. 4.2(b)). At constant $d/L_0 = 0.5$,

$$BC = b_0/\cos \alpha_0 = b/\cos \alpha \quad \text{or} \quad b = b_0 \frac{\cos \alpha}{\cos \alpha_0}, \tag{4.5}$$

where "b_0" is the distance between the same orthogonal in deep waters.

A plot showing the coordinate positions of the orthogonal is for a given location with defined bathymetry (depth contours) is called as a *refraction diagram*. Figure 4.3 shows a typical view of understanding of wave refraction. By drawing a refraction diagram, the following interpretation can be made.

(i) Convergence of orthogonals (e.g., towards a headland)
(ii) Divergence of orthogonal (e.g., approaching a bay between two headlands).

Convergence of orthogonals leads to increase in wave heights or energy, resulting in shore erosion. Divergence of orthogonals leads to reduction in wave heights or energy, resulting in accession or deposition.

The basic assumptions for construction of refraction diagram are as follows:

• Wave energy between the wave rays or orthogonal is conserved.
• The direction of the advance of waves is given by the direction of orthogonal.

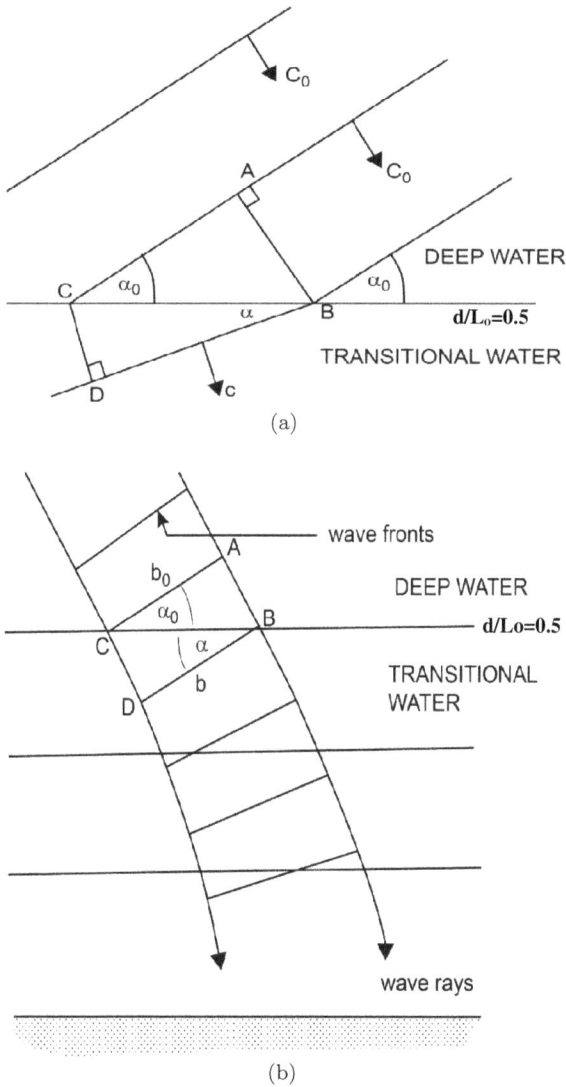

(a)

(b)

Fig. 4.2   (a) Wave deformation definition sketch. (b) Schematic representation of wave deformation.

- The speed of a given wave is a function of only depth.
- The bottom slope is gradual.
- Small amplitude waves of constant period.
- Effects of currents, winds, and reflection from beaches are negligible.

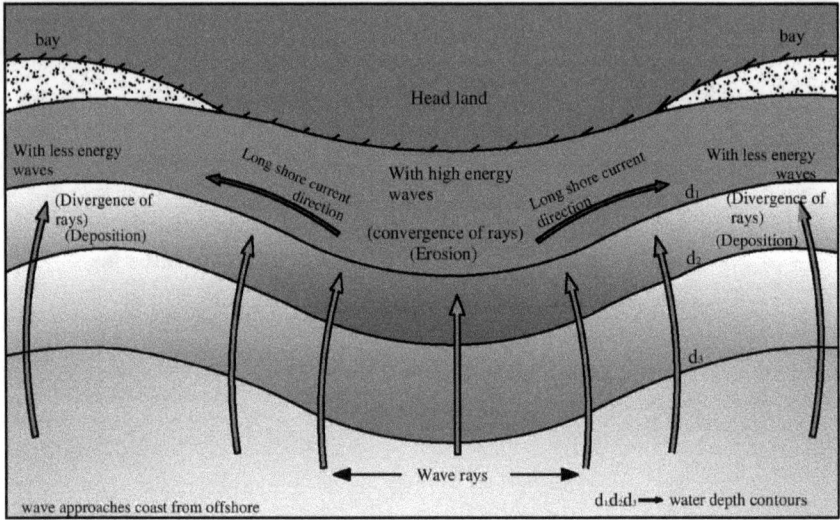

Fig. 4.3   Refraction diagram.

Graphical methods or computer program may be used for drawing the refraction diagrams. In the analyses of refraction phenomenon, it is assumed that for an advancing wave approaching the shore, there is an absence of energy flow along the wave crest in the lateral direction. It implies that the energy transmitted remains a constant between orthogonal.

The average power transmitted by a wave is given by

$$\overline{P} = nb\overline{E}C \quad \text{where} \quad n = \frac{1}{2}\left[1 + \frac{2kd}{\sinh 2kd}\right]. \tag{4.6}$$

For deep water conditions, $(2kd/\sinh 2kd) \to 0$ and hence

$$\overline{P}_0 = \frac{1}{2}b_0\overline{E}_0C_0. \tag{4.7}$$

Equation (4.5) leads to

$$\frac{\overline{E}}{E_0} = \frac{1}{2}\left(\frac{1}{n}\right)\left(\frac{b_0}{b}\right)\left(\frac{C_0}{C}\right). \tag{4.8}$$

We know

$$\frac{\overline{E}}{\overline{E}_0} = \frac{\rho g H^2/8}{\rho g H_0^2/8} = \frac{H^2}{H_0^2}. \tag{4.9}$$

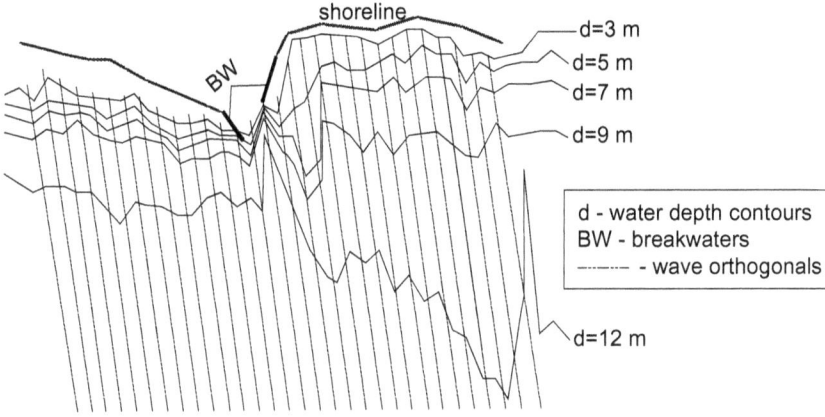

Fig. 4.4   Refraction diagram for Paradip Port.

Hence

$$\frac{H}{H_0} = \sqrt{\frac{1}{2}\left(\frac{1}{n}\right)\left(\frac{C_0}{C}\right)}\sqrt{\frac{b_0}{b}}, \tag{4.10}$$

where

$$\sqrt{\frac{1}{2}\left(\frac{1}{n}\right)\left(\frac{C_0}{C}\right)} = K_s = \text{Shoaling coefficient},$$

$$\sqrt{\frac{b_0}{b}} = K_r = \text{Refraction coefficient}.$$

K_S is given in the wave tables as a function of d/L (see Shore Protection Manual, 1984). Using Eq. (4.10), one can evaluate the wave heights in shallow as well as transitional water depths. If K_r is greater than unity, the wave energy is concentrated leading to an increase in the wave height in the near shore and if K_r is lesser than unity it is vice versa. A typical refraction diagram for Paradip Port due to waves of period 8 s approaching from south is shown in Fig. 4.4.

4.5   Wave Diffraction

When the energy from water waves is transferred laterally along the wave crest, i.e., normal to the wave direction, the phenomenon is called wave diffraction.

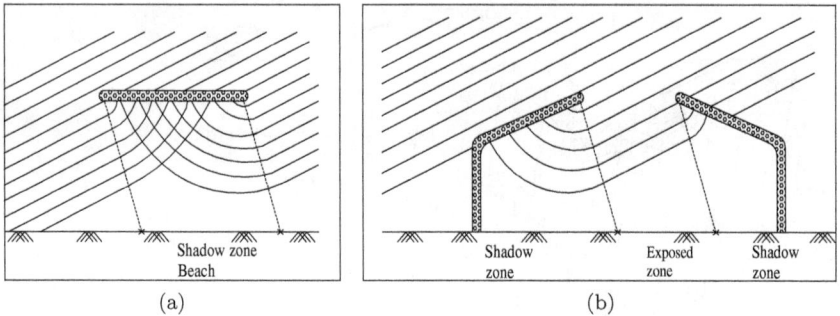

Fig. 4.5   Diffraction of waves around (a) single breakwater and (b) a pair of breakwaters.

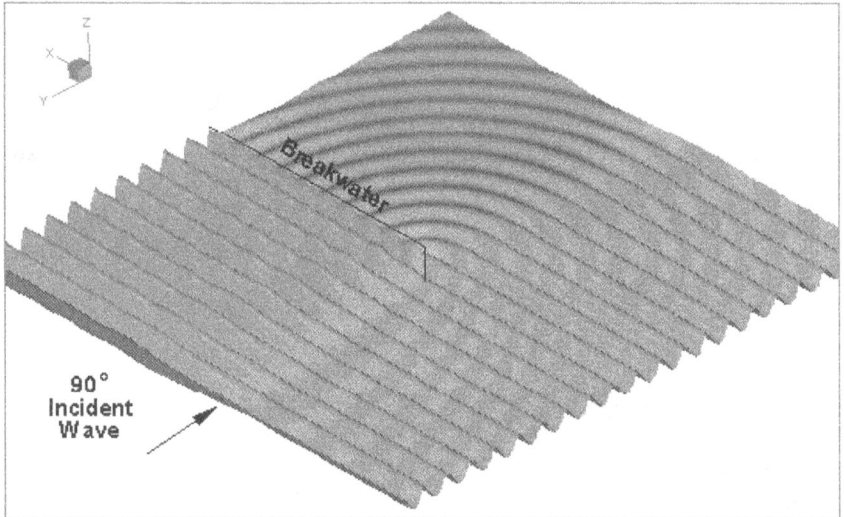

Fig. 4.6   Diffraction pattern past semi-infinite breakwater in constant water depth.

It is dominant around natural barriers or man-made structures such as breakwaters, groins, training walls, etc. Any of these structures could interrupt an otherwise regular train of waves, in which case, the waves curve around the barrier penetrates into the sheltered area. Sample wave diffraction patterns for a single and a pair of breakwaters are illustrated in Figs. 4.5(a) and 4.5(b), respectively. A further explanation to the diffraction phenomena presented in Fig. 4.6, where a "geometric shadow region" is

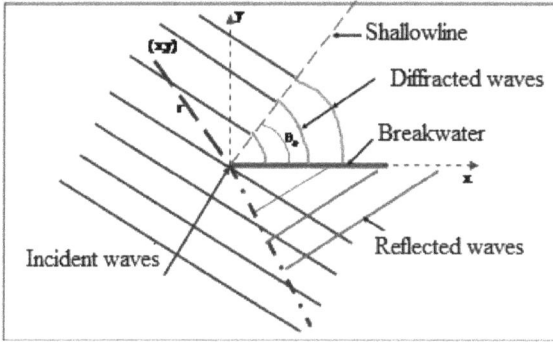

Fig. 4.7   Wave diffraction characteristics.

observed on the protected side of the breakwater of finite length, when the waves propagate normal to it.

In order to understand more clearly, we refer to Fig. 4.7. The origin of the coordinate system is taken to be at the tip of the breakwater with the x and y-axes running parallel and normal to the breakwater as shown in the figure. The regions may be idealized as follows.

(1) The region $0 < \theta < \theta_0$ is the shadow zone of the breakwater in which the solution consists of only the scatter waves. In this area, the wave crests form circular arcs centered at the origin.
(2) The region $\theta_0 < \theta < (\theta_0 + \pi)$ is the one in which the scattered waves and the incident waves are combined. It is assumed that the wave crests are undisturbed by the presence of the break water.
(3) The region $(\theta_0 + \pi) < \theta < 2\pi$ is the region in which the incident and the reflected waves are superimposed to form an oblique incidence and a partial standing wave for normal incidence.

4.5.1   *Importance of calculation of diffraction effects*

The wave height distribution can be predetermined to a certain extent of accuracy in a harbor or a sheltered bay by means of the diffraction characteristics of both natural and artificial structures offering protection from the incident waves. Hence, it is vital in finalizing the layout of harbors.

In certain harbors, after certain year of its existence, it might call for its rehabilitation or may require additional absorbing of wave energy by providing suitable dampers. This could be finalized through diffraction plots.

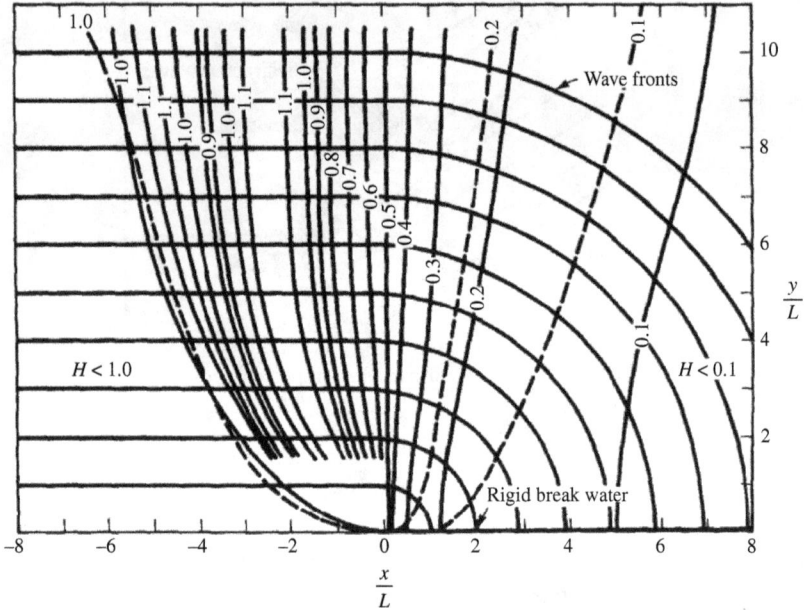

Fig. 4.8   Wave fronts and isolines.

A sound knowledge in wave diffraction is a prerequisite for an effective design and location of harbor entrances so as to mitigate siltation and harbor resonance.

The wave diffraction is governed by the Helmholtz equation given by

$$\frac{\partial^2 F}{\partial x^2} + \frac{\partial^2 F}{\partial y^2} + k^2 F(x, y) = 0. \tag{4.11}$$

$F(x, y)$ is complex, containing both amplitude and phase distributions. Figure 4.8 represents wave fronts and isolines of relative wave height for $y > 0$. The solution to the Helmholtz equation was found by Sommerfeld in 1986 who applied it to diffraction of light as detailed by Dean and Dalrymple (1994).

According to Penney and Price (1952), the solution for $F(x, y)$ for engineering applications can be obtained as follows.

First for large y, the relative wave height approaches one half on a line separating the geometric shadow and illuminated regions (see Fig. 4.9).

Second, for $y/L > 2$, isolines of wave height behind a breakwater may be determined in accordance with the parabolic equation:

$$\frac{x}{L} = \sqrt{\frac{\beta_r^4}{16} + \frac{\beta_r^2}{2}\frac{y}{L}}, \tag{4.12}$$

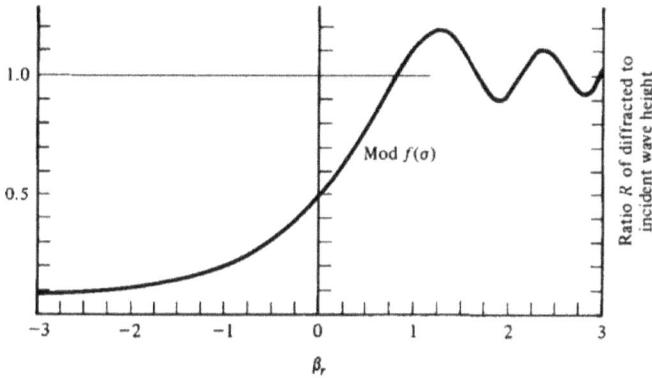

Fig. 4.9 Variation of K_d with distance parameter β_r.

in which β_r is obtained from Fig. 4.9 for any value of relative wave height $R = H/H_I$. The dashed lines in Fig. 4.8 compare several isolines obtained by Eq. (4.12) with those from the complete solution. Typical results on the variation of diffraction coefficient, K_D = wave height inside the sheltered zone, H_d = incident wave height, H_i = past breakwater gaps as a function of relative gap ratio, gaps between breakwaters and wavelength are reported in Wiegel (1964).

4.6 Combined Refraction and Diffraction

Although the waves undergo refraction alone (open coast with gradual varying bathymetry with no obstructions) or diffraction alone (wave penetration in to a harbor of constant water depth), they often undergo combined refraction and diffraction. This may in fact include or exclude reflection from the structure. A schematic representation of the above phenomena and a combined refraction and diffraction results due to a pair of training walls are shown in Figs. 4.10 and 4.11, respectively.

4.7 Breaking of Waves

4.7.1 Breaking criteria

The two breaking criteria are "steepness limited" that holds good in offshore and the "depth limited". In the offshore, the waves break due to excessive transfer predominantly or only from the wind. As they propagate towards the coast, they undergo shoaling leading to an increase in their height and reduction in length and ultimately break near the shore. Ocean wave would break leading to the dissipation of energy under the conditions listed below.

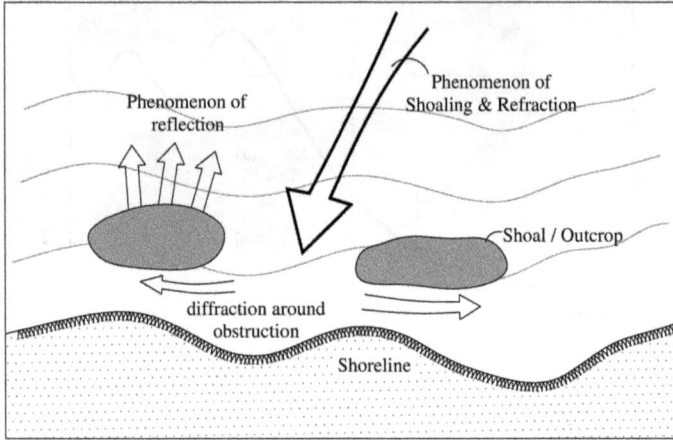

Fig. 4.10   Schematic representation of combined reflection, refraction and diffraction.

- When horizontal particle velocity at the crest exceeds the celerity of the wave.
- When vertical particle acceleration is greater than acceleration due to gravity.
- When crest angle is less than 120°.
- When the wave steepness, $H/L > 0.142$ tanh kd, and for deep waters tanh kd will become 1.
- When wave height is greater than $0.78d$.

4.7.2   Wave height at the breaking point

An approximate estimate of breaking wave height can be determined under the assumption that linear shallow water wave theory holds good at the point of wave breaking.

We know $L_0 = \frac{g}{2\pi}T^2$ and at point of breaking $H_b/d_b = 0.8$.

The dissipation of energy during transition from deep water to the point of wave breaking is assumed to be negligible. Thus the shore normal energy flux is equal for/at both deep water (E_0) and breaking point (E_b).

$$E_b = \frac{1}{8}\rho g H_b^2 C_b = \frac{1}{8}\rho g H_b^2 \sqrt{gd_b},$$

$$\frac{H_0}{H_b} = \left\{2\sqrt{H_0}\frac{2\pi}{gT}\sqrt{1.25g}\right\}^{2/5} \quad \text{since } E_0 = E_b \text{ and } d_b = H_b/0.8.$$

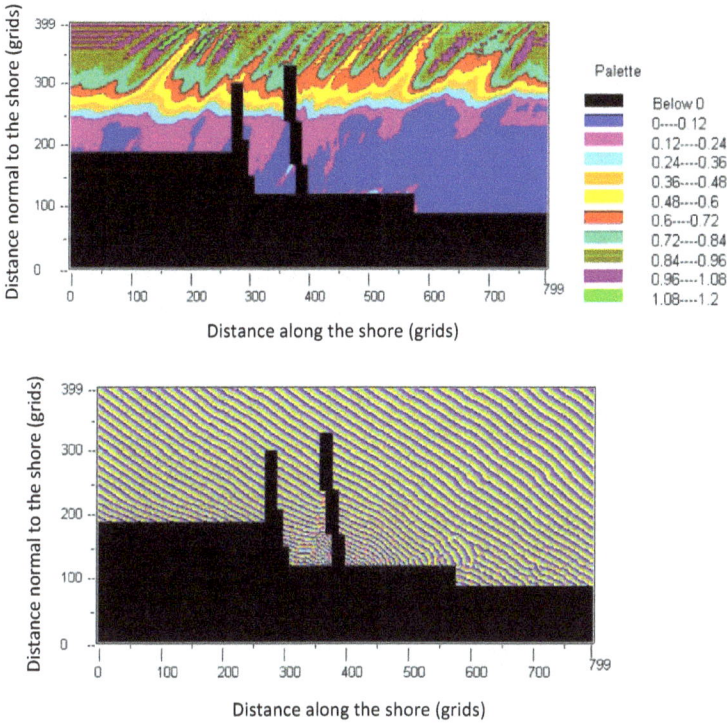

Fig. 4.11 Typical wave height and wave phase distribution due to combined refraction and diffraction.

Hence,

$$\frac{H_0}{H_b} = 0.50 \left(\frac{L_0}{H_0}\right)^{1/5}. \qquad (4.13)$$

Empirical formulas for the breaker depth proposed by several investigators are discussed in detail by Rattanapitikon and Vivattanasirisak (2000).

4.7.3 Types of breakers

The breaker type dictates the extent of mixing of sediments in the breaker zone. The breaking of waves may be classified as follows:

Spilling breakers: Waves of low steepness waves break over beaches of mild slopes. The breaking is by continuous spilling of foam down the front face sometimes called as "white water". Breaking is gradual. The type of breaking can be classified according to the parameter

$$N_1 = \frac{tan\beta}{\sqrt{\frac{H_0}{L_0}}} \leq 0.5, \quad N_2 = \frac{tan\beta}{\sqrt{\frac{H_b}{L_0}}} \leq 0.4.$$

Plunging breakers: Waves of medium steep breaking over beach of medium steepness fall under this category. The waves at breaking curl over. The breaking is instant and $N_1 = 0.5$–3.3 or $N_2 = 0.4$–2.0.

(a) Spilling breaker

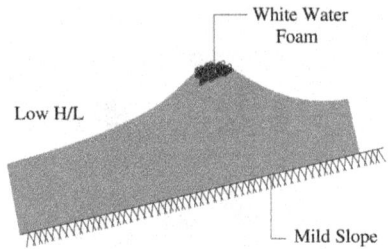

(b) Plunging breaker

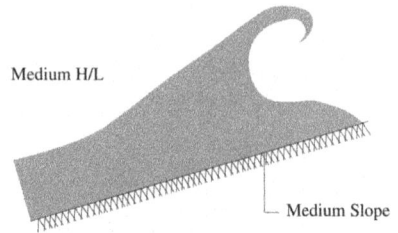

(c) Surging breaker

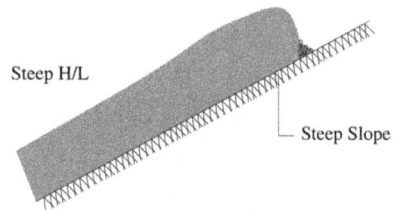

Fig. 4.12   View and schematic representation and types of breakers.

Surging breakers: These occur with the steepest waves breaking over steep slopes. The base of the wave surges up the beach generating considerable foam. It builds up as if to form the plunging type. For this type of breaker, $N_1 > 3.3$ or $N_2 > 2.0$. Galvin (1969) also describes a fourth category of breaking waves called as "*Collapsing breakers*" that would be in between the plunging and surging breakers. Views and schematic representation of the different breakers are projected in Fig. 4.12.

4.8   Wave Reflection

For a wave of elevation, that is incident, η_i on a vertical wall is $\eta_i = a \sin(kx - \sigma t)$. Let the reflected wave elevation, η_r, be $\eta_r = a \sin(kx + \sigma t)$. The composite wave elevation, η_τ by adding the components η_i and η_r is given by

$$\eta_T = \eta_i + \eta_r = 2a \sin kx \cos \sigma t. \tag{4.14}$$

When waves reach a vertical wall, the maximum vertical displacement of wave elevation on its surface will be twice the incident wave amplitude, with these maxima occurring in stationary patterns separated by stationary patterns of zero water surface elevations. In such cases, the waves are termed as "standing waves" and the pressures and forces on walled structures due to such waves will be high and further the orbital velocity gradient over the wall will be high resulting in scour and thereby dictating the stability of the structure. Hence, this factor is very important to be considered in the design of coastal structures and in harbor design. It is wise to avoid such vertical face structures inside harbor basins to be exposed to waves as the tranquility will be of great concern and more so at locations exposed to long waves. Partial reflection is an important factor to consider in partially submerged structures. In most problems in the marine environment, partial reflection occurs due to the absorption of incident or diffracted waves to an extent, which would be significant in the case of permeable structures.

Let a_i, a_r, a_t and a_{ab}, respectively, be the incident, reflected, transmitted, and absorbed wave amplitude for the case of a wave propagating over an obstacle. Since the wave energy is proportional to the square of the wave amplitude, it can be stated that

$$a_i^2 = a_r^2 + a_t^2 + a_{ab}^2. \tag{4.15}$$

In case the structure is emerging through the free surface and if impermeable, a_t will be zero, and hence

$$a_i^2 = a_r^2 + a_{ab}^2. \tag{4.16}$$

On solving for a_r/a_i, the reflection coefficient K_r can be obtained as

$$K_r = \frac{a_r}{a_i} = \sqrt{1 - \left(\frac{a_{ab}}{a_i}\right)^2}. \tag{4.17}$$

It is mentioned that coastal structures like breakwaters, seawalls, groins, etc. will be a function of its permeability and its slope. For flatter slope as well as for structures formed with a higher degree of permeability, the K_r will be less. This is because energy loss by bottom friction also plays an important role in dissipating the incident wave energy. For waves past structures, wherein energy is not absorbed but transmitted past it, Eq. (4.15) becomes

$$a_i^2 = a_r^2 + a_t^2. \tag{4.18}$$

On solving for a_t/a_i, the transmission coefficient K_t can be obtained as

$$K_t = \frac{a_t}{a_i} = \sqrt{1 - \left(\frac{a_r}{a_i}\right)^2}. \tag{4.19}$$

The transmission coefficient depends on the size, shape, porosity and position of the structures from the MSL being the most important factors.

4.9 Waves on Currents

Waves would propagate in the absence or presence of currents that flow at an angle to its direction. This current will influence some of the wave characteristics, and, in their turn, the waves will influence certain current characteristics. If the wave current and the wave propagation are unidirectional, a reduction in wave height and an increase in wavelength are observed. On the contrary, if they are not unidirectional, a reversal of the same is observed, i.e., the wave height increases and wavelength decreases. If the current is in the opposite direction, the wave height will increase and the wavelength will decrease. The wave celerity, in the absence of current, is given by

$$C = \sigma/k. \tag{4.20}$$

If the celerity of the wave in presence of current of magnitude, U is C^*, we could define, $C^* = C + U$.

The frequency will change to σ^* assuming wavelength as constant and represented as

$$\sigma^* = C^*k = (C + U)k = \sigma + k^*U, \tag{4.21}$$

where σ^* is wave frequency in the case of waves and a current and this component, kU is often referred to as the *Doppler effect*. Let us consider waves propagating over a current, U in the same direction and on the assumption of the conservation of the number of wave crests

$$k^*(U + C^*) = kC, \tag{4.22}$$

where k and k^* are the wave number in absence and presence of the current, respectively. C and C^* refer to the wave celerity in the absence and presence of the current, respectively. If the waves are assumed to be in deep waters as a simple case,

$$(C^*)^2 = g/k^* \quad \text{and} \quad (C)^2 = g/k, \tag{4.23}$$

Eq. (4.22) can be written as quadratic equation in C^*/C with the solution

$$\frac{C^*}{C} = \frac{1}{2}\left[1 + \left(1 + \frac{4U}{C}\right)^{1/2}\right]. \tag{4.24}$$

When the current is flowing in the same direction as that of waves (traveling) $U > 0$, the velocity C^* and the wavelength L^* increase compared to their respective values in absence of current. If the waves propagate over an opposing current, $U < 0$, C^* and L^* would then decrease. Further, for waves against current, for $U < -C/4$, no solution for Eq. (4.24) is possible. At this critical point, $U = -C/4$ and $C^* = C/2$, or $C_G = C^*/2 = -U$. Hence, when the local C_G is equal and opposite to the current velocity, no wave energy can propagate against the current, and at this stage the waves will steepen and break.

The change in wave height can be obtained from the principle of conservation of wave action, defined by E/σ. In the present case, the principle leads to the equation

$$E^*\left(U + \frac{1}{2}C^*\right)C^* = \text{const} = \frac{1}{2}EC_0^2. \tag{4.25}$$

where "E" and "a" are the energy per unit area of the waves and wave amplitude in absence of current, respectively. The "*" as suffix indicated the variables in the presence of current. Since $E^* = 0.5\rho g(a^*)^2$ and $E = 0.5\rho g(a)^2$, it can be shown that

$$\frac{a^*}{a} = \frac{C_0}{[C^*(C^* + 2U)]^{1/2}}. \tag{4.26}$$

With an opposing current, $U < 0$, the amplitude increases and approaches infinity as the velocity U approaches $-C/2$ the limit found above.

Worked Out Examples

Problem 4.1

The coast is oriented 24° with respect to north. The period of the wave is 12 s, while its direction in deep waters with respect to north is 160°. Find the shallow water wave angle at water depth of 8 m.

Solution

$d/L_0 = 0.036$, $d/L = 0.07867$ and hence, $L = 101.69\,\mathrm{m}$,
$C = L/T = 8.474\,\mathrm{m/s}$.
As $\sin\theta_0/\sin\theta = C_0/C$, we have $\sin 44°/\sin\theta = 18.72/8.474$.
Hence $\theta = 18°$.

Problem 4.2

A wave with a period of 12 s and a height of 2 m propagates from deep to shallow waters. Assuming that the bottom contours are parallel to each other, compute the wave height at a water depth of 6 m.

Solution

$L_0 = 1.56, T^2 = 1.56 * 144 = 224.64\,\mathrm{m}$, $d/L_0 = 0.027$ and corresponding $d/L = 0.06747$,
$C_0 = L_0/T = 18.72\,\mathrm{m/s}$, $L = 88.93\,\mathrm{m}$, $C = 88.93/12 = 7.41\,\mathrm{m/s}$,
$k = 2\pi/L = 0.07$, $2kd = 2 \times 0.07 \times 6 = 0.84$,

$$n = \frac{1}{2}\left[1 + \frac{2kd}{\sinh 2kd}\right] = 0.95,$$

$$\frac{H}{H_0} = \sqrt{\frac{C_0}{C}\frac{1}{2n}} = \sqrt{\frac{18.72}{7.41}\frac{1}{2*0.95}} = 1.1531.$$

Therefore, wave height, H, at $d = 6\,\mathrm{m}$ is $1.1531 * H_0 = 2.3 > H_0$.

Problem 4.3

A deep water wave of height 3.2 m and period 12 s is refracted so that the distance between the orthogonal is reduced by 50% in a depth of 8 m and reduced by 10% at 4 m water depth. What will be the height of the wave here assuming no energy loss?

Solution

(a) H in depth of 8 m,

$$\frac{H}{H_0} = K_s K_r, \quad K_s = \sqrt{\frac{1}{2}\frac{1}{n}\frac{C_0}{C}}, \quad K_r = \sqrt{\frac{b_0}{b}} = \sqrt{2} \text{ since, } \frac{b_0}{b} = 2,$$

$$L_0 = 1.56T^2 = 224.64 \text{ m}, \quad d/L_0 = 0.036,$$

$$d/L = 0.07867, \quad \text{and} \quad L = 101.69 \text{ m},$$

$$C_0 = L_0/T = 18.72 \text{ m/s}, \quad \text{and} \quad C = L/T = 8.474 \text{ m/s}$$

$$k = 2\pi/L = 0.062, \quad 2kd = 2 \times 0.062 \times 8 = 0.992$$

$$n = \frac{1}{2}\left[1 + \frac{2kd}{\sin h2kd}\right] = 0.93$$

$$k_s = \sqrt{\frac{1}{2n}\frac{C_0}{C}} = 1.09$$

$$\frac{H}{H_0} = 1.07 * \sqrt{2} = 1.54.$$

Since $H_0 = 3.2$, $H = 3.2 * 1.54 = 4.93$ m which is greater than H_0 of 3.2 m in $d = 8$ m. In the above problem, even if refraction is not considered,
$H/H_0 = K_s = 1.07$ and hence $H = 1.07* H_0 = 3.424$ m $> H_0$ of 3.2 m.

(b) H in depth of 4 m

$$K_r = \sqrt{\frac{1}{0.9}}, \quad \frac{H}{3.2} = \sqrt{\frac{1}{2 * 0.963}\frac{18.72}{6.11}}\sqrt{\frac{1}{0.9}}$$

$$H = 3.2 * 1.33 = 4.25 \text{ m}.$$

In this case, as per the given problem, H in $d = 4$ m is 4.25, which is greater than the breaking wave height in this depth of $0.78 \times 4 = 3.12$ m. This only suggests that the 10% reduction in 4 m water depth would not have wave of height greater than 3.12 m.

Problem 4.4

For a wave with $T = 10$ s, $H_0 = 3$ m, evaluate the variation in the wave height when the wave approaches shallow waters. Also determine the depth of water in which the wave would break theoretically.

Solution

Given $H_0 = 3\,\text{m}$, $T = 10\,\text{s}$, $L_0 = 156\,\text{m}$. The variation of wave height as the wave approaches shallow waters is given in the table below. The method involves the computation of d/L_0 values for progressive waves. With this information, it is possible to obtain the d/L values and the shoaling coefficient. The depth in which the wave would break is computed from the Miche steepness equation, i.e.,

$$\frac{H}{L} \geq 0.142 \tanh kd.$$

$d(\text{m})$	d/L_0	d/L	L	H/H_0	H	H/L	$0.142 *$ $\tanh kd$
70	0.45	0.4531	154.491	0.9847	2.9541	0.0191	0.1410
60	0.385	0.3907	153.57	0.9728	2.9184	0.0190	0.1399
50	0.32	0.3302	151.423	0.9553	2.8659	0.0189	0.1376
45	0.288	0.3014	149.303	0.9449	2.8347	0.0190	0.1357
35	0.224	0.2455	142.566	0.9242	2.7726	0.0194	0.1296
25	0.160	0.1917	130.412	0.9130	2.739	0.0210	0.1186
15	0.0964	0.1380	108.696	0.9358	2.8074	0.0258	0.0994
6	0.0386	0.08175	73.395	1.072	3.216	0.0438	0.0671
4	0.0257	0.06613	60.487	1.159	3.477	0.0575	0.0558

The wave would break at $d \cong 4\,\text{m}$.

Problem 4.5

A deep water wave has a period of $10\,\text{s}$, a height of $3\,\text{m}$ and is traveling at $35°$ to the shoreline. Assuming that the seabed contours are parallel, find the height, depth, celerity, and angle of wave attack when it breaks.

Solution

$$C_0 = gT/2\pi = 15.60\,\text{m/s},$$

$$b = b_0 \frac{\cos \alpha}{\cos \alpha_0} \quad \text{and} \quad K_r = \sqrt{\frac{b_0}{b}}.$$

d/L_0	$d\,(\text{m})$	C/C_0	$C(\text{m/s})$	K_s	α (degrees)	K_r	H/H_0	$H(\text{m})$	d_b (m)
0.1	15.6	0.71	11.06	0.933	30	0.97	0.90	2.71	2.12
0.05	7.8	0.53	8.28	1.023	22	0.94	0.96	2.88	2.24

The above table can be prepared for various d/L_0. Only the final results are the ones shown for $d/L_0 = 0.05$ and 0.1.

Problem 4.6

For a wave with period $T = 12$ s, deep water wave height $H_0 = 4.0$ m and beach slope, $m = 1/25$ calculate the breaking wave height using various formulae.

Solution

Sl.No	Formula for Breaking wave height	Breaker height H_b in (m)
1	Le Mehaute and Koh (1967) $H_b = 0.76 * H_0 * (H_0/L_0)^{(-0.25)} * m^{(1/7)}$	5.254
2	Komar and Gauhgen (1972) $H_b = 0.56 * H_0 * (H_0/L_0)^{(-0.2)}$	5.013
3	Sumura and Horikowa (1974) $H_b = H_0 * m^{0.2} * (H_0/L_0)^{(-0.25)}$	5.752
4	Singamsetti and Wurd (1980) $H_b = 0.575 * H_0 * m^{0.031} * (H_0/L_0)^{(-0.254)}$	5.791
5	Ogawa and Shuto (1984) $H_b = 0.68 * H_0 * m^{0.09} * (H_0/L_0)^{(-0.25)}$	5.573

Problem 4.7

Given wave height $(H) = 1.5$ m, $d = 3.5$ m, wave period $T = 10$ s, current velocity $U = 0.9$ m/s. Calculate the amplitude and length of the waves when currents are in the same direction and in opposite direction. Also calculate the critical velocity at which wave would tend to break.

Solution

(a) *Wave amplitude and length when currents in the same direction*
   We have

$$\frac{C^*}{C} = 1/2[1 + (1 + 4U/C)^{1/2}],$$

where C^* is the wave velocity in the presence of currents and U the velocity of current, $L_0 = 1.56T^2 = 156$ m, $d = 3.5$ m, $d/L_0 = 0.0224$, $d/L = 0.0606$ and $L = 57.76$ m, $C = L/T = 57.76/10 = 5.776$ m/s. Substituting these values in the above equation, we get $C^* = 5.776/2[1 + (1 + 4*0.9/5.776)^{1/2}]$ and $C^* = 6.57$ m/s.

   Further we have $a^*/a = C_0/[C^*(C^* + 2U)]^{1/2}$, $C_0 = L_0/T = 15.6$ m/s, $a = H/2 = 0.75$ m and by substituting the values of C^*, C_0, U we get the value of a^*.

$$a^* = 1.58\,\text{m},$$

$$L^*/T = C^*, \quad L^* = 6.57 \times 10 = 65.7\,\text{m}$$

Thus, we see the length of the waves increase if it travels along with a following current.

(b) *Wave amplitude and length when currents in opposite direction*

We have $C^*/C = 1/2[1 + (1 + 4U/C)^{1/2}]$, where $U = -0.9\,\text{m/s}$ and $C = 5.776\,\text{m/s}$.

We get $C^* = 4.66\,\text{m/s}$. Then $a^*/a_0 = C_0/[C^*(C^* - 2U)]^{1/2}$. Therefore $a = 3.2\,\text{m}$.

$L^* = 2\pi/k^*$ but $k^* = g/(C^*)^2$. Therefore, $k^* = 0.452$ and $L^* = 2 \times 3.14/0.452 = 13.89\,\text{m}$.

Herein, we see the length of the waves decreases when it propagates over an opposing current.

At $U = C/4$, the waves would steepen and tends to break, then it would be the critical velocity for breaking. Hence, the critical velocity, $U = C/4 = 5.776/4 = 1.44$ m/s.

Chapter 5

Finite Amplitude Wave Theories

5.1 General

The small amplitude wave theory developed by Airy in 1845 also regarded as first-order Stokes (1847) theory and discussed in detail in Chapter 3 in general can be applied for all ranges of d/L. Small amplitude wave theory was based on the premises that motions are sufficiently small to allow the free surface boundary condition to be linearized, i.e., the terms involving wave amplitude to the second order and higher orders are considerably negligible. If the wave amplitude is large, it is must to retain higher order terms to obtain an accurate representation of wave motion. Finite amplitude wave theories or Stokes' higher order theories are similar as per the concept in which additional higher order terms that were neglected in the Airy's theory are included. The velocity potential function will be more complicated as it would include terms of $(H/L)^n$ and $\cos(n\theta)$ or $\sin(n\theta)$, where "n" indicates the order of the wave theory. Refer to Section 3.2 in which the higher order terms $(u^2+v^2+w^2)$ were neglected that can no longer be ignored. In general, the higher the order of the wave theory, the higher will be the limiting wave height for which it would be valid. They are complicated due to the relative importance of the additional parameters namely H/d and H/L. Keulegen (1950) classified finite amplitude wave theories depending on d/L as shown in Fig. 5.1(a).

5.2 Classification of Finite Amplitude Wave Theories

The classification of different waves and their shapes according to Wilson (1963) is shown in Fig. 5.1(b).

(a)

small amplitude waves
(one theory)

1/2 ——————————————————— 1/20 ——————— d/L

Deep water ——→|←—— Intermediate depth ——————→|←—— Shallow water
water
Final amplitude waves
(atleast 3 thoeries)

1/10　　　1/50

d/L

Stokian theory ————————————————→ Cnoidal ←— solitary
(also Gosetner) ←———————————————— theory —→wave theory

L_T　$\dfrac{H}{L}$ important ——————→　　$H = \dfrac{H}{L} = \dfrac{H}{d}$ important　　$\dfrac{11}{d} \cdot d$ important

(b)

STILL WATER LEVEL

η_0　H

WAVE LENGTH, L
PERIOD, T

d

URSELL'S Dimensionless parameter $(\eta_0 L^2/d^3)$

Airy's Wave : $\eta_0/H < 0.505$; $\eta_0 L^2/d^3 < 1$, $C < \sqrt{gd}$

Stoke's Wave : $\eta_0/H < 0.635$; $\eta_0 L^2/d^3 < 30$, $C \leq \sqrt{gd}$

Cnoidal Wave : $0.635 < \eta_0/H < 1$; $\eta_0 L^2/d^3 > 10$, $C \geq \sqrt{gd}$

Solitary Wave : $\eta_0/H = 1$; $\eta_0 L^2/d^3 \to \infty$　$C = 1.33\sqrt{gd}$

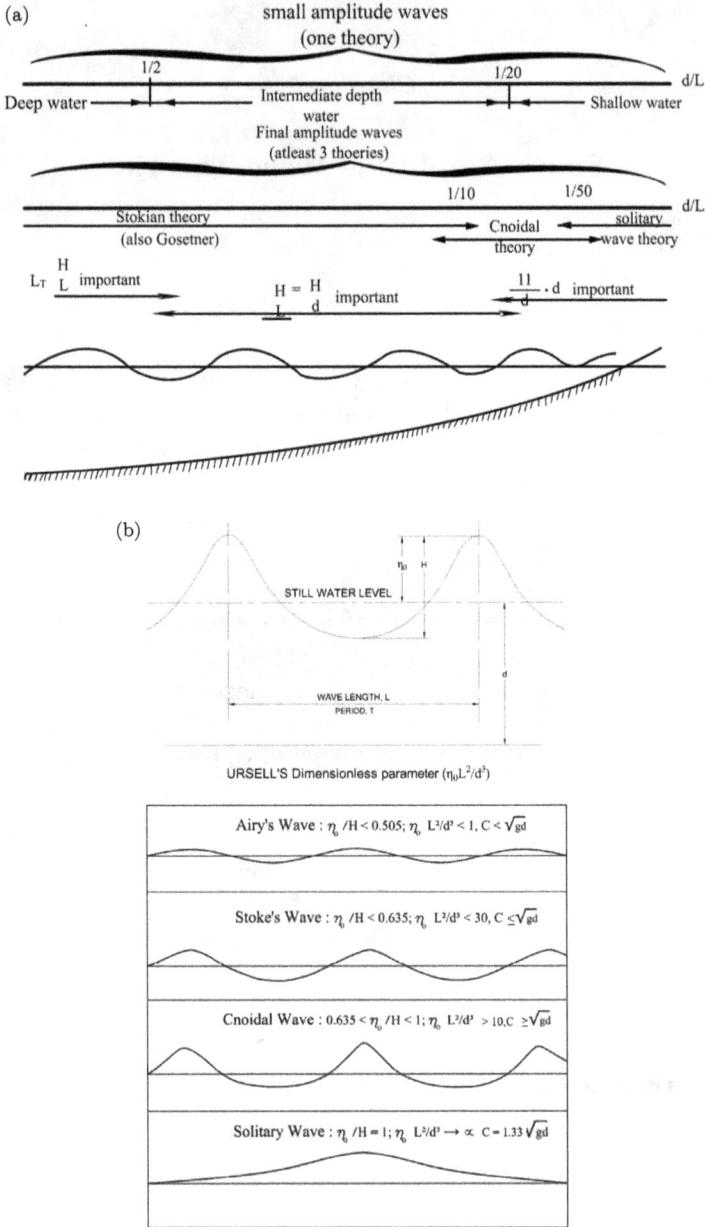

Fig. 5.1　(a) Classification of wave theories (Keulegen, 1950). (b) Profile shapes of Gravity waves (Wilson, 1963).

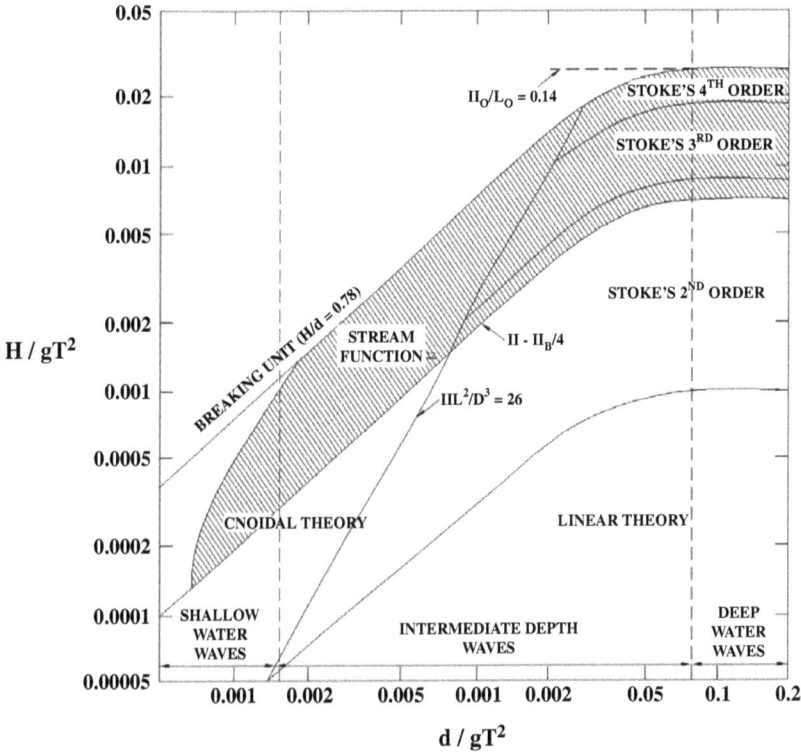

Fig. 5.2   Regions of validity for various wave theories (Le Mehaute, 1976).

5.2.1   Stoke's wave theories

Nonlinear theories have been developed for various surface profiles. The form was expressed as a series of terms of $\cos n(kx - \sigma t)$, where $n = 0, 1, 2, 3, \ldots$, and such waves are irrotational. In deep waters, if terms involving $(H/L)^3$ and higher powers are neglected form the Fourier series relating to such an analysis.

As the ratio H/L increases, the crest becomes sharper and the trough becomes flatter than Airy wave. This has the effect of raising the median height relative to the SWL. The regions of validity for the various wave theories are given in Fig. 5.2.

The expressions for wave elevation, orbit velocities, displacements, mass transport velocity and pressure based on Stokes's second-order wave theory are given below.

For first order,

$$\eta = \frac{H}{2}\cos(kx - \sigma t) = \frac{H}{2}\cos\theta = a\cos\theta.$$

A more general expression would be

$$\eta = a\cos(\theta) + a^2 B_2(L, d)\cos(2\theta) + a^3 B_3(L, d)\cos(3\theta)$$

$$+ \cdots + a^n B_n(L, d)\cos(n\theta). \tag{5.1}$$

If $a < \frac{H}{2}$, for orders higher than 2nd, $B_2, B_3, \ldots$ are specified functions of L and d.

$$\eta = \frac{H}{2}\cos(kx - \sigma t) + \left(\frac{\pi H^2}{8L}\right)\frac{\cosh kd}{\sinh^3(kd)}[2 + \cosh 2kd]\cos 2\theta. \tag{5.2}$$

For deep water $\frac{d}{L} > \frac{1}{2}$, the above equation becomes

$$\eta = \frac{H_0}{2}\cos(k_0 x - \sigma t) + \frac{\pi H_0^2}{4L_0}\cos 2(k_0 x - \sigma t). \tag{5.3}$$

Particle Kinematics
Orbital velocities

$$u = \frac{HgT}{2L}\frac{\cosh[k(z+d)]}{\cosh kd}\cos(kx - \sigma t)$$

$$+ \frac{3}{4}\left(\frac{\pi H}{L}\right)^2 C\frac{\cosh[2k(z+d)]}{\sinh^4 kd}\cos 2(kx - \sigma t), \tag{5.4}$$

$$w = \frac{\pi H}{L}C\frac{\sinh[k(z+d)]}{\sinh kd}\sin(kx - \sigma t)$$

$$+ \frac{3}{4}\left(\frac{\pi H}{L}\right)^2 C\frac{\sinh[2k(z+d)]}{\sinh^4 kd}\sin 2(kx - \sigma t). \tag{5.5}$$

Horizontal water particle displacement is

$$\delta_x = \frac{-HgT^2}{4\pi L}\frac{\cosh k(d + z)}{\cosh kd}\sin(kx - \sigma t)$$

$$+ \frac{\pi H^2}{8L}\frac{1}{\sinh^2 kd}\left\{1 - \frac{3}{2}\frac{\cosh 2k(d+z)}{\sinh^2 kd}\right\}\sin 2(kx - \sigma t)$$

$$+ \left(\frac{\pi H}{L}\right)^2 \frac{Ct}{2}\frac{\cosh 2k(d + z)}{\sinh^2 kd}. \tag{5.6}$$

Vertical displacement is

$$
\delta_y = \frac{HgT^2}{4\pi L} \frac{\sinh[k(d+z)]}{\cosh kd} \cos(kx - \sigma t)
$$

$$
+\frac{3}{16} \frac{\pi H^2}{L} \frac{\sinh[2k(d+z)]}{\sinh^4 kd} \cos 2(kx - \sigma t). \tag{5.7}
$$

Mass transport velocity

$$
\overline{u}(z) = \left(\frac{\pi H}{L}\right)^2 \frac{C}{2} \frac{\cosh[2k(z+d)]}{\sinh^2 kd}. \tag{5.8}
$$

Subsurface pressure

$$
p = \rho g \frac{H}{2} \frac{\cosh[k(d+z)]}{\cosh kd} \cos(kx - \sigma t)
$$

$$
+\frac{3}{8}\rho g \frac{\pi H^2}{L} \frac{\tanh kd}{\sinh^2 kd} \left\{ \frac{\cosh 2k(z+d)}{\sinh^2 kd} - \frac{1}{3} \right\} \cos 2(kx - \sigma t)
$$

$$
-\rho g z - \frac{1}{8}\rho g \frac{\pi H^2}{L} \frac{\tanh kd}{\sinh^2 kd} \{\cosh 2k(d + z) - 1\}. \tag{5.9}
$$

Relationship for particle velocities and the wave forms become complex as the number of terms retained increases. The use of digital computers has enabled rapid solutions to be obtained for higher order formulations. The second-order solutions have been developed by Biesel (1952). Skjelbreia (1959) has developed the third-order theory, whereas Skjelbreia and Hendrickson (1960) presented fifth-order theory. In practice, the fifth-order stokes wave is frequently used in estimating wave forces on structures, since many of the experimental constants derived from field measurements have been correlated with this theory. The complete expressions for the fifth-order theory is provided in Appendix B. A typical comparison of second-order Stokes' profile with linear wave profile is shown in Fig. 5.3.

5.2.2 *Solitary wave theory*

In very shallow waters wave crests become peaked and trough flattened. The surface profile is entirely above the SWL in the case of solitary waves. The wave is therefore not periodic and has no definite wavelength. Boussinesq (1872) derived the characteristics of the solitary wave, in shallow water depth, directly from the general equation for steady flow. Solitary wave has proved useful in engineering problems such as the study of very long waves

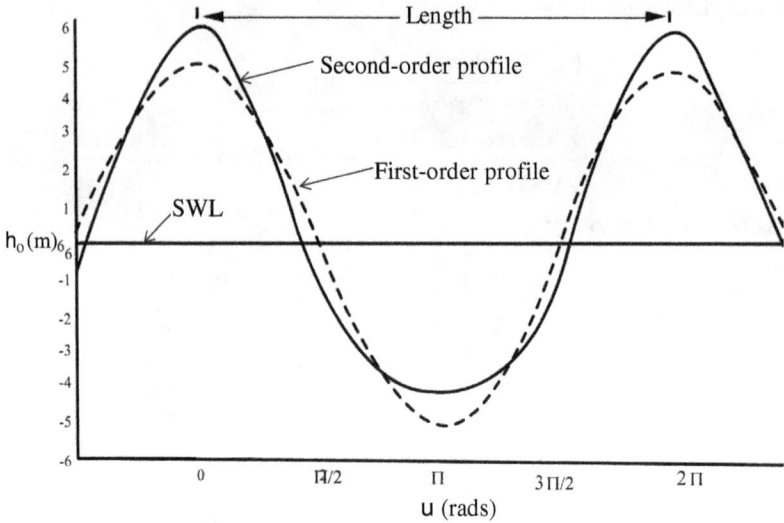

Fig. 5.3   Comparison of second order with linear theory.

like tsunamis and in determining wave properties near breaking in shallow water and for studying waves of maximum steepness in deep waters. Under such conditions, the wave characteristics are independent of L and T and depend only on H and d.

Let

$$H_b = 0.78d, \tag{5.10}$$

$$C = \sqrt{2g(H + d)}, \tag{5.11}$$

$$\eta = H \operatorname{sech}^2 \sqrt{\frac{3}{4}\frac{H}{d^3}}(x - Ct). \tag{5.12}$$

Here 'x' being origin at the crest. Then

$$u = \frac{CN\left[1 + \cos(M\left[\frac{z+d}{d}\right])\cosh(M\frac{x}{d})\right]}{\left\{\cos(M\left[\frac{z+d}{d}\right]) + \cosh(M\frac{x}{d})\right\}^2}, \tag{5.13}$$

$$w = \frac{CN\left[\sin(M\left[\frac{z+d}{d}\right])\sinh\left(M\frac{x}{d}\right)\right]}{\left\{\cos(M\left[\frac{z+d}{d}\right]) + \cosh\left(M\frac{x}{d}\right)\right\}^2}. \tag{5.14}$$

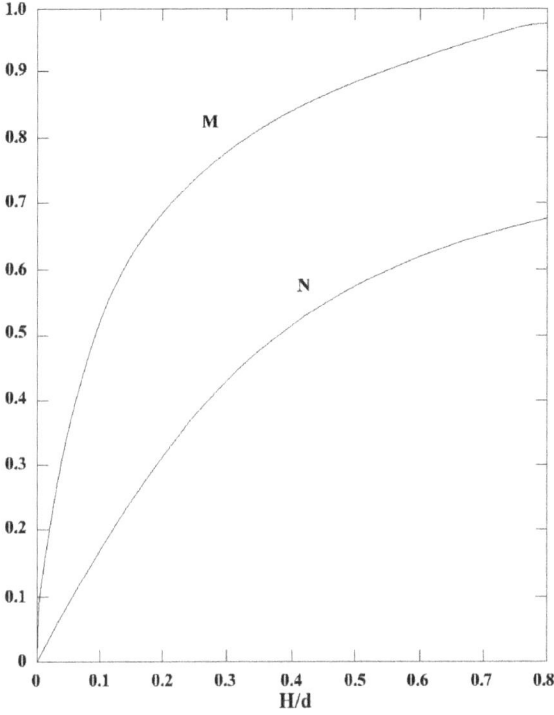

Fig. 5.4   Variation of M and N with H/d.

where m, n are functions of H/d as shown in Fig. 5.4 and can also be determined from

$$\frac{H}{d} = \frac{N}{M} \tan \frac{1}{2}\left[M\left(1 + \frac{H}{d}\right)\right],$$

$$N = \frac{2}{3}\sin^2\left[M\left(1 + \frac{2H}{3d}\right)\right]. \qquad (5.15)$$

Now $u_{\max}$ will occur when $x = t = 0$ which is given by

$$u_{\max} = \left[CN \bigg/ \left(1 + \cos\left\{M\left(\frac{z+d}{d}\right)\right\}\right)\right]. \qquad (5.16)$$

In a solitary wave, the kinetic and potential energies are distributed evenly among the total energy.

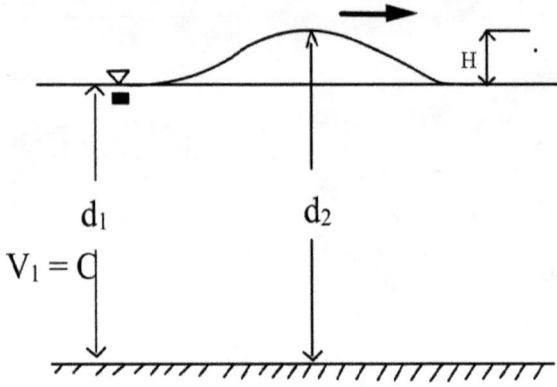

Fig. 5.5   Solitary wave profile.

The total energy per unit crest width is

$$E = \frac{8}{3\sqrt{3}} \rho g H^{3/2} d^{3/2}. \tag{5.17}$$

However, the local fluid velocity influences the pressure beneath a solitary wave and so does the pressure under a Cnoidal wave. It may be approximated as

$$p = \rho g (y_s - y). \tag{5.18}$$

A simplified derivation for the celerity for solitary wave fronts is given below. Consider the steady-flow case where the wave form (Fig. 5.5) shown remains solitary and water enters with a velocity $V_1 = C$. Applying Bernoulli's equation,

$$d_1 + \frac{V_1^2}{2g} = d_2 + \frac{V_2^2}{2g}. \tag{5.19}$$

From continuity equation, $d_1 V_1 = d_2 V_2$ for width perpendicular to the paper. Substituting and rearranging $V_1 = V_2 = C$,

$$\frac{V^2}{2g} = \frac{d_1 + d_2}{1 - (d_2/d_1)^2}. \tag{5.20}$$

Substitute $d_2 = H + d_1$ and dropping higher terms,

$$C = \sqrt{gd} \left[1 + \frac{3}{2} \cdot \frac{H}{d} \right]^{1/2}. \tag{5.21}$$

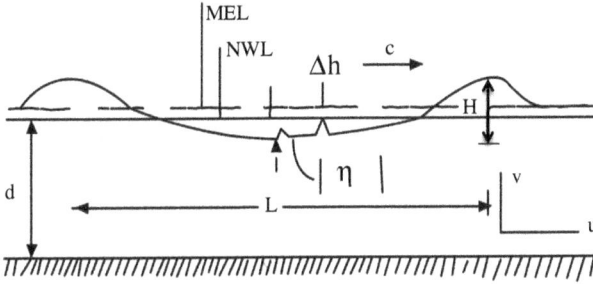

Fig. 5.6   Cnoidal wave profile.

Note: When H becomes small with respect to depth, the limit is $C = \sqrt{gd}$. For extremely shallow water, $C = \sqrt{g(d + H)}$.

5.2.3   *Cnoidal wave theory*

The Cnoidal wave theory first described by Korteweg and Devries (1895) is applicable over the range $\frac{1}{50} < \frac{d}{L} < \frac{1}{10}$. However, both the Stokes wave theory and the Solitary wave theory are valid over a portion of this same range. The theory considers terms of second order and therefore gives more accurate results which may be necessary under special circumstances. Further, Cnoidal wave theory is valid for Ursell's parameter, $\frac{L^2 H}{d^3} > 26$. A typical Cnoidal wave profile is shown in Fig. 5.6.

As the wavelength increases towards infinity, the Cnoidal wave theory approaches solitary wave theory. Wave characteristics are described in parametric form in terms of the modules "k" of the elliptic integrals. While "k" as such has no physical significance, it is used to describe the relationships between the various wave parameters. The ordinate of the water surface y_s measured above the bottom is given by

$$y_s = y_t + HCn^2 \left[2K(k) \left(\frac{X}{L} - \frac{t}{L} \right), k \right], \qquad (5.22)$$

where y_t is the distance from the bottom to the wave trough, C_n the elliptic cosine function, $K(k)$ the complete elliptic integral of the first kind, and K the modulus of the elliptic integrals.

As the argument of C_n^2 is often denoted simply by ( ), Eq. (5.22) can be written as

$$y_s = y_t + HCn^2( ). \qquad (5.23)$$

The elliptic cosine is a periodic function where

$$Cn^2 \left\{ 2K(k) \left[\left(\frac{x}{l} \right) - \left(\frac{t}{T} \right) \right] \right\}$$

has a maximum amplitude equal to unity. The modulus "k" is defined over the range between 0 and 1. When $k = 0$, the wave profile becomes a sinusoid, as in the linear theory; when $k = 1$, the wave profile becomes that of a solitary wave.

The distance from the bottom to the wave through y_t, as used in Eqs. (5.22) and (5.23) is given by

$$\frac{y_t}{d} = \frac{y_c}{d} - \frac{H}{d} = \frac{16d^2}{3L^2} K(k) \left[K(k) - E(k) + 1 - \frac{H}{d} \right], \tag{5.24}$$

where y_c is the distance from the bottom to the crest, $E(k)$ the complete elliptic integral of the second kind, and

$$L = \sqrt{\frac{16d^3}{3H} k K(k)}. \tag{5.25}$$

The wave period is given by

$$T\sqrt{\frac{g}{d}} = \sqrt{\frac{16y_t}{3H}} \frac{d}{y_t} \left[\frac{kK(k)}{1 + \frac{H}{y_t k^2} \left(\frac{1}{2} - \frac{E(k)}{K(k)} \right)} \right].$$

For any elevation y, pressure under a Cnoidal wave is complex that depends upon the local fluid velocity which may be written approximately as

$$p = \rho g (y_s - y). \tag{5.26}$$

A linear variation for pressure distribution is assumed from ($\rho g y_s$) at the bed to zero at the surface. Shoaling computations are done using Cnoidal theory, which best describes wave motion in relatively shallow waters. Simple and completely satisfactory procedures are not available for applying Cnoidal wave theory.

Figures 5.7(a) and 5.7(b) show the dimensionless Cnoidal wave surface profiles for various values of the square of the modulus of the elliptic integrals k^2. Figures 5.8–5.12 present dimensionless plots of the parameters, which characterize Cnoidal waves.

In Figs. 5.8 and 5.9, it is the exponent of $k^2 = 1 - 10^{-\alpha}$ that varies along the vertical axis. The values of k^2 should be read with care as they are close to 1.0. For complete description of Cnoidal wave theory, one can refer to Coastal Engineering Manual (2002).

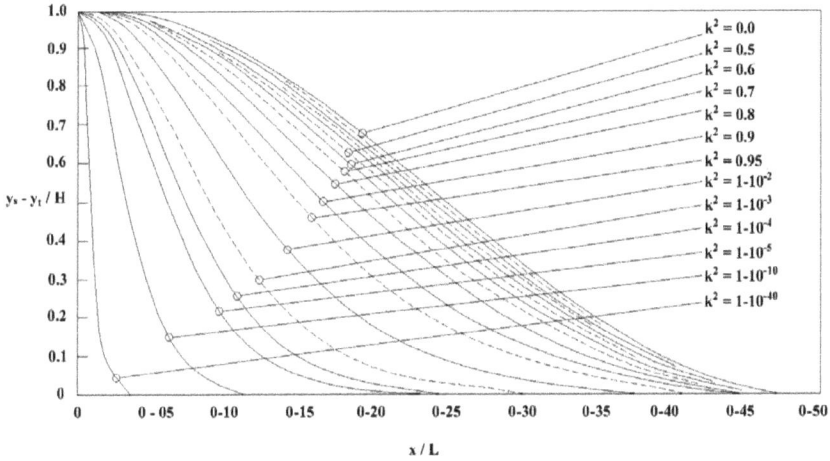

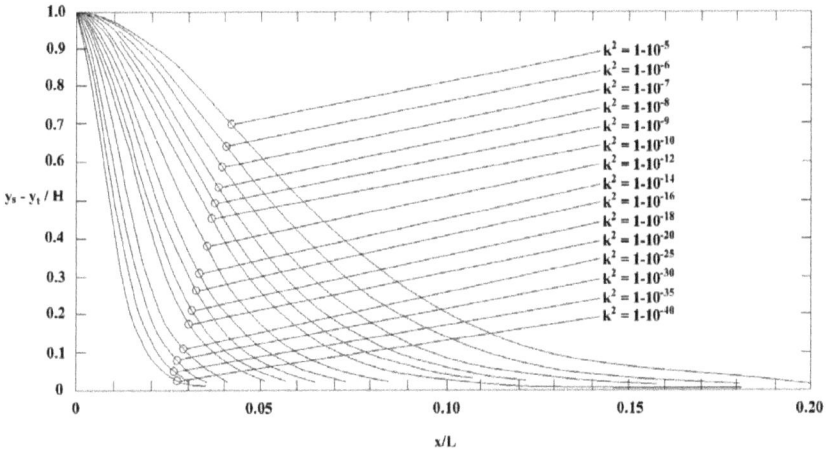

Fig. 5.7   Cnoidal wave profiles as a function of k^2 and x/L (Wiegel, 1960).

5.2.4   *Stream function theory*

Although the stokes waves satisfy the basic Laplace equation and the seabed boundary conditions, the free surface boundary conditions are not fully satisfied. The components of flow at the surface are not necessarily in accordance with the shape of the surface and its motions. Further, not a restriction placed on the pressure immediately below the free surface.

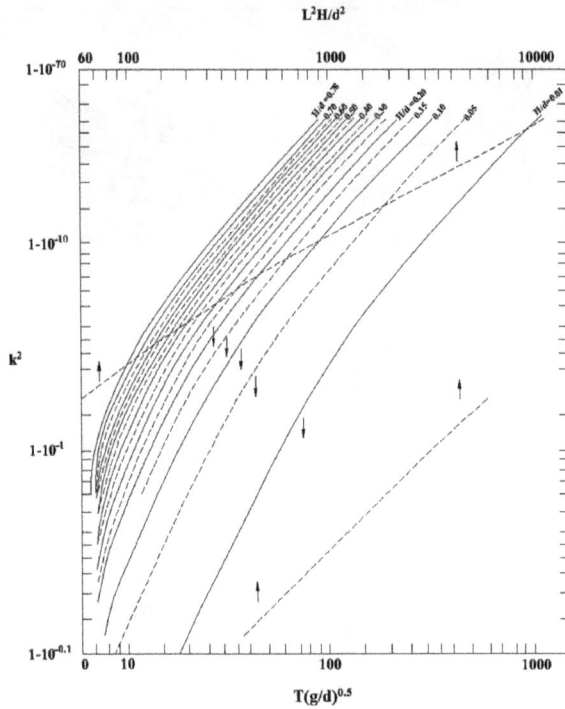

Fig. 5.8   Relationship between k^2, H/d and $T\sqrt{g/d}$ (Wiegel, 1960).

The dynamic free surface boundary condition is

$$\eta + \frac{1}{2g}[(u - C)^2 + w^2] - \frac{C^2}{2g} = \text{const.} \qquad (5.27)$$

The stream function solution may be expressed as

$$\psi(x, z) = \frac{L}{T}z + \sum_{n=1}^{N} X(n)\sinh\left[\frac{2\pi n}{L}(d + \eta)\right]\cos\left[\frac{2\pi n}{L}x\right] \qquad (5.28)$$

and evaluated by setting $z = \eta$ to give a surface

$$\eta = \frac{T}{L}\psi_n - \frac{T}{L}\sum_{n=1}^{N} X(n)\sinh\left[\frac{2\pi n}{L}(d + \eta)\right]\cos\left[\frac{2\pi n}{L}x\right]. \qquad (5.29)$$

For a particular water depth, wave height, and period, the function exactly satisfies the Laplace equation, the seabed, and the surface flow boundary conditions for arbitrary values of the constants L, ϕ_n and $X(n)$,

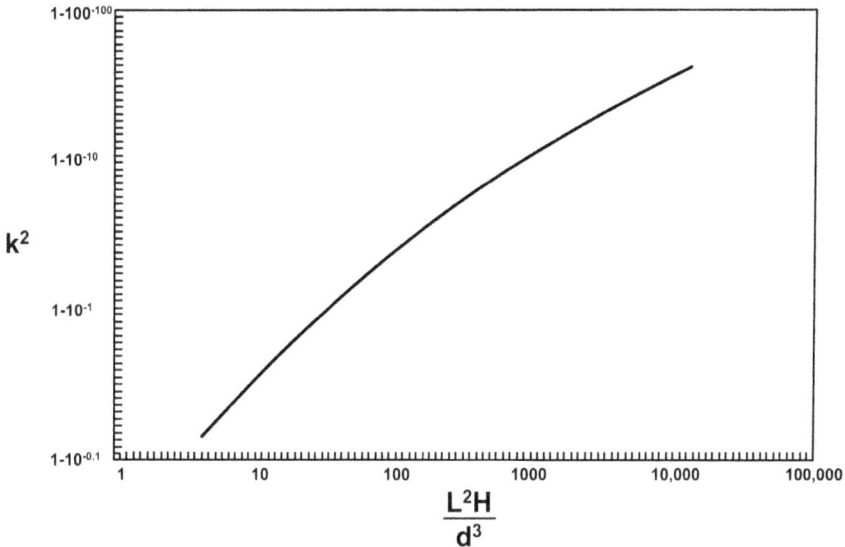

Fig. 5.9   Relationship between k^2 and L^2H/d^3.

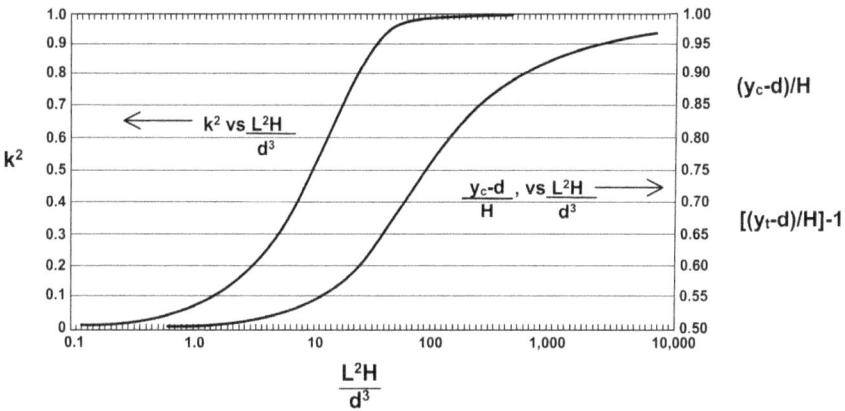

Fig. 5.10   Relationship between k^2, $T\sqrt{g/d}$ and L^2H/d^3 and between $[(y_c - d)/H]$ and $[(y_t - d)/H] - 1$.

respectively. These values can be obtained numerically so that the dynamic free surface boundary condition has been published by Dean (1965).

Comparison between the different wave theories and laboratory measurements (Dean, 1965), as shown, for example, in Figs. 5.13 and 5.14,

$$T\sqrt{\frac{g}{d}} = \sqrt{\frac{16d}{3H}\frac{k\,K(k)}{1+\frac{H}{dk^2}\left(\frac{1}{2}-\frac{E(k)}{K(k)}\right)}}$$

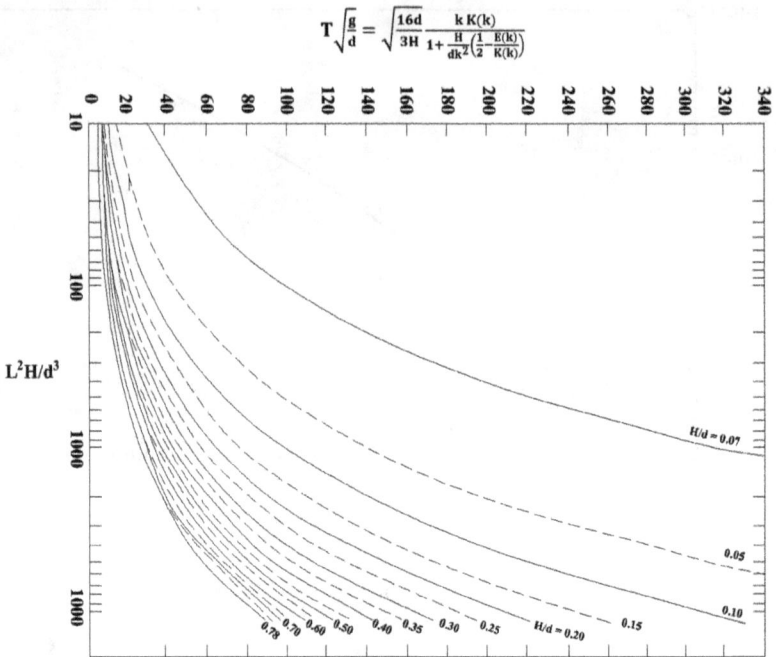

Fig. 5.11   Relationship between and L^2H/d^3 and $T\sqrt{g/d}$ for different H/d.

demonstrates the wide divergence of results which can arise from the various assumptions made the different wave theories.

Worked Out Examples

Problem 5.1

The following example best illustrates the use of the above figures.

Given a wave traveling in 3 m water depth with a period of 15 s and height 1 m.

(a) Using Cnoidal wave theory, find the wavelength L and compare this length with the length determined using Airy theory.
(b) Determine the celerity C. Compare this celerity with the celerity determined using Airy theory.
(c) Determine the distance above the bottom of the wave crest y_c and wave through y_t.
(d) Determine the wave profile.

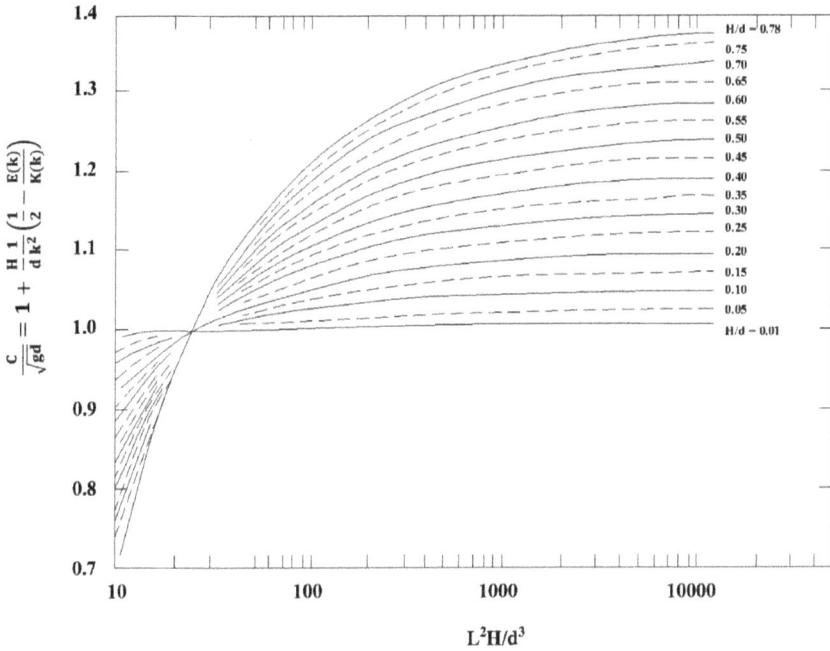

Fig. 5.12   Relationship between Cnoidal wave velocity and L^2H/d^3.

Solution

Given $d = 3\,\text{m}$, $T = 15\,\text{s}$, $H = 1\,\text{m}$.

(a) $H/d = 1/3 = 0.33$,

$$T\sqrt{\frac{g}{d}} = 15\sqrt{\frac{9.8}{3}} = 27.11.$$

Using Fig. 5.8 we can determine k^2 (square of the modulus of the complete elliptic-integral)

We get $k^2 = 1 - 10^{-5}$.

Figure 5.9 gives the relationship between k^2 and $L^2H/d^3 = 290$, i.e.,

$$L = 290\frac{d^3}{H}, \quad L = \frac{290(3)^3}{1}, \quad \text{i.e., } L = 88.5\,\text{m}.$$

By the Airy theory,

$$L = \frac{gT^2}{2}\tanh\frac{(2\pi d)}{L} = 80.6\,\text{m}.$$

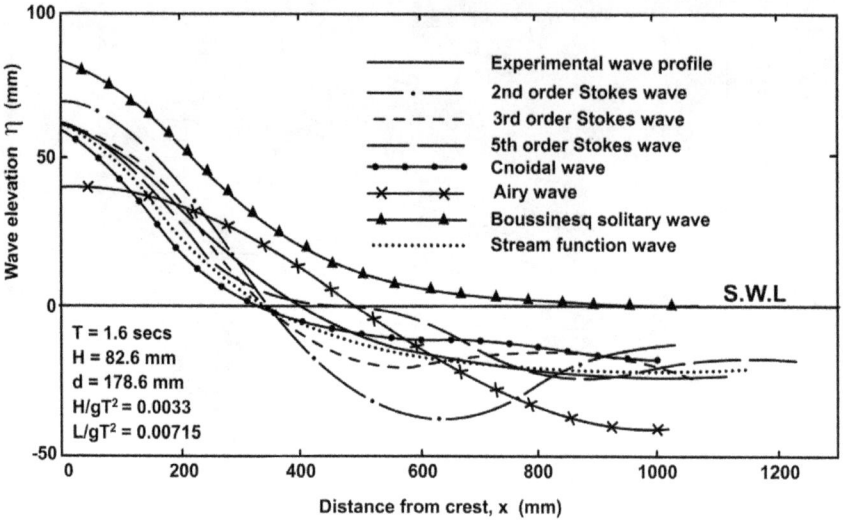

Fig. 5.13   Comparison between surface profiles (Dean, 1965).

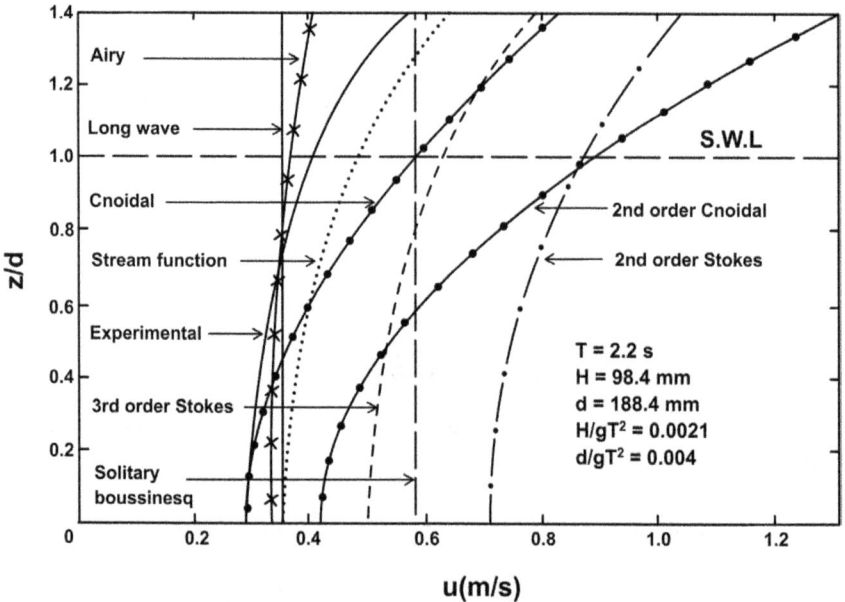

Fig. 5.14   Comparison between maximum horizontal velocity under the wave crest and along the depth (Dean, 1965).

To check the validity of Cnoidal wave theory for the given wave condition, we have

$$d/L = 3/88.5 = 0.0339 < 1/8.$$

Ursell or stoke's parameter,

$$L^2H/d^3 = 1/(d/L)^2(H/d) = 290 > 26.$$

Hence the Cnoidal theory is applicable.

(b) Wave celerity is given by

$$C = L/T = 88.5/15 = 5.9 \, \text{m/s}.$$

By Airy's theory,

$$C = L/T = 80.6/15 = 5.27 \, \text{m/s}.$$

Thus, if it is, assumed that the wave period is the same for both Cnoidal and Airy theories, we can say that

$$\frac{C_{\text{cnodial}}}{C_{\text{Airy}}} = \frac{L_{\text{cnoidal}}}{L_{\text{Airy}}} \cong 1.$$

(c) Figure 5.10 gives the percentage of the wave height above SWL. For $L^2H/d^3 = 290$,

$$(y_c - d)/H \text{ is found to be } 86.5\%.$$

Hence $y_c = 0.865H + d = 0.865(1) + 3$ and $y_c = 3.865 \, \text{m}.$
Similarly, from Fig. 5.10,

$$(y_t - d)/H + 1 = 0.865,$$

$y_t = (0.865 - 1)(1) + 3 = 2.865 \, \text{m}.$

(d) The dimensionless wave profile is shown in Fig. 5.8.
For $k^2 = 1 - 10^{-5}$, it can be seen that the SWL is approximately 0.14 H above the wave trough or 0.86 H below the wave crest.

Problem 5.2

A wave of height 1 m and length 60 m propagates in a water depth of 6 m. Which theory could be adopted for the evaluation of the water particle kinematics? For the wave with the same frequency, find the range of wave heights in which case stream function theory could be applied. Take care of the breaking criteria.

Solution

$d = 6\,\mathrm{m}$, $H = 1\,\mathrm{m}$, $L = 60\,\mathrm{m}$.

$d/L = 6/60 = 0.1$ and corresponding $d/L_0 = 0.056$, $L_0 = 107$, Hence, $T = 8.2\,\mathrm{s}$.

$H/gT^2 = 1/(9.81 * 8.2 * 8.2) = 0.0015$, $d/gT^2 = 0.009$.

Hence the Airy theory is valid for kinematics.

For stream function theory, H/gT^2 should lie between 0.017 and 0.006.

Using these limits, $H = 4\text{--}11.48\,\mathrm{m}$. However, the maximum wave height is $0.78 * 6 = 4.68\,\mathrm{m}$.

Hence H should range from $4\,\mathrm{m}$ to $4.68\,\mathrm{m}$.

Problem 5.3

Given water depth $= 20\,\mathrm{m}$, $T = 5$ to $15\,\mathrm{s}$. Find the range of wave heights in which case (a) stream function theory is applicable, (b) stokes theory is applicable and which order take T in intervals of $1\,\mathrm{s}$ and tabulate the results (remember the wave breaking).

Solution

Breaking wave height $= 0.78d = 0.78 \times 20 = 15.6\,\mathrm{m}$.

(a) Stream function theory

T	d/gT^2	H/gT^2	$H(\mathrm{m})$	Valid range of wave $H(\mathrm{m})$
5	8.15×10^{-2}	–	Not applicable	
6	5.66×10^{-2}	–	Not applicable	
7	4.16×10^{-2}	–	Not applicable	
8	3.18×10^{-2}	–	Not applicable	
9	2.15×10^{-2}	$1 \times 10^{-2}\text{--}2 \times 10^{-2}$	7.97–5.89	7.97–15.60
10	2.04×10^{-2}	$9.5 \times 10^{-3}\text{--}1.0 \times 10^{-2}$	9.31–9.83	9.31–9.83
11	1.68×10^{-2}	$7 \times 10^{-3}\text{--}0.5 \times 10^{-3}$	5.94–8.31	5.94–8.31
12	1.415×10^{-2}	$5 \times 10^{-3}\text{--}9 \times 10^{-3}$	7.06–12.71	7.06–12.71
13	1.21×10^{-2}	$5 \times 10^{-3}\text{--}1 \times 10^{-2}$	6.63–16.58	8.28–15.60
14	1.04×10^{-2}	$4 \times 10^{-3}\text{--}9 \times 10^{-3}$	7.69–17.304	7.69–15.60
15	9.06×10^{-3}	$2.3 \times 10^{-3}\text{--}6.8 \times 10^{-3}$	5.08–15.01	5.08–15.01

(b) Stokes wave theory

T	d/gT^2	H/gT^2 Second order	Third order	Fourth order	H Second order	Third order	Fourth order
5	8.15×10^{-2}	10^{-3}– 7×10^{-3}	8.5×10^{-3}– 1.7×10^{-2}	1.7×10^{-3}– 3×10^{-2}	0.245–1.716	2.084–4.169	4.17–7.36
6	5.66×10^{-2}	10^{-3}– 6.5×10^{-3}	8.5×10^{-3}– 1.6×10^{-2}	1.6×10^{-2}– 3×10^{-2}	0.353–2.295	2.825–5.65	5.65–10.59
7	4.16×10^{-2}	9×10^{-4}– 6×10^{-3}	7×10^{-3}– 1.2×10^{-2}	1.2×10^{-2}– 2.5×10^{-2}	0.4326– 2.884	3.364–5.768	5.77–12.01
8	3.18×10^{-2}	8×10^{-4}– 5×10^{-3}	6×10^{-3}– 0.8×10^{-2}	0.8×10^{-2}– 1.5×10^{-2}	0.4561– 2.851	3.421–4.56	4.56–8.15
9	2.516×10^{-2}	7×10^{-4}– 4.5×10^{-3}	5.7×10^{-3}– 0.5×10^{-2}	0.5×10^{-2}– 1×10^{-2}	0.556–3.575	4.529–3.97	3.97–7.94
10	2.04×10^{-2}	6×10^{-4}– 4×10^{-3}	5×10^{-3}– 9×10^{-3}	Not Applicable	0.5886– 3.924	4.905–8.829	Not Applicable
11	1.68×10^{-2}	5.2×10^{-4}– 3.8×10^{-3}	4.3×10^{-3}– 7×10^{-3}	Not Applicable	0.617–4.51	5.104–8.309	Not Applicable
12	1.415×10^{-2}	4.9×10^{-4}– 3×10^{-3}	4×10^{-3}– 7×10^{-3}	Not Applicable	0.69–4.238	5.65–9.89	Not Applicable
13	1.21×10^{-2}	4.3×10^{-4}– 2.8×10^{-3}	3.5×10^{-3}– 5×10^{-3}	Not Applicable	0.713–4.62	5.803–8.29	Not Applicable
14	1.04×10^{-2}	4.1×10^{-4}– 2.5×10^{-3}	3.2×10^{-3}– 4.5×10^{-3}	Not Applicable	0.788–4.806	6.152–8.652	Not Applicable
15	9.06×10^{-3}	3×10^{-4}– 1.7×10^{-3}	Not Applicable	Not Applicable	0.662–3.752	Not applicable	Not Applicable

Chapter 6

Characteristics of Random Waves

6.1 Generation of Ocean Waves

If one blows over a container filled with still water, its surface oscillates according to the intensity of blowing air, its duration as well as the distance over which the blowing is effective. Similarly, wind blowing the disturbance of the ocean surface takes place due to the action of tangential and normal stresses of the wind induced on the sea surface and hence waves are generated. Initially, ripples, short period or high frequency waves are generated. Gradually, the contribution of energy towards the lower frequency waves takes place. The energy continuously transfers from high frequency to low frequency until the phase velocity or the celerity of the wave C is equal to the wind velocity V, i.e., $C = V$ and in more general, the group celerity, $C_g = V$ is considered. When $C_g > V$, the transfer of energy from the wind does not contribute any more towards the growth of the waves. This means that waves reach equilibrium at certain extent. This equilibrated sea state is called *fully developed sea* (FDS). The wind wave development and generation thus depend significantly on the ratio of C and V, which is termed *wave age*. Before reaching FDS, wave age may range from 0.1 to 2.0.

If we notice the ocean surface for a few minutes, the height and length of the wave will not be found to be a constant. The water surface will be seen to vary at random in both space and time and the statistical properties of the waves such as mean height averaged over say, hundred waves change from day to day. In general, the higher the wind velocity, the longer the fetch over which it blows; and the longer will be the average waves. Waves still under the action of the winds (within Fetch) that created them are called wind waves, or a sea. If the waves propagate out of the fetch region, the waves try to evolve in length. The period of swells that propagate outside the fetch area can be 8 s to 25 s even though the general perception is that swells are very long waves.

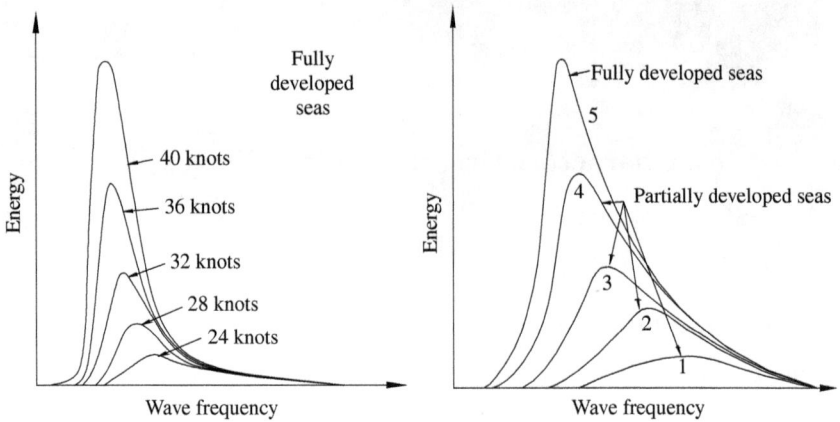

Fig. 6.1   Growth of wind-generated ocean waves.

The growth of waves with the generation of longer waves, the shifting of the energy peak to the low frequency zone due to the continuous transfer of energy from as well as due to the increase in the wind speed are represented in Fig. 6.1. The FDS as well as the partially developed sea, PDS are also indicated. Prior to the waves attaining their equilibrium, if the wind stops blowing, the sea state is *underdeveloped* compared to its potential. This sea state is called *duration limited*. On the other hand, if there is a constraint in the space for the wave propagation or for them to grow, the sea state then is called "fetch limited" sea.

Wind is the exciting force in this case and gravity is the restoring force. Hence, the sea is developed as a vertically oscillating component influenced by the gravity, g, because of which, the wind-waves are also called gravity waves.

From the foregoing discussion, FDS $= f(V, g)$

$$\text{Duration limited Sea} = f(V, t_d, g)$$

$$\text{Fetch limited Sea} = f(V, F, g)$$

Once it is known that any sea state has an equilibrium sea state for a particular wind climate, it is obvious to seek a solution to arrive at the wave characteristics such as significant wave height and mean wave period reasonably well directly through several of the theories in existence. Standard spectral model such as PM (Pierson–Moskowitz) spectrum evolves from the FDS state.

6.2 Collection of Wave Data

The measuring techniques now practically applied can be divided into two types. The first one measures a physical quantity which is closely related to the sea surface fluctuation, and is called the indirect measuring technique. The second one measures directly the fluctuation in the sea surface by some means, and is called the direct measuring technique. These two types can also be subdivided further according to the measuring principle. Representative techniques among these are explained briefly below.

The modest way is to observe the sea and compose a visual assessment, nevertheless a substantial experience is essential to acquire reliable data. The second method involves observation of surface water against a vertical scale erected over a pier in case of shallow water, or on a float, which is damped to limit short period vertical oscillations in deep water. Yet another method that is often adopted is to house a pressure-sensor at an elevation below the MSL or closer to the seabed. The pressure sensor continuously records the water level fluctuations from which the wave elevation time history can be derived after adopting a few corrections. This method is frequently used for measurements in the coastal zone. The response of the pressure wave gauge is not fast enough for short period waves. Therefore, it is commonly said that the present type of wave gauge cannot be used for waves shorter than about 3 s. In the strict sense, the lower limit of the wave period to be measured depends upon the water depth and the elevation of the pressure gauge above the sea bottom. In order to measure the pressure fluctuation, numerous pressure gauges are used, such as the variable resistance type, direct recording type, strain-gauge type, pressure difference detector type, differential transformer type.

As seawater is an electrolyte, information on height variations of the sea surface can be obtained using electrical devices by measuring the changes in resistance or capacitance of a vertical conductor placed in seawater, as the water surfaces move up and down. A step resistance wave gauge operates by switching an electric circuit on and off through electrodes attached to a vertical pole, corresponding to the sea surface variation. In the case of an ultrasonic wave gauge, an ultrasonic pulse is emitted toward the sea surface from a transceiver held under the water or in the air, and the time, t, required for the pulse to return from the sea surface is measured. The distance from the transceiver to the sea surface, 1, is calculated by the equation $l = ct/2$, where c is the sonic speed. This type of wave gauge is not suitable for use in places where the water conditions are complex, or where there is heavy navigation.

A ship-borne wave recorder measures the wave characteristics in deep water by measuring the pressure at a point on the hull to give the wave height relative to the ship.

There are two types: one is the Tucker type which consists of a pressure gauge and accelerometer placed on the ship's bottom, and the other is the Mark type which is an ultrasonic wave gage hanging off the bow. The principle of this type is to measure the vertical acceleration of the buoy due to wave motion and to twice integrate this measurement electronically to obtain the vertical displacement of the buoy. This method is now used extensively.

All the above methods provide information on waves only at a point. Ocean waves are spatial and time-dependent phenomena; hence the surface elevation should be expressed by $\eta(x, y, t)$. All the foregoing methods measure the surface fluctuation time history $\eta(t)$. Therefore, the data obtained by these methods are not adequate to obtain a detailed picture of ocean waves. To compensate for their inadequacy, several trials are made to obtain a surface elevation spatial distribution, $\eta(x, y)$, at a certain time. Stereo photographs are taken by using cameras installed separately in two airplanes. In recent years, wave information over a wide area in a short time has been obtained by radar or laser mounted on an airplane, or by microwave sensor from a satellite.

Spatial information on waves can be obtained from stereo photographs or by taking records on the sea surface using altimeters mounted on aircraft or satellite. Satellite measurements or microwave scattering give very nearly correct information on both wave and wind speed and direction.

Directional information of the ocean waves can be obtained by recording waves at different locations and examining their corresponding phase relationships simultaneously through a method of analysis. The alternate method is to measure the waves with a pressure sensor, velocity fields in two directions in the horizontal plane simultaneously usually referred to as single point measurement and later carry out a rigorous analysis campaign. The above-stated information can be also through stereo photographs or from measurements made from aircrafts or satellites. The most common wave direction measurement method is to use a transit compass from an elevated position such as a coastal cliff, coastal dune, or tower. Wave direction is clearly discerned on aerial photographs. The Rayleigh disk has been developed to measure the direction of incoming waves based on the principle that a lightweight disk supported in the vertical plane is at rest perpendicular to the direction of oscillatory flow. Another way is to use two electromagnetic current meters set at cross directions to each other. The principle of this instrument is to continuously measure the two horizontal components of wave force acting on a sphere, and to determine the predominant wave direction by analyzing the records.

6.3   Analysis of Ocean Waves

6.3.1   *General*

Although, the ocean waves are assumed to be monochromatic long-crested waves, their striking feature is its irregularity. The said feature characterizes them into long-crested irregular waves and short-crested irregular waves. The ideal arrangement would be to record the waves continuously as shown in Fig. 6.2(a), but this would result in far too much data to be analyzed. No doubt, the accuracy of any statistical analysis of the data or a random variable such as the wave height increases with the sample size (number of waves). Cartwright (1958) has mathematically analyzed this aspect and his results are shown in Fig. 6.2(b). It can be seen that standard error in the statistical averages worked out on the basis of 100 consecutive waves

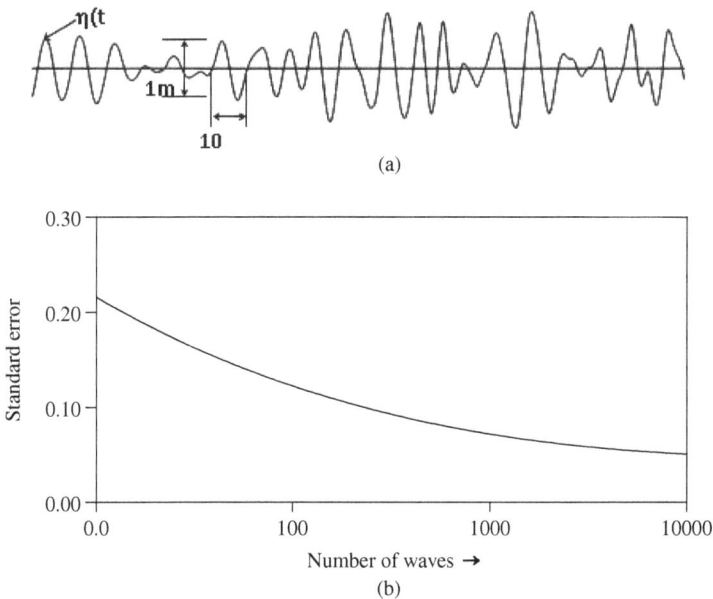

Fig. 6.2   (a) Typical sample of wave record. (b) Standard error vs. number of waves (Cartwright, 1958).

is about 12% and if number of consecutive waves considered is 10,000, the standard error is about 6.5%. Hence, considering a very large sample size does not appreciably increase the accuracy. It is a standard practice of record once in every hour for a period of 20 min such that at least about 100 waves are recorded each time.

It is suggested that the reader has a basic knowledge on statistical analysis and random process, which is also provided in Appendix C.

6.3.2   *Sea waves as a stationary ergodic process*

It is possible to generate in the laboratory a number of wave records, $\eta_1(t), \eta_2(t), \ldots$ each of which having the same significant wave height, H_s and peak period, T_p. This cannot occur in the real ocean. The above simulated wave records may be expressed as a stochastic process as

$$\eta(t) = \{\eta_1(t), \eta_2(t), \eta_3(t), \ldots, \eta_j(t), \ldots\}. \tag{6.1}$$

The braces $\{\}$ in the above equation indicate as an ensemble which would mean that the sample records are considered to have come from the ensemble. Furthermore, we assume that the probability density function of $\eta(t)$ is given not for one sample $\eta_j(t)$ but for the ensemble itself at a particular time t. For an ensemble to be a stationary process, it would require that all the statistical properties as ensemble means are time invariant. For instance, the stationary condition applied to the arithmetic mean is time invariant and autocorrelation function gives

$$E[\eta(t)] = E[\eta(0)] = \eta, \quad -\infty < t < \infty, \tag{6.2}$$

$$E[\eta(t+\tau)\eta(t)] = E[\eta(\tau)\eta(0)] = R(\tau), \quad -\infty < t < \infty \tag{6.3}$$

in which

$$E[\eta(t)] = \lim_{n \to \infty} \frac{1}{N} \sum_{j=1}^{N} \eta_j(t), \tag{6.4}$$

$$E[\eta(t+\tau)\eta(t)] = \lim_{N \to \infty} \frac{1}{N} \sum_{j=1}^{N} \eta_j(t+\tau)\eta_j(t). \tag{6.5}$$

When a process satisfies the conditions of Eqs. (6.2) and (6.3), it is good enough to state that it is stationary stochastic process in the strict sense, and all other statistics become stationary. If the time-averaged statistics for a particular sample $\eta_j(t)$ are equal to those of the ensemble average, the process can be termed as ergodic; that is,

$$E[\eta(t)] = \bar{\eta}_j(t) = \lim_{t_0 \to \infty} \frac{1}{t_0} \int_0^{t_0} \eta_j(t)dt, \tag{6.6}$$

$$E[\eta(t+\tau)\eta(t)] = \overline{\eta_j(t+\tau)\eta_j(t)} = \lim_{t_0 \to \infty} \frac{1}{t_0} \int_0^{t_0} \eta_j(t+\tau)\eta_j(t)dt. \tag{6.7}$$

A stochastic process that has ergodicity is always a stationary process, but the reverse need not be true. A Gaussian process indicates that the probability density of wave profile will follow a Gaussian distribution that will be discussed later. For further details, the reader is suggested to refer Goda (2010). As we know that the basic wave characteristics like H_s and T_p from measurements in the ocean will not be the same and vary with time, it is not expected to be a stationary process. However, the sea state over a short duration of several minutes may be assumed to be stationary for short wave records. Based on the foregoing discussion, we apply the theory of stationary stochastic process for the analysis of random sea waves.

6.4   Detailed Analysis

6.4.1   General

The different methods of analysis of ocean waves are shown in Fig. 6.3. Two methods of analysis of waves are the statistical procedure and the spectral method. In the case of the statistical procedure, individual waves are defined by zero up crossings or zero down crossings. The unit of analysis

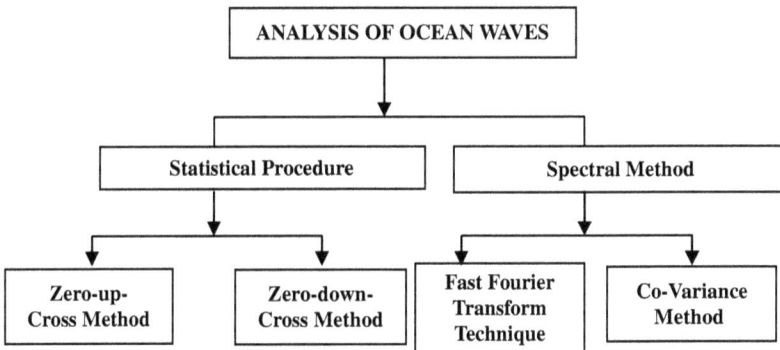

Fig. 6.3   Methods of ocean wave analysis.

in the case of spectral method is $\eta(t)$ which is read at a fixed and a very small time interval Δt.

The time series $\eta(t)$, wave record is digitized at a constant time interval Δt. The amplitude of the time series is read from base line, which should be well below the lowest trough. All such data points are subtracted with the distance between the mean line and the base line. The mean of the data points (amplitude of the time series) is found and then the time series is converted to mean zero process. The $\eta(t)$ has to be normalized as $\eta(t) = (\eta(t)/\bar{\eta})$ where $\bar{\eta}$ is the mean. The initial exercise would be to check if $\eta(t)$ follows a normal distribution or is a Gaussian process. The probability density function, pdf of normal distribution is given by

$$P(x_i) = \frac{1}{\sigma\sqrt{2\pi}} e^{[-(x_i-\bar{x})^2/2\sigma^2]}, -\infty < x_i < \infty, \tag{6.8}$$

where "σ" is the standard deviation.

6.4.2   Statistical procedure

6.4.2.1   General

The individual waves in the record are evaluated based on either zero up cross (ZUC) or zero down cross (ZDC) methods. A portion of the time series through which the individual waves are derived based on both methods is projected in Fig. 6.4. The H_{u*}, T_{u*} are through ZUC, whereas H_{d*}, T_{d*} waves are defined as ZDC. It has been proved by researchers that difference in wave heights based on these two methods is small. However, wave periods, defined by ZUC method is slightly more than that of ZDC method. There could be a slight difference in the probability density estimated through both methods like the ZDC method might yield a bimodal function, whereas the ZUC might exhibit a unimodal function for the same time series.

6.4.2.2   Short-term wave statistics

On arriving at the individual wave heights, H and periods, T the observed pdf of these two variables should be compared with the theoretical Rayleigh distribution function given as

$$P_R(H_i) = \frac{\pi H_i}{2\bar{H}^2} e^{-(\pi/4)[H_i/\bar{H}]^2} \quad \text{for } H >= 0. \tag{6.9}$$

If $\eta(t)$ is a Gaussian process, it is usual that H would follow the Rayleigh distribution. If H does not follow the Rayleigh distribution, certain other

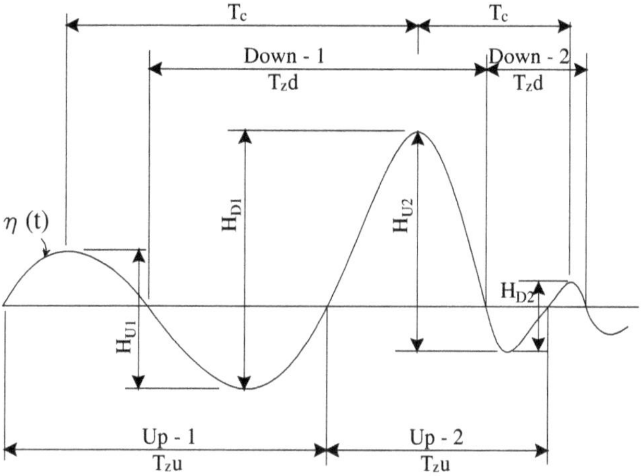

Fig. 6.4 Definition of waves with ZUC and ZDC methods.

standard distributions could be tried and Weibull distribution is another distribution which is commonly applied and is given by

$$P(H_i) = \alpha\beta H_i^{(\beta-1)} e^{(-\alpha H_i^\beta)}. \tag{6.10}$$

The peakedness coefficient β is given by

$$\beta = 4\int_0^\infty H_i f(H_i)dh \quad \text{and} \quad \alpha = [\Gamma(1+1/\beta)/\bar{H}]^\beta,$$

where $\bar{H}$ is the mean wave height and Γ is gamma function. It should be noted that the above distribution becomes Rayleigh if $\beta = 2$ and $\alpha = 1/H_{rms}^2$.

With the individual wave heights, one can derive the statistical wave characteristics like $H_{1/3}$ (Average of highest 1/3 waves termed as *significant wave height*), $H_{1/10}$ (Average of highest 1/10th waves), $H_{1/100}$ (Average of the highest of 1/100th waves), H and H_{rms}. With the wave periods of individual waves, the respective wavelengths could be evaluated using the dispersion relationship discussed in Chapter 3.

The root mean square wave height H_{rms} of a wave time history apart from its definition can also be evaluated by the method of Tucker (1963) which is quite simple and on the assumption it is a narrowband spectra (discussed later in this chapter) and can be determined as follows:

$$H_{\mathrm{rms}} = \sqrt{2}H_1(2\theta)^{-1/2}(1+0.289\theta^{-1}-0.248\theta^{-2})^{-1} \qquad (6.11a)$$

$$= \sqrt{2}H_2(2\theta)^{-1/2}(1-0.289\theta^{-1}-0.103\theta^{-2})^{-1}, \qquad (6.11b)$$

where "H_1" is the distance between the highest crest and the lowest trough, "H_2" the distance between the second highest crest to the second lowest trough and $\theta = \log N_z$, where N_z is the number of zero up-crosses in the selected wave record. The other wave characteristics like $H_{\max}$, H_s and $H_{1/10}$ as a function of H_{rms} can be determined through the relationships given below.

$$H_{\max} = H_{\mathrm{rms}}2(2\theta)^{1/2}(1+0.298\theta^{-1}-0.247\theta^{-2}),$$

$$H_s = 4H_{\mathrm{rms}},$$

$$H_{1/10} = 5.09H_{\mathrm{rms}}. \qquad (6.12a)$$

The average zero crossing period T_z and the average crest period T_c are evaluated as follows:

$$T_z = (\text{Duration of the record/No. of zero up or down crosses})$$

$$T_c = (\text{Duration of the record/No. of crests}).$$

If we adopt Rayleigh distribution as an approximation to the distribution of individual wave heights, according to Longuet-Higgins (1952)

$$H_{1/10} = 1.27H_{1/3} = 2.03\bar{H}, H_{1/3} = 1.60\bar{H}$$

$$\text{or,} \quad \bar{H} = 0.885H_{\mathrm{rms}}, \quad H_{1/3} = 1.416H_{\mathrm{rms}},$$

$$H_{1/10} = 1.8H_{\mathrm{rms}}, \quad H_{\max} = 2.172H_{\mathrm{rms}}. \qquad (6.12b)$$

Further the ratio between H_{max} and $H_{1/3}$ was proposed to vary between 1.6 and 1.8. In the case of random wave studies, if the above ratio exceeds 2.0 it is termed *Freak waves*. Freak or Rogue waves are unusually large, which when occurs can lead to catastrophe for vessels exposed to such waves (Sundar *et al.*, 1999).

These results represent the mean values of wave records taken together. Individual wave records containing less than 100 waves may result in deviations from the above mean relations.

The above relations are valid for the narrowband spectrum. For the broadband spectrum Cartwright and Longuet-Higgins (1956) suggested that Eq. (6.12b) be multiplied by $(1-\varepsilon^2)^{1/2}$ where, the spectral width parameter, ε indicating the shape of the spectrum or the range of frequencies over which the energy is spread is given by

$$\varepsilon = \sqrt{1 - (T_c/T_z)^2}. \tag{6.13}$$

6.4.3   *Wave–wave spectrum relationship*

One of the fundamental premises of the spectral approach is that irregular waves are the results of the super-position of an infinite number of simple sine waves of small amplitudes that have a continuous frequency distribution. This process can be approximated with a finite number of small amplitude sine waves having discrete frequencies. Under these conditions, the mean total wave energy per unit surface area is given by

$$E = \frac{\rho g}{8}(H_1^2 + H_2^2 + \cdots + H_n^2), \tag{6.14}$$

where H_n is the wave height associated with a given discrete angular frequency, $\omega_n = 2\pi f_n$ and f_n is the linear frequency in Hz of the nth component. The frequency distribution of energy is called the energy spectrum. The reader is suggested to see Problem 6.1 for a clear understanding.

The useful information that can be observed from a spectral density curve are as follows:

(i) The range of frequencies those are important for the contribution to the seaway.
(ii) The frequency at which the maximum energy occurs or supplied.
(iii) The content of energy at different frequency bands.
(iv) The existence of a swell at low frequencies.

6.4.4   *Spectral method*

6.4.4.1   *Through autocorrelation*

In order to perform the spectral analysis of the random wave record, the number of lags is determined $(0, 1, 2, \ldots, m)$ where $m = 10\%$ of the number of data points (say).

The autocorrelation function R_r is estimated for these lags:

$$R_r = \frac{1}{N-r} \sum_{i=1}^{N-r} x_i x_{i+r,r=0,1,2,\ldots,m}. \tag{6.15}$$

The smoothed spectral density function estimates are computed for these lags, $s_\eta(f)$:

$$S_\eta(f) = S_\eta\left(\frac{Rf_c}{m}\right) = 2\Delta t\left[R_0 + 2\sum_{r=11}^{m-1} D_r R_r \cos\left(\frac{\pi rk}{m}\right)\right], \qquad (6.16)$$

where $k = 0, 1, 2, \ldots, m$, f_c is cut-off frequency $= 1/2\Delta t$, and

$$D = 1/2(1 + \cos(\pi r/m)) \quad \text{for } r = 0, 1, 2, \ldots, m$$

$$= 0 \quad \text{for } (r > m),$$

where $S_\eta(f)$ is linear spectral density estimates at different frequencies, f.

The spectral density estimates can be checked by determining the area under the spectral density curve ($S_\eta(f)$ vs. f) which is equivalent to zero spectral moment (m_0) and comparing this with the autocorrelation function at zero log (R_0) and the square of the standard deviation of η, (σ_η^2) originally estimated, i.e.,

$$m_0 = R_0 = (\sigma_\eta^2). \qquad (6.17)$$

Once "m_0" is evaluated, the significant wave height, H_s or H_{m0} can be obtained as $4\sqrt{m_0}$. The different wave characteristics from a spectrum can be evaluated from the nth order of spectral moment given by

$$m_n = \int_0^\infty f^n S_\eta(f)df. \qquad (6.18)$$

The different characteristics of the measured wave record as per the frequency-domain analysis can be derived as given in Table 6.1.

Note that the above characteristics should have to be multiplied with a correction factor $(1 - \varepsilon^2)^{1/2}$ in case the observed spectrum is broadband.

Table 6.1   Seaway characteristics based on spectral method.

Characteristic	Relationship
ε^2	$1 - m^2/m_0 m^4$
$\bar{H}$	$2.5\sqrt{m_0}$
H_s	$4\sqrt{m_0}$
$H_{1/10}$	$5.09\sqrt{m_0}$
$H_{1/100}$	$6.67\sqrt{m_0}$
T_c	$\sqrt{m_2/m_4}$
T_{av}	$\sqrt{m_0/m_2}$

6.4.4.2 *Through fast Fourier transformation method*

The faster method of deriving the spectrum from a time series is through the fast fourier transformation (FFT) technique. The $\eta(t)$ is represented in the form of Fourier series with a_n and b_n as the components of the time series

$$\eta(t) = \sum_{n=1}^{N/2} \left[a_n \cos \frac{2\pi i \Delta t}{T} + b_n \sin \frac{2\pi n i \Delta t}{T} \right]. \tag{6.19}$$

Herein $\eta(t)$ is the water surface elevation to have zero mean. The components a_n and b_n are available at the fundamental frequency $(1/T)$, the integral multiples theory up to Nyquist frequency, $1/2\,\Delta t$, and there are $N/2$ components of which a_0, the mean, is one, but b_0 and $b_{n/2}$ being identically zero are not included. Our best estimation of the variance in the process at the frequency $n/T (= f_n)$ is associated with frequency interval of $1/T = \Delta f$. Thus, we can define our elementary sample estimate of the spectrum by

$$\hat{S}_\eta(f_n)\Delta f = 1/2(a_n^2 + b_n^2), \tag{6.20}$$

i.e.,

$$\hat{S}_\eta(f_n) = 1/2\Delta f(a_n^2 + b_n^2) = T/2(a_n^2 + b_n^2). \tag{6.21}$$

6.5   Presentation of Wave Characteristics

On evaluating the H_s and T_z from a number of wave records, the summary of the wave characteristics according to months, seasons or annual represents the weather window that helps in the planning of offshore operations. This is prepared in the form of a scatter diagram similar to the one projected in Fig. 6.5(a), which gives the number of occurrences of a particular combination of H_s and T_z.

   The wave characteristics are reported as percentage of occurrence which is a slight extension of the scatter diagram, a sample of which for Chennai harbor based on an analysis of wave data for the duration April 1974 to March 1984 (Sundar (1986)) is projected in Fig. 6.5(b). To include the wave direction, the data in the form of wave rose plots are presented as reported in Fig. 1.7(c). For most of the operations in the marine environment, it would be essential to have information on the duration of the time over which the wave heights are less than a critical value. For example, the limiting value of wave height for a vessel operating in the offshore. This information

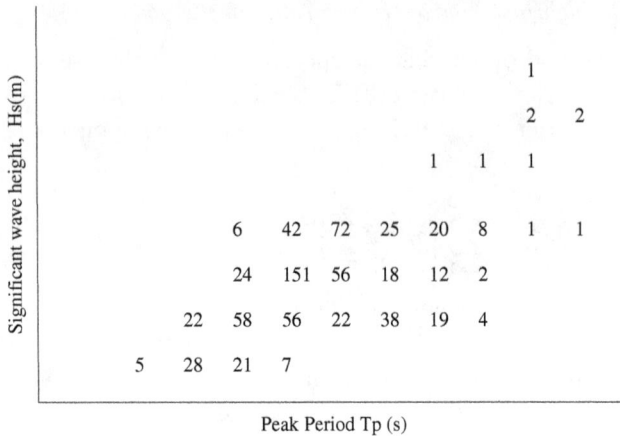

Significant wave height, Hs(m)

```
                                                              1

                                                        2     2

                                            1     1     1

              6     42    72    25    20    8     1     1

              24    151   56    18    12    2

        22    58    56    22    38    19    4

  5     28    21    7
```

Peak Period Tp (s)

(a)

Wave height groups in (m)	Wave period groups in sec						Total
	4–6	6–8	8–10	10–12	12–14	14–16	
0.4–0.6	–	0.79	9.71	12.97	3.79	0.11	27.37
0.6–0.8	–	0.86	10.12	7.91	0.86	–	19.75
0.8–1.0	–	0.22	5.74	4.09	0.22	–	10.27
1.0–1.2	–	2.55	13.80	5.81	0.86	–	23.02
1.2–1.4	–	1.50	3.56	2.47	0.11	–	7.64
1.4–1.6	–	2.32	4.84	0.94	0.07	–	8.17
1.6–1.8	–	0.52	0.41	0.34	–	–	1.27
1.8–2.0	–	0.30	0.26	0.45	0.04	–	1.05
2.0–2.2	0.07	0.26	0.49	0.22	–	–	1.05
2.2–2.4	0.04	0.04	–	–	–	–	0.08
2.4–2.6	–	0.15	0.15	–	–	–	1.30
Total	0.11	9.51	49.08	35.20	5.95	0.11	100

(b)

Fig. 6.5 (a) Wave scatter diagram. (b) Percentage of occurrence of wave height and period for Chennai harbor.

may be prepared either month-wise or season-wise as stated earlier. Typical persistence diagram reporting the number of occurrences, the wave height is greater than a critical value as well as its duration of its occurrence is shown in Fig. 6.6(a). Similarly, typical persistence diagram reporting the number of occurrences, the wave height is lesser than a critical value as well as its duration of its occurrence is shown in Fig. 6.6(b) (Haskins and Leggett, 1983). The above kinds of information help in the planning of offshore operations.

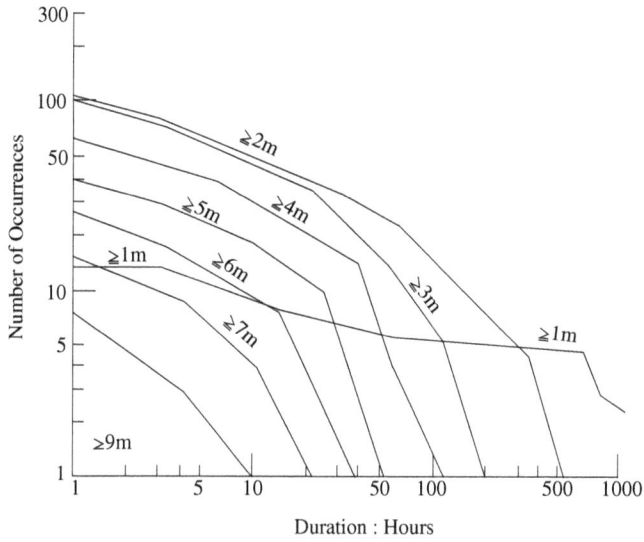

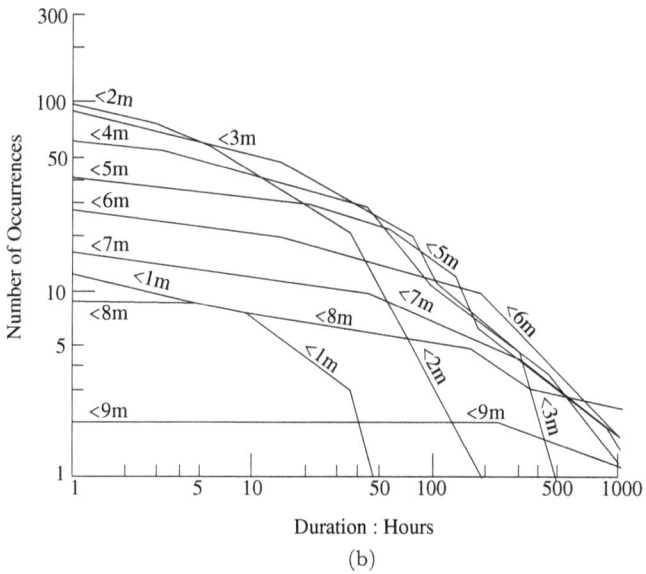

Fig. 6.6   Persistence diagram (a) wave height $> H$ and (b) wave height $< H$ (Haskins and Leggett, 1983).

6.6   Ocean Wave Prediction Models

6.6.1   *Pierson–Moskowitz spectrum*

For a FDS, following the formulation of Pierson and Moskowitz (1964), the estimates for significant wave height (H_s or H_{m0}) and peak wave frequency (f_0) are derived as follows:

$$H_{m0} = 0.21V^2/g, \tag{6.22}$$

$$f_0 = 0.87g/(2\pi V), \tag{6.23}$$

where "V" the wind speed is in m/s, and g is the gravitational constant. The above formulas were derived for wind speeds between 10 and 20 m/s and it was assumed that the sea was neither fetch limited nor duration limited.

6.6.2   *SMB wave prediction curves*

A wave prediction procedure based on wave energy growth concepts with empirical calibration using a limited amount of field data was developed by Sverdrup and Munk (1946) which was further improved by Bretschneider (1952, 1958) by calibrating using vast field data. The method is known as the SMB method after the three authors. Consider a dimensional analysis of the basic wave prediction relationship,

$$H_s, T_s = f(V, F, t_d, g). \tag{6.24}$$

Depending on whether the wave generation is fetch or duration-limited, the fetch or the duration term on the right side would control the estimation:

$$\frac{gH_s}{V^2} = 0.283\tanh\left[0.0125\left(\frac{gX}{V^2}\right)^{0.42}\right], \tag{6.25}$$

$$\frac{gT_s}{2\pi V} = 1.2\tanh\left[0.077\left(\frac{gX}{V^2}\right)^{0.25}\right], \tag{6.26}$$

$$\frac{gt_d}{V} = Ke^{\left[\left\{A\left(1n\left(\frac{gX}{V^2}\right)^2\right)-B1n\left(\frac{gX}{V^2}\right)+C\right\}^{0.5}+D1n\left(\frac{gX}{V^2}\right)\right]}, \tag{6.27}$$

where $K = 6.5882$, $A = 0.0161$, $B = 0.3692$, $C = 2.2024$ and $D = 0.8798$.

The above relation has been presented in the form of empirical equations and dimensional plots and is shown in Fig. 6.7 (USACE, 2006).

For a fetch-limited wave condition, the solid lines can be used to predict the significant wave height and period. For a duration-limited wave

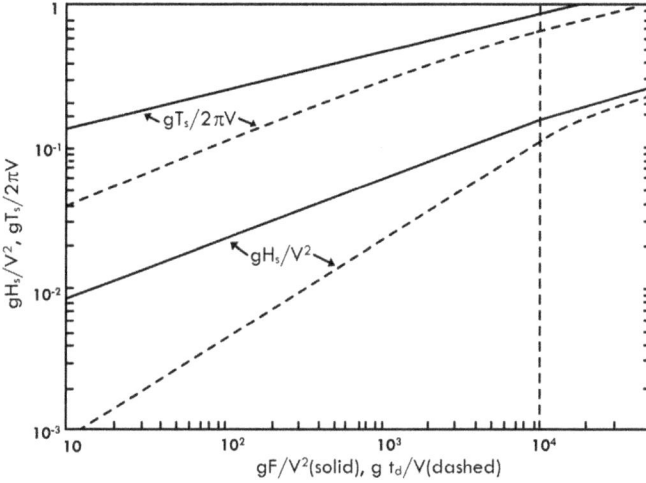

Fig. 6.7   SMB wave prediction curves (Bretschneider, 1952, 1958).
Note: W is the wind speed, t_d is the duration of wind and F is the fetch.

condition, the dashed line can be used. Note that the parameters, fetch, duration, significant wave height and wave period were dimensionless in terms of wind speed. The curves tend to become asymptotic to each other and horizontal lines on the right-hand edge. This limit is the fully developed sea condition.

6.6.3   *SPM: Deep water wave prediction*

A parametric model based on the Joint North Sea Wave Project (JON-SWAP) studies has estimated wave characteristics under both the fetch limited and duration limited wind conditions.

For fetch limited condition:

$$\frac{gH_{m0}}{V_A^2} = 0.0016 \left(\frac{gF}{V_A^2} \right)^{1/2}, \tag{6.28}$$

$$\frac{gT_p}{V_A} = 0.286 \left(\frac{gF}{V_A^2} \right)^{1/3}. \tag{6.29}$$

For duration limited condition:

$$\frac{gt_d}{V_A} = 68.8 \left(\frac{gF}{V_A^2} \right)^{2/3}. \tag{6.30}$$

Here, the wind is adjusted to V_A from V_{10} as, $V_A = 0.71V_{10}^{1.23}$. It has to be noted that the above expressions have empirical coefficients and hence it is sensitive to units. Wind speed is given in terms of m/s.

6.7   Standard Representation of Wave Spectra

6.7.1   *General*

Similar to the representation of short-term wave statistics through theoretical probability distribution as discussed under Section 6.4.2, the measured spectrum can also be checked if it follows an established standard spectrum. Further, it is preferred to generate $\eta(t)$ of pre-defined spectral characteristics for the testing of structures in the laboratory. A number of mathematical spectrum models have been proposed by researchers in the past. These have been tested with measured wave elevations and are generally a function of significant wave height and/or wave period and in certain types a shape factor is associated. The widely adopted single-parameter spectrum is the Pierson and Moskowitz (1964) model based on the significant wave height or wind speed. Bretschneider (1969), Scott (1965), ISSC (1964) and ITTC (1966) have proposed two parameter spectra that are also widely adopted. Joint North Sea Wave Project (JONSWAP) spectrum described by Hasselman (1973, 1976) is a five-parameter spectrum. A double peak spectrum like the swell or sea-dominated spectrum was suggested by Ochi and Hubble (1976). A few of the above spectral models are briefly discussed below.

6.7.2   *Pierson–Moskowitz (P–M) spectrum*

The Pierson–Moskowitz (P–M) spectrum is a one-parameter model describing a fully developed sea, in which fetch and duration are assumed to be infinite. This model has been found to be useful in representing a severe storm wave in offshore structure design.

The P–M spectrum model in terms of the frequency of the spectral peak is written as

$$S_\eta(\omega) = \alpha g^2 \omega^{-5} e^{[-1.25(\omega/\omega_0)^{-4}]}, \qquad (6.31)$$

where $\alpha = 0.0081$. Herein, ω is the angular frequency $= 2\pi f$ and ω_0 is the angular frequency of spectral peak. The P–M spectrum in terms of the linear frequency, $f(=\omega/2\pi)$ is given by

$$S_\eta(f) = \frac{\alpha g^2}{(2\pi)^4} f^{-5} e^{[-1.25(f/f_0)^{-4}]} \tag{6.32}$$

where $f_0 = \omega_0/2\pi$ and $S_\eta(\omega) = s_\eta(f)/2\pi$.

6.7.3   Bretschneider spectrum

This is a two-parameter spectral model on the assumption that the wave is narrow banded with the wave characteristics following the Rayleigh distribution. Bretschneider (1959, 1969) proposed this spectral model as follows:

$$S_\eta(\omega) = 0.2107 H_s^2 \frac{\omega_0^4}{\omega^5} e^{-0.8429(\omega_0/\omega)^4}, \tag{6.33}$$

where H_s is the significant wave height and T_s is the significant wave period defined as the average period of the significant waves.

6.7.4   ISSC spectrum

The International Ship Structures Congress (1964) suggested slight modification of the Bretschneider spectrum:

$$S_\eta(\omega) = 0.3123 H_s^2 \frac{\omega_0^4}{\omega^5} e^{-1.2489(\omega_0/\omega)^4}. \tag{6.34}$$

6.7.5   ITTC spectrum

The P–M spectrum was modified in the International Towing Tank Conference (1966, 1969, 1972):

$$S_\eta(\omega) = \alpha g^2 \omega^{-5} e^{[-4\alpha g^2 \omega^{-4}/H_s^2]}, \tag{6.35}$$

where $\alpha = \frac{0.0081}{k^4}$ and $k = 0.5649 \frac{\sqrt{g/H_s}}{\omega_z}$.

6.7.6   JONSWAP spectrum

The JONSWAP spectrum was developed by Hasselman *et al.* (1973) during a Joint North Sea Wave Project. The formula for the JONSWAP spectrum can be written by modifying the P–M formulation as follows:

$$S_\eta(\omega) = \alpha g^2 \omega^{-5} e^{[-1.25(\omega/\omega_0)^{-4}]} \gamma e^{\left[-(\omega-\omega_0)^2/2\tau^2\omega_0^2\right]}, \tag{6.36}$$

in which γ = peakedness parameter, and τ = shape parameter (τ_a for $\omega \leq \omega_0$ and τ_b for $\omega > \omega_0$) given by

$$\gamma = 3.30 \quad \text{may vary from 1 to 7},$$

$$\tau_a = 0.07,$$

$$\tau_b = 0.09, \quad \alpha = 0.0081.$$

6.7.7 Scott spectrum

The Scott (1965) spectral formula is independent of the wind speed, fetch, or duration, and, as such, should represent a fully developed sea spectrum. The Scott spectrum is a two-parameter model given by

$$S_\eta(\omega) = \begin{cases} 0.214 H_s^2 e^{[-(\omega-\omega_0)/0.065(\omega-\omega_0+0.26)]^{1/2}} & \text{for } -0.26 < (\omega - \omega_0) \\ & \quad < 1.65, \\ 0 & \text{elsewhere.} \end{cases} \tag{6.37}$$

6.7.8 Ochi–Hubble spectrum

Ochi and Hubble (1976) developed a six-parameter spectrum model, consisting of the effect of swell (low-frequency dominated) and wind (high-frequency dominated), which enables us to know which of these two spectra dominates the seaway. This spectral model is given as

$$S_\eta(\omega) = \frac{1}{4} \sum_{j=1}^{2} \frac{\left(\frac{4\lambda_j+1}{4}\omega_{0j}^4\right)}{\Gamma(\lambda_j)} \frac{H_{sj}^2}{\omega^{4\lambda_j+1}} e^{[-(4\lambda_j+1/4)(\omega_0/\omega)^4]}, \tag{6.38}$$

where H_{s1}, ω_{01}, and λ_1 are the significant wave height, peak frequency, and shape factor of the low-frequency components, respectively. Similarly, H_{s2}, ω_{02}, and λ_2 represent the high-frequency components. In the above expression, if the significant wave height, peak frequency, and shape factor values are constant, the factor/ parameter λ_j dictates the shape of the spectral peak, particularly the sharpness. If $\lambda_1 = 1$ and $\lambda_2 = 0$, the above spectrum reduces to modified P–M spectral model. In the general formulation of Eq. (41), the equivalent significant height, H_s is obtained from

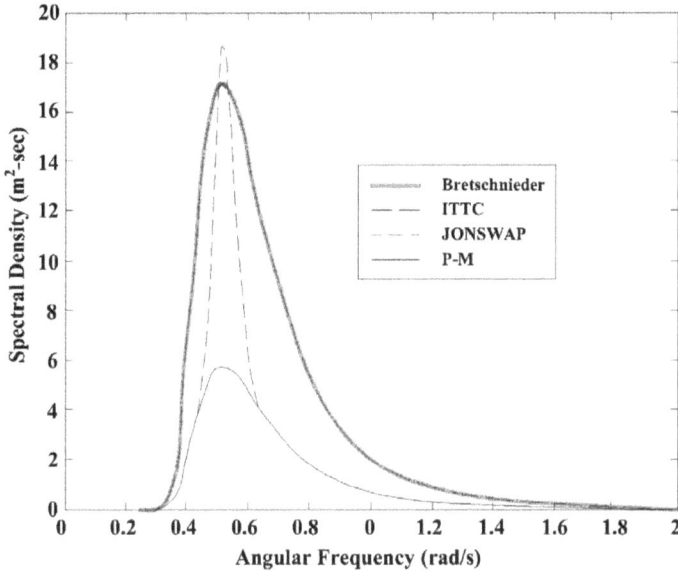

Fig. 6.8   Variation of different standard spectra for $H_s = 10\,\mathrm{m}$ and $T_p = 12\,\mathrm{s}$.

$$H_s = \sqrt{H_{s1}^2 + H_{s2}^2}. \tag{6.39}$$

For additional information on standard spectra, the reader can refer to Chakrabarti (1987).

The different standard spectra as discussed above with $T_p = 12\,\mathrm{s}$ and $H_s = 10$ m superposed in Fig. 6.8 which reveals the following conditions.

- The JONSWAP ($g = 3.3$) exhibits the highest peak value and is narrow banded.
- The ITTC and Bretschneider spectra have the same energy levels.
- The P–M spectrum has the lowest energy, since only the peak frequency is given as input and not the H_s.

6.7.9   Swell and wind-dominated spectrum

There are two types of double-peaked spectra namely wind-dominated spectrum and swell-dominated spectrum. Wind-dominated, low-frequency swell system after traveling considerable distance and losing energy meets a wind-wave system that gives rise to a wind-dominated spectrum. If the energy

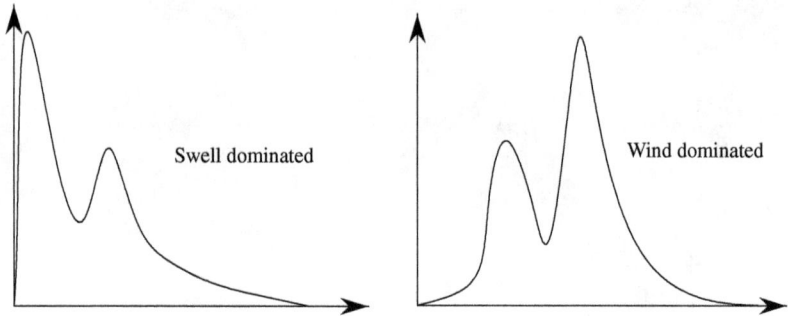

Fig. 6.9   A representation of wind and swell-dominated spectrum.

in a seaway is dominated by the low-frequency components the spectrum is termed a swell-dominated spectrum. These two typical types of double-peaked spectra are shown in Fig. 6.9.

6.8   Simulation of Time Series

6.8.1   *Simulation of waves with pre-defined spectral characteristics*

Herein we discuss unidirectional waves. Although it is usual to derive the spectrum for a measured wave elevation time history, it is also necessary to simulate the time history of $\eta(t)$ with pre-defined spectral characteristics or to calculate the wave height as a function of frequency from a standard spectrum. This is illustrated with the help of a mathematical spectral curve shown in Fig. 6.10. If the energy density at a frequency f_1 be $S_\eta(f_1)$, the wave height can then be defined as

$$H(f_1) = 2\sqrt{2S_\eta(f_1)\Delta f}, \tag{6.40}$$

where H and T together define the wave characteristics. A phase angle associated with each H and T is chosen uniformly distributed in the range of $(0, 2\pi)$ associated by a random number generator, R_n as

$$\varepsilon(f_1) = 2\pi R_n. \tag{6.41}$$

As η is a function of both space x and time t, for a particular x, where the wave elevation be defined, as a function of t, the wave profile is as

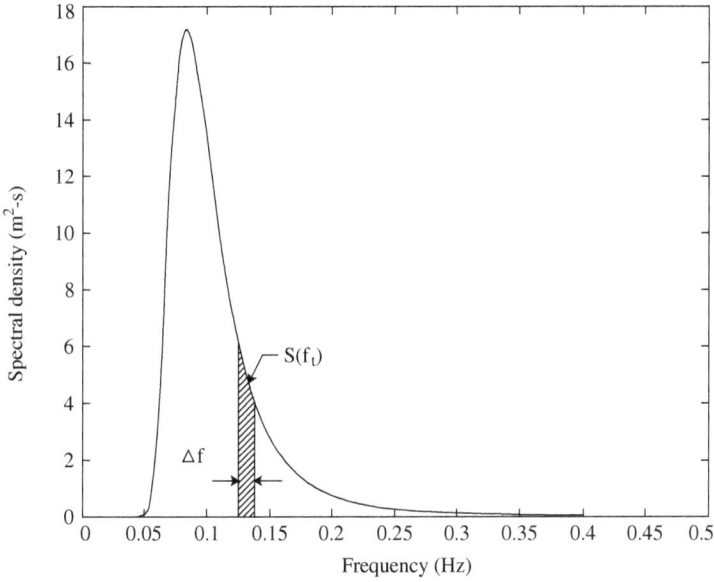

Fig. 6.10   Mathematical spectral curve.

follows:

$$\eta(x,t) = \sum_{n=1}^{N} \frac{H(n)}{2} \cos[k(n)x - 2\pi f(n)t + \varepsilon(n)], \qquad (6.42)$$

where $k(n) = 2\pi/L(n)$ and $L(n)$ corresponds to the wavelength of the nth frequency $f(n)$, "N" is the total number of frequency components, and Δf being the frequency bandwidth. If more randomness is deemed necessary, $f(n)$ can be assigned randomly within each frequency interval. To simulate, the wave height or amplitude $(H/2)$ from (6.40) will be through a standard spectrum or from a measured energy density spectrum. The spectrum curve is divided into several frequency components. Further, if enough component is adopted, it could be enough to apply equal frequency interval, Δf. The number of components should be at least 100 for exhibiting randomness and simulating with more number of components, say, 200 components will duplicate the spectrum accurately. Since the amplitudes of the individual frequency components in this principle of superposition are deterministic, it is referred to as a deterministic spectral amplitude (DSA) model. If only the phases of individual frequency components are chosen at random, it is then called the random phase method. Tucker et $al.$ (1984) have stated that this method of representing the wave simulation does not necessarily satisfy the condition that $\eta(t)$ is Gaussian unless N tends to infinity.

An alternative scheme is to represent $\eta(t)$ in terms of two Fourier coefficients as

$$\eta(x,t) = \sum_{n=1}^{N} \{a_n \cos[k(n)x - 2\pi f(n)t] + b_n \sin[k(n)x - 2\pi f(n)t]\}.$$

(6.43)

Here, a_n and b_n are independent, Gaussian distributed random variables with zero mean and variance of $S_\eta(f_n)\Delta f$. The amplitude and a phase hence are equivalently replaced by two amplitudes, being the coefficients of the sine and cosine components of the wave elevation. The above equation represents that of a Gaussian sea for which large value for N is necessary for generating a real random sea. This is referred to as a random coefficient scheme. Since the amplitudes are random in the case, it is also termed as a non-deterministic spectral amplitude (NSA) model. The main difference between the DSA and NSA models is the satisfaction (or lack of it) of the Gaussian sea. In both the methods, many Fourier components greater than 1000 is used, the difference between would be a minimum.

6.8.2   *Simulation of water particle kinematics*

The method to simulate particle kinematics for random wave time history $\eta(t)$ as per Reid (1957) is discussed below.

A Fourier integral representation of a random wave record $\eta(t)$ is given as follows:

$$\eta(t) = \int_0^\infty A_n \cos(\theta_n - \omega t)d\omega,$$

(6.44)

$$a_n = A_n \cos\theta_n = \frac{1}{\pi} \int_0^{Tr} \eta(t)\cos\omega_n t dt,$$

(6.45a)

$$b_n = A_n \sin\theta_n = \frac{1}{\pi} \int_0^{Tr} \eta(t)\sin\omega_n t dt,$$

(6.45b)

where a_n and b_n are Fourier coefficients of nth wave component.

Here A_n, θ_n, and ω_n are the amplitude, phase, and angular frequency of the nth component, respectively, whereas a_n and b_n are the

Fourier coefficients. The A_n and θ_n are given by

$$A_n = \sqrt{a_n^2 + b_n^2}, \quad \theta_n = \tan^{-1}(b_n/a_n). \tag{6.46}$$

For the expressions for particle kinematics u, $\dot{u}$, w and $\dot{w}$, one can refer to Section 3.5 and for p refer Section 3.7. With these expressions, it is possible to generate the particle time series provided we know the corresponding frequency response function:

$$u(t) = \int_0^\infty A_n R_u(w_n) \sin(\theta_n - w_n t) dw, \tag{6.47a}$$

$$\dot{u}(t) = -\int_0^\infty A_n R_{\dot{u}}(w_n) \cos(\theta_n - w_n t) dw, \tag{6.47b}$$

$$w(t) = -\int_0^\infty A_n R_w(w_n) \cos(\theta_n - w_n t) dw, \tag{6.47c}$$

$$\dot{w}(t) = -\int_0^\infty A_n R_{\dot{w}}(s_n) \sin(\theta_n - w_n t) dw, \tag{6.47d}$$

$$p(t) = \rho g \int_0^\infty A_n R_p(w_n) \sin(\theta_n - w_n t) dw. \tag{6.47e}$$

The R_* represents the frequency response function of the corresponding particle kinematics based on the linear wave theory and, for u, w and p, we define

$$R_u(w_n) = w_n \frac{\cosh k_n(d+z)}{\sinh k_n d}, \ R_{\dot{u}}(w_n) = w_n R_u(w_n), \tag{6.48a}$$

$$R_w(w_n) = w_n \frac{\sinh k_n(d+z)}{\sinh k_n d}, \quad R_{\dot{w}}(w_n) = w_n R_w(w_n), \tag{6.48b}$$

$$R_p(w_n) = \rho g \frac{\cosh k(d+z)}{\cosh k d}. \tag{6.48c}$$

The R_u can be derived as follows:

$$u = \frac{\pi H}{T} \frac{\cosh k(d+z)}{\sinh k d} \sin \theta = \frac{\pi H}{T} G_z \sin \theta,$$

where

$$G_z = \frac{\cosh k(d+z)}{\sinh k d},$$

$$u = \frac{2\pi}{T} G_z \left(\frac{H}{2} \sin \theta \right)$$

$$= w G_z \eta.$$

If the system is linear, the output "u" is related to the input $\eta(t)$:

$$u \propto \eta \text{ or } u = C_u \eta, C_u = \omega G_z \text{ and therefore } R_u = C_u = \omega \frac{\cosh k(d+z)}{\sinh kd}.$$

Similarly, for R_p,

$$p = \rho g \frac{\cosh k(d+z)}{\cosh kd} \frac{H}{2} \sin \theta = \rho g G_z \eta,$$

$$p \propto \eta \text{ or } p = C_p \eta \text{ and therefore } R_p = C_p = \rho g \frac{\cosh k(d+z)}{\cosh kd}.$$

6.9   Directional Waves

6.9.1   *General*

The foregoing discussion pertains to unidirectional random waves often referred to as two-dimensional (2D) random waves. In real ocean, in particular, in the offshore, the waves propagating from different directions are multi-directional or short crested, and hence the procedure of determining the loads with unidirectional waves is conservative. The 2D spectrum provides the energy distribution as a function of frequency for a particular direction of the waves. However, the energy is also spread over different directions around a mean direction. If the mean direction changes over time, the waves are called long-term directional waves. Due to the energy being distributed over different directions, the forces on structures due to short crested waves lead to a reduction in forces as compared to that due to unidirectional waves. The concept of directional spectrum has evolved to describe the sea state due to superposition of not only the frequency components but also its directional components. The directional spectrum $S_\eta(f, \theta)$ can be represented as shown in Fig. 6.11. The difference between 2D spectrum and 3D spectrum are illustrated in Figs. 6.12(a) and 6.12(b), respectively.

6.9.2   *Simulation of directional waves*

The $S_\eta(f)$ is generally written as

$$S_\eta(f, \theta) = S_\eta(f) \cdot D(f, \theta). \tag{6.49}$$

While a 2D spectrum provides the energy distribution over different frequencies without its variations with respect to the directions being considered, the directional spectrum provides the energy of the waves (per

$$S_\eta(f,\theta) = S_\eta(f).D(f,\theta)$$

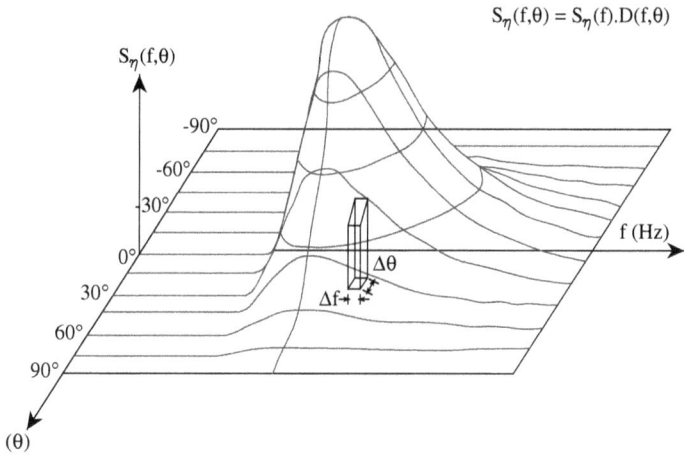

Fig. 6.11   A directional wave spectrum.

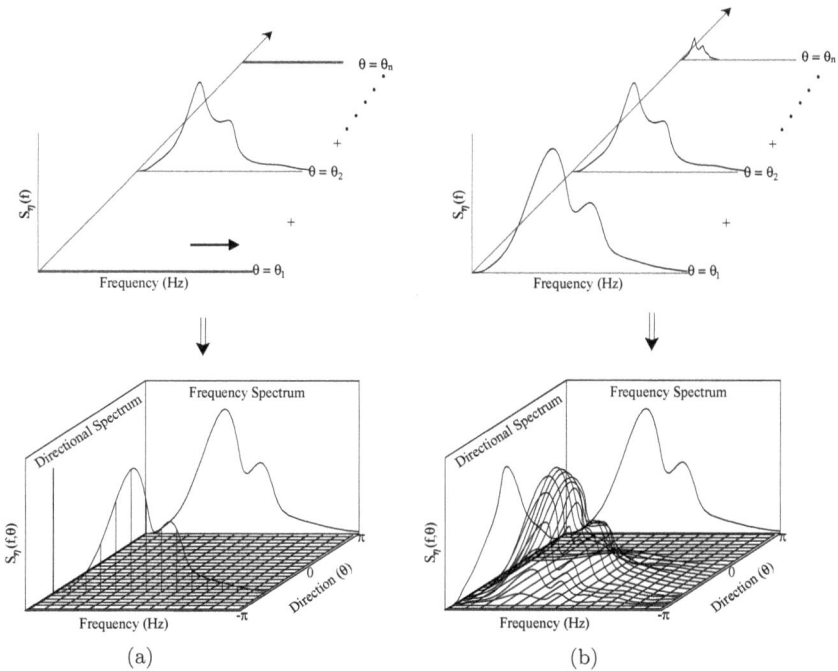

Fig. 6.12   (a) Unidirectional 2D spectrum and (b) multi-directional 3D spectrum.

unit plan area) against various frequencies as well as directions. The average wave energy per unit horizontal surface area contributed by waves of frequency f and a bandwidth df propagating with a direction θ to the dominant wind direction and a spread interval $d\theta$ can be represented by the volume $S_\eta(f, \theta)$. df. $d\theta$. The $S_\eta(f, \theta)$ is termed as directional wave spectral density function, directional wave spectrum, spreading function, angular spreading function or as directional distribution.

The directional spreading function $D(f, \theta)$ has no dimension and is normalized as $\int_{-\pi}^{\pi} D(f, \theta), d\theta = 1$.

The $S_\eta(f)$ is the absolute value of the wave energy density, while $D(f, \theta)$ is the relative magnitude of directional spreading of wave energy. If "m" is the number of frequency components and "n" is the number of directions, the water surface elevation $\eta(x, y, t)$ can be represented as

$$\eta(x, y, t) = \sum_{m=1}^{\infty} \sum_{n=1}^{\infty} a_{m,n} \cos(k_m x \cos \theta_n + k_m y \sin \theta_n - 2\pi f_m t + \varepsilon_{m,n})$$

(6.50)

or,

$$\eta(x, y, t) = \frac{H_{m,n}}{2} \cos(k_m x \cos \theta_n + k_m y \sin \theta_n - 2\pi f_m t + \varepsilon).$$ (6.51)

Here, $a_{m,n}$ is the amplitude of component wave, k_m is the wave number corresponding to the frequency f_m, while θ_n represents the direction of the wave propagation and $\varepsilon_{m,n}$ is the random phase angle distributed uniformly between 0 and 2π. The horizontal coordinates of the point of consideration of the surface elevation are x, y. The wave number k_m is related to the frequency f_m by the dispersion relationship given by

$$(2\pi f_n)^2 = g k_m \tanh k_m d.$$ (6.52)

The amplitude $a_{m,n}$ as a function of directional spectral density $S(f, \theta)$ may be

$$\sum_{f_m}^{f_m + \delta f_m} \sum_{\theta_n}^{\theta_n + \delta \theta_n} \frac{1}{2} a_{m,n}^2 = S_\eta(f_m, \theta_n) \delta f_m \delta \theta_n.$$ (6.53)

In practice, $\eta(x, y, t)$ from Eqs. (6.50) and (6.53) is difficult to obtain as we need to deal with an infinitely large number of wavelets distributed at infinitely small intervals. Hence, an approximate solution is to adopt sufficiently large number of wavelets in Eq. (6.50) and of sufficiently small intervals of frequency and direction in Eq. (6.53). Hence, Eq. (6.50) can be rewritten as

$$\eta(x, y, t) = \sum_{m=1}^{M} \sum_{n=1}^{K} a_{m,n} \cos(k_m x \cos \theta_n + k_m y \sin \theta_n$$

$$- 2\pi f_m t + \varepsilon_{m,n}), \tag{6.54}$$

$$\frac{1}{2} a_{m,n}^2 = S_\eta(f_m, \theta_n) \Delta f_m \Delta \theta_n \quad (n = 30, \; m = 200 \text{ say}). \tag{6.55}$$

The directional spectrum $S_\eta(f, \theta)$ is expressed as in Eq. (6.49).
If the relationship of Eq. (6.49) is used, Eq. (6.55) can be written as

$$a_{m,n} = \sqrt{2S_\eta(f_m)\Delta f_m} \sqrt{D(f_m, \theta_n)\Delta \theta_n}. \tag{6.56}$$

For a location, x and y are fixed, and hence Eq. (6.54) can be further transformed as

$$\eta(t) \text{ at } (x, y) = \sum_{m=1}^{M} A_m \cos(2\pi f_m t - \varphi_m), \tag{6.57}$$

where $A_m = \sqrt{C_m^2 + S_m^2}$, $\varphi_m = \tan^{-1}(S_m/C_m)$.
In the above expressions, C_m and S_m are defined as follows:

$$C_m = \sum_{n=1}^{K} a_{m,n} \cos(k_m x \cos \theta_n + k_m y \sin \theta_n + \epsilon_{m,n}), \tag{6.58a}$$

$$S_m = \sum_{n=1}^{K} a_{m,n} \sin(k_m x \cos \theta_n + k_m \sin \theta_n + \epsilon_{m,n}). \tag{6.58b}$$

In the case of numerical simulation of 3D waves, the above equation is sometimes employed, where the wavelet A_m is determined as

$$A_m = \sqrt{2S_\eta(f_m)\Delta f_m}. \tag{6.59}$$

From the point of view of wave variability, the evaluation of A_m through the above equation is incorrect which could be corrected by letting x and y be zero in Eq. (6.58) for a fixed location without the generality being lost.

Hence,

$$C_m = \sum_{n=1}^{K} a_{m,n} \cos \varepsilon_{m,n}, \qquad (6.60a)$$

$$S_m = \sum_{n=1}^{K} a_{m,n} \sin \varepsilon_{m,n}. \qquad (6.60b)$$

As $\varepsilon_{m,n}$ is uniformly distributed between 0 and 2π, the C_m and S_m are zero mean process, the variances of which can be expressed as

$$\text{Var}(C_m) = \text{Var}(S_m) = S(f_m)\Delta f_m. \qquad (6.61)$$

The η_{rms} of the simulated wave profile of Eq. (6.57) using the normal distribution of C_m and S_m can be derived as

$$\eta_{rms} = \sqrt{\overline{\eta^2}} = \sqrt{\frac{1}{2} \sum_{m=1}^{M} A_m^2} = \sqrt{\frac{1}{2} \sum_{m=1}^{M} (C_m^2 + S_m^2)}. \qquad (6.62)$$

The expected value of η_{rms}^2 can also easily be derived as

$$E[\eta_{\text{rms}}^2] = \sum_{m=1}^{M} S_\eta(f_m) \Delta f_m = m_0. \qquad (6.63)$$

6.9.3 Directional spreading functions

Due to the difficulty in reliable field measurements of directional waves, the knowledge on its spreading function is rather limited. The commonly adopted spreading functions are given below.

(a) A circular normal distribution of Mobarek (1965) and Fan (1968) is given by

$$D(f, \theta) = \frac{e^{\alpha \cos(\theta - \alpha)}}{2 l_0(\alpha)}, \qquad (6.64)$$

where "α" or "θ_0" is the central direction of wave travel, α is reciprocal of variance and $I_0(\alpha)$ is modified. Bessel function is second kind and zeroth order.

(b) Wrapped — around Gaussian model: Borgman (1969)
This model is an exponential model based upon a normal distribution and can be expressed as

$$D(f, \theta) = \sum_{k=-\infty}^{k} \frac{e^{-(\theta - 2\pi k - \theta_0)^2 / 2\sigma^2}}{\sigma \sqrt{2\pi}}, \tag{6.65}$$

where θ_0 is the mean and σ^2 is the variance.

(c) A finite Fourier series is expressed as

$$D(f, \theta) = \frac{a_0}{2} = \sum_{n=1}^{N} [a_n \cos n\theta + b_n \sin n\theta]. \tag{6.66}$$

(d) Cosine squared is expressed as

$$D(f, \theta) = \begin{cases} (2/\pi) \cos^2 \theta & \text{for } |\theta| < \pi/2, \\ 0 & \text{otherwise.} \end{cases} \tag{6.67}$$

(e) Mitsuyasu type
Mitsuyasu *et al.* (1975) have proposed the following function on the basis of their detailed field measurements:

$$D(f, \theta) = D_0 \cos^{2s} \left(\frac{\theta}{2} \right), \tag{6.68}$$

where "θ" is the azimuth measured counterclockwise from the principle wave direction. D_0 is a constant introduced to satisfy the following condition:

$$\int_{-\pi}^{\pi} D(f, \theta) d\theta = 1,$$

i.e.,

$$D_0 = \int_{\theta_{\min}}^{\theta_{\max}} \left[\cos^{2s} \left(\frac{\theta}{2} \right) d\theta \right]^{-1} \tag{6.69}$$

where "s" is a parameter related to the frequency which will be discussed later. If the range of the angle is such that $\theta_{\min} = -\pi$, $\theta_{\max} = \pi$, the constant D_0 becomes

$$D_0 = \frac{1}{\pi} 2^{2s-1} \frac{\Gamma^2(s+1)}{\Gamma(2s+1)}, \tag{6.70}$$

where Γ is the Gamma function. By setting $s = 10$, D_0 becomes about 0.9033 and the directional spreading function is calculated as shown by the solid line of Fig. 6.13.

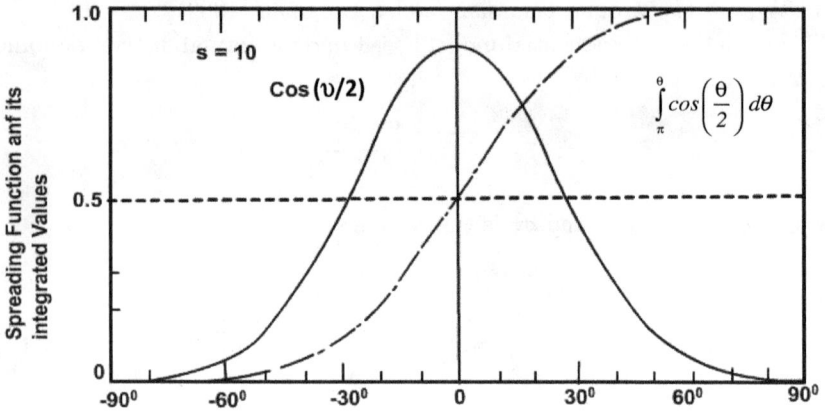

Fig. 6.13   Example of directional spreading function.

The cumulative value of $D(f,\theta)$ from $\theta = -\pi$ is also shown in this figure as the dash-dot line. From this cumulative distribution of $D(f,\theta)$, it is observed that about 85% of the wave energy is contained with the waves spread between $\pm 30°$.

The parameter "s" of the directional spreading function of Mitsuyasu et al. (1975) represents the degree of directional energy concentration and has a peak value around the frequency of the spectral peak. The value of "s" decreases towards the higher and lower frequency zones on either side of the peak frequency.

(f) Spreading function of Goda and Suzuki (1975)

The peak of the parameter "s" of the earlier formulation of Mitsuyasu et al. (1975) was denoted as s_{max} by Goda and Suzuki (1975) and

$$S = \begin{cases} S_{max}(f/f_0)^5 & \text{for } f \le f_0, \\ S_{max}(f/f_0)^{-2.5} & \text{for } f \ge f_0. \end{cases} \tag{6.71}$$

The frequency of spectral peak f_0 is related to $T_{1/3}$ as $f_0 = 1/1.05T_{1/3}$.

The directional spreading function as a function of dimensionless frequency, $f_* = f/f_0$ for $s_{max} = 20$, is illustrated in Fig. 6.14. The range of the energy distribution is set as $-\pi/2 \le \theta \le \theta/2$ and the normalization constant D_0 is evaluated by numerical integration for each directional spreading function.

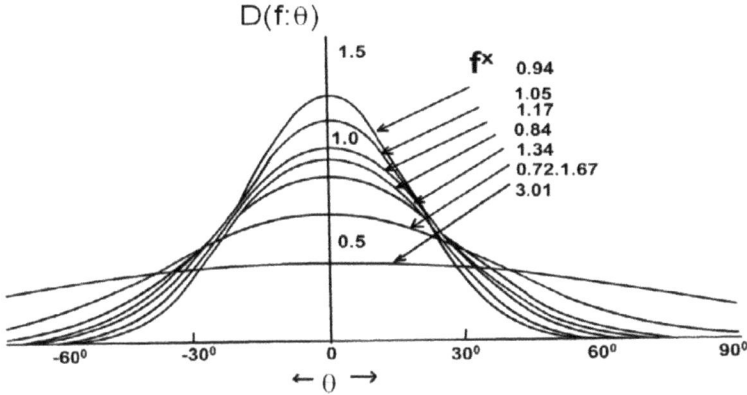

Fig. 6.14 Example of Mitsuyasu-type spreading function for $s_{max} = 20$.

The values of s_{max} for practical engineering applications are given below.

(i) For wind waves $\qquad s_{max} = 10$

(ii) Swell with short-decay distance

(with relatively large-wave steepness) $\qquad s_{max} = 25$

(iii) Swell with long-decay distance

(with relatively small-wave steepness) $\qquad s_{max} = 75$

6.9.4 *Particle kinematics*

The particle kinematics under three-dimensional (3D) waves can be derived as given below and these are essential in evaluating wave forces on structures in multi-directional waves.

The particle velocities, u, v and w in the x, y, z directions are given as follows:

$$u\left(x, y, z, t\right) = \sum_{m=1}^{M} \sum_{n=1}^{K} A_{mn} \frac{\cosh k_m \left(d + z\right)}{\sinh k_m d} \cos \theta_n \cos \psi_{mn}, \quad (6.72a)$$

$$v\left(x, y, z, t\right) = \sum_{m=1}^{M} \sum_{n=1}^{K} A_{mn} \frac{\cosh k_m \left(d + z\right)}{\sinh k_m d} \sin \theta_n \cos \psi_{mn}, \quad (6.72b)$$

$$w\left(x, y, z, t\right) = \sum_{m=1}^{M} \sum_{n=1}^{K} A_{mn} \frac{\sinh k \left(d + z\right)}{\sinh k d} \sin \theta_n \cos \psi_{mn}. \quad (6.72c)$$

where

$$\psi_{mn} = k_m \left(x \cos \theta_n + y \sin \theta_n \right) - 2\pi f_m t + \epsilon_{mn}. \qquad (6.73)$$

6.9.5   Analysis of directional wave fields

6.9.5.1   General method

The directional spreading function can be estimated from measured data using different methods. Input data for evaluating the directional spread usually should be available in the form of the time series of different wave properties such as water surface elevation, slopes, or velocities. The amplitude ratio or the phase difference between different measured quantities can be used to determine the directional distribution of the waves. The analysis techniques are based on the principle of cross-spectral analysis between pairs of signals that include, for example, the wave elevations and two components of horizontal velocities. The directional characteristics of the wave field can then be expressed in terms of the cross-spectra.

In order to obtain information about wave directionality, the measured time series are first Fourier transformed to yield a frequency-dependent, cross-spectral density matrix. Most current methods of analysis then utilize the relationship between the cross-spectral density matrix of the measured data and the directional wave spectrum to estimate the directional spreading function. This relationship can be expressed in a general form as per Isobe *et al.* (1984) given by

$$S_{mn}(\omega) = \int H_m(k, \omega) exp\left\{-ik(x_n - x_m)\right\} S(k, \omega) dk. \qquad (6.74)$$

Here, the subscripts m and n represent any two measured quantities, $H(k, \omega)$ is a complex transfer function that relates a measured quantity to the water surface elevation, and x_m and x_n are the location of the measurements, respectively. The above relationship is valid only for a spatially homogeneous wave field where, there is no correlation of the wave components traveling in different directions. The transfer function can be expressed as

$$H_m(k, \omega) = (\cos \theta)^\alpha (\sin \theta)^\beta R_m(\omega), \qquad (6.75)$$

where $R(\omega)$ is a direction-independent transfer function. The values of $R(\omega)$, α, and β take the values of 1, 0, and 0, respectively for the quantity water surface elevation; $R(\omega), 1$ and 0 for water particle velocity u; and $R(\omega)$, 0, and 1 for water particle velocity v, respectively. The value of $R(\omega)$ for the water particle velocity is the transfer function of velocities with water surface elevation. Instead of using linear wave theory, the transfer function $R(\omega)$ can be estimated from the co-spectra of the measured quantities as

$$R(\omega) = \sqrt{\frac{C_{uu} + C_{vv}}{C_{\eta\eta}}} \tag{6.76}$$

as

$$S_{mn}(\omega) = C_{mn}(\omega) + iQ_{mn}(\omega) \tag{6.77}$$

where $S_{mn}(\omega)$, $C_{mn}(\omega)$ and $Q_{mn}(\omega)$ denote spectral densities of cross-spectra, co-spectra and quadrature spectra, respectively. The use of above equation to estimate the transfer function minimizes the effect of the errors introduced in the calibration of current meters.

For the measurements of the water surface elevation and the two orthogonal components of the horizontal velocity at a single location, Eq. (6.74) reduces to

$$\int_{-\pi}^{\pi} D(\omega, \theta) qi(\theta) d\theta = \psi_i(\omega), \quad i = 1, \dots, 5 \tag{6.78}$$

where

$$
\begin{aligned}
\psi_1(\omega) &= 1, & q_1 &= 1, \\
\psi_2(\omega) &= C_{\eta u}/[R(\omega)\, S\eta(\omega)], & q_2 &= \cos\theta, \\
\psi_3(\omega) &= C_{\eta v}/[R(\omega)\, S\eta(\omega)], & q_3 &= \sin\theta, \tag{6.79}\\
\psi_4(\omega) &= (C_{uu} - C_{vv})/[R2(\omega)\, S\eta(\omega)], & q_4 &= \cos 2\theta, \\
\psi_5(\omega) &= 2C_{uv}/[R2(\omega)\, S\eta(\omega)], & q_5 &= \sin 2\theta.
\end{aligned}
$$

The salient aspects of the three analysis methods is discussed in the following sections.

6.9.5.2   The Fourier series approach

The directional spreading function can be expanded as a Fourier series in direction θ as

$$D(\omega, \theta) = a_0 + \sum_{n=1}^{\infty} (a_n \cos n\theta + b_n \sin n\theta) \tag{6.80}$$

where a_n and b_n are Fourier coefficients. Because of the limited number of measured quantities for a wave probe-current meter array, the expansion can only be carried out to obtain the first five Fourier coefficients. Substituting Eq. (6.80) into Eq. (6.78) and using the orthogonality of the cosine and sine functions gives the Fourier coefficients as

$$a_o = \psi_1/(2\pi), \quad a_1 = \psi_2/\pi, \quad a_2 = \psi_4/\pi,$$

$$b_1 = \psi_3/\pi, \quad b_2 = \psi_5/\pi \tag{6.81}$$

There are no constraints on the values of the Fourier coefficients so unrealistic negative values of the directional spreading function can appear. Thus, smoothing functions are required to ensure a positive spreading function but this however results in a further decrease in directional resolution.

6.9.5.3 *Maximum likelihood method*

The maximum likelihood method (MLM) is normally used in probability theory to estimate parameters of a probability distribution which maximize the likelihood of data set.

In directional wave analysis, the MLM estimate of the spreading function minimizes the error of the weighted sum of Fourier coefficients, with the weighting function dependent on the direction of wave propagation. The MLM is described in detail by Isobe *et al.* (1984). The MLM estimate of the directional spreading function can be expressed as

$$D(\omega, \theta) = \frac{a_0}{v^T \Psi^{-1} v}, \tag{6.82}$$

where a_0 is a normalizing coefficient that ensures Eq. (6.54) is satisfied, and $\{v\}$ is the weighting function given by

$$\{v\} = \left\{ \begin{array}{c} 1 \\ \cos\theta \\ \sin\theta \end{array} \right\}, \tag{6.83}$$

and $[\psi]$ is the normalized cross-spectral density matrix of η, u and v, and is given by

$$[\Psi] = \left[\begin{array}{ccc} & \Psi_1 \Psi_2 \Psi_3 & \\ \Psi_2 & \dfrac{1+\Psi_4}{2} & \dfrac{\Psi_5}{2} \\ \Psi_3 & \dfrac{\Psi_5}{2} & \dfrac{1+\Psi_4}{2} \end{array} \right]. \tag{6.84}$$

The cross-spectral density matrix at each frequency, obtained from an FFT analysis of the measured time series can be substituted into Eq. (6.83) to yield a discrete directional distribution.

6.9.5.4 *Maximum entropy method*

The maximum entropy method (MEM) is based on a similar concept used in probability theory, where probability distributions are estimated from a limited amount of information. The probability approach was first applied to directional wave analysis by Kobune and Hashimoto (1986). The application of the MEM to directional wave analysis is detailed here. The spreading function can be the probability distribution of wave energy at a given frequency over direction. The problem of estimating the spreading function thus becomes one of estimating, a probability distribution from a limited amount of data. The concept of entropy is introduced as an index of uncertainty of the directional function. Information is available in the form of integral equation, Eq. (6.78), in which the kernel functions $q_j(\theta)$ and the cross spectra components $\psi_j(f)$ are known, and the directional spreading function is unknown. There are different directional distributions that can satisfy the integral equation. Based on maximum entropy ideas, the least biased distribution is the one that maximizes the entropy associated with the directional distribution. The entropy, I, is defined as

$$I = -\int_{-\pi}^{\pi} D(\omega, \theta) \ln D(\omega, \theta) d\theta. \tag{6.85}$$

Since there is information provided by the relationship between the measured cross-spectral density matrix and the directional distribution, the maximum entropy solution is chosen to satisfy that relationship (Nwogu *et al.*, 1987). Maximizing the entropy subject to the constraints given in Eq. (6.54) produces the solution

$$D(\omega, \theta) = \exp\left[-1 \sum_{-i=1}^{5} \mu_i q_i(\theta)\right] \tag{6.86}$$

where μ_i are the Lagrange multipliers chosen such that the estimate of the spreading function is consistent with the measured cross-spectral density matrix.

Since $q_1(\theta) = 1$, the multiplier μ_1 is just a constant and can be modified to absorb the extra constant "-1" in the above equation. Substituting the above in Eq. (6.78) results in a set of nonlinear equations:

$$\int_{-\pi}^{\pi} \exp\left[\sum_{-i=1}^{5} \mu_i q_i(\theta)\right] q_i(\theta) d\theta = \psi_i(\omega), \quad i = 1, \ldots, 5. \tag{6.87}$$

The Newton–Raphson procedure is used to solve for μ_i. A suitable initial guess for the parameter μ_i can be determined from the MLM solution of $D(\theta)$.

6.9.6 *Directional parameters*

The directional characteristics of the directional spectrum elude quantitative definition. One possible definition of wave direction is the predominant wave direction at the peak frequency at which the power spectrum shows a maximum or to define a single-wave direction for a whole-wave group.

The various parameters that are adopted to define the nature of the directional waves are discussed by Kobune *et al.* (1985), Kondo and de Koning (1986) and Nwogu *et al.* (1987).

Worked Out Examples

Problem 6.1. Evaluate the distribution of energy for the following wave characteristics.

Wavelength (m)	684	170	76	42	26	18
Wave height (m)	3	3.5	5	4	3	2

Solution

From

$$C = \sqrt{gL/2\pi} \quad \text{(Deep water Condition)},$$

$$\sigma_i \text{ or } w_i = \sqrt{2\pi g/L_i}, \quad g = 9.81$$

$$w_1 = \sqrt{\frac{2\pi \times 9.81}{684}} = 0.3.$$

Similarly, $w_2 = 0.6$, $w_3 = 0.9$, $w_4 = 1.2$, $w_5 = 1.5$, $w_6 = 1.8$.

The total energy per square meter of the wave surface is given by

$$E_T = \frac{\rho g}{2}(a_1^2 + a_2^2 + a_3^2 + a_4^2 + a_5^2 + a_6^2)$$

$$= \frac{1025}{2}(1.5^2 + 1.75^2 + 2.5^2 + 2^2 + 1.5^2 + 1^2)$$

$$= 94582\,\text{N/m}.$$

The distribution of this total energy is shown in Fig. 6.15. In this figure, the ordinates are obtained by dividing the individual energy content by the bandwidth of 0.3 rad/s in the present case.

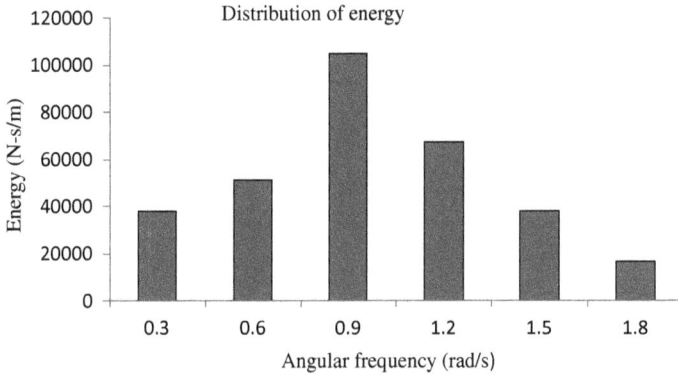

Fig. 6.15   Distribution of energy as a function of frequency.

Note: The dimension of energy is kg/m. Since the area under the total curve should give this dimension, the ordinates represent kg-s/m, since the abscissa has the dimension s^{-1}. We have thus seen in the problem that the energy in waves and that the area under the curve yield the same quantity. Now, instead of drawing the spectrum like before one can draw a different in which the ordinates represent

$$\frac{1}{2}(a_1^2 + a_2^2 + a_3^2 + \cdots + a_n^2).$$

Note: ρg is divided throughout and the area under the curve of the new figure, generally denotes m_0, is later multiplied by to obtain the energy. Figure 6.16 is wave spectrum and the ordinates that are represented are called the spectral density of the wave energy.

Hence

$$\frac{1}{2}(1.5^2 + 1.75^2 + 2.5^2 + 2^2 + 1.5^2 + 1^2) \quad \text{and} \quad \text{Total area} = 9.406\,\text{m}^2.$$

For a wave spectrum, the above values corresponding to different waves are divided by the bandwidth (0.3) to get the ordinates. The total area under the wave spectrum in Fig. 6.15 would have a value 9.406 which when multiplied by $\rho g(1025 \times 9.81\,\text{N/m}^3)$ gives the energy, which is 94,579.68 N/m.

Problem 6.2.

The following wave heights were recorded.

Height(m)	0–1 m	1–2 m	2–3 m	3–4 m	4–5 m
No. of waves	4200	5400	1440	720	240

Solution

The histogram data are shown in the following table.

Wave H_t (m) (1)	Mean of H_i/H (2)	No. of waves (3)	% occurrence = col(3)/sum × 100 (4)	% of occurrence per m of wave H_t from record = col(4)/ΔH (5)
0–1	0.5	4200	35	35
1–2	1.5	5400	45	45
2–3	2.5	1440	12	12
3–4	3.5	720	6	6
4–5	4.5	240	2	2
SUM = 12,000				

Note: The values of col (5) are obtained by dividing col (4) by 2 which is interval of H. This gives the ordinates for the wave histogram, i.e., [pdf = prob $(x_i / \Delta x)$].

Sample calculation for Rayleigh distribution:

$$H_{rms}^2 = \frac{\sum_i H_i^2 f_i}{\sum_i f_i}$$

$$= \frac{\begin{matrix} 0.5^2 \times 4200 + 1.5^2 \times 5400 + 2.5^2 \times 1440 + 3.5^2 \\ \times 720 + 4.5^2 \times 240 \end{matrix}}{12000}$$

$$= \frac{35880}{12000}$$

$$= 2.99\,\mathrm{m}^2$$

$$H_{rms} = 1.73\,\mathrm{m}.$$

Now, according to Rayleigh distribution,

$$p(H) = \frac{2H_i}{H_{rms}^2} \exp -[H_i^2/H_{rms}^2],$$

$$p(H = 1m) = \frac{2 * 0.5}{2.99} e^{-(0.5^2)/2.99} = 0.307 = 30.7\%.$$

Similarly

$$p(H = 1.5) = 0.473, \quad \text{in terms of } \% = 47.3,$$

$$p(H = 2.5) = 0.207, \quad \text{in terms of } \% = 20.7,$$

$$p(H = 3.5) = 0.0389, \quad \text{in terms of } \% = 3.89,$$

$$p(H = 4.5) = 0.0034, \quad \text{in terms of } \% = 0.34.$$

Note: The total area under the Rayleigh distribution curve should be equal to 1 (i.e., the total probability should be 1).

Area from $H = 0.5$ to $H_i = 4.5$ (Simpson rule)

$$= \frac{1}{3}h \sum \text{product, i.e., first and last ordinates should be}$$

multiplied by I and II, IV, i.e., even ordinate by 4 and odd

ordinates III, V, etc. by 2,

$h = $ interval,

$$\text{Area(from } H = 0.5 \text{ to } 4.5) = \frac{1}{3} \times [0.307 \times 1 + 0.473 \times 4 + 0.206 \times 2$$

$$+0.0389 \times 4 + 0.0034 \times 1]$$

$$= 0.923,$$

$$\text{Area(from } H = 0 \text{ to } 0.5) = \frac{1}{2} \times 0.5 \times 0.307 = 0.07675,$$

Total area $= 0.923 + 0.07675 \cong 1.0$.

The comparison of the observed probability density function with the theoretical Rayleigh distribution is shown in Fig. 6.16.

Therefore, if the area under the histogram curve is known, one can directly relate $\bar{H}$ or H_{rms} to Rayleigh distribution formula. The probability of H being greater than a specific value of H_i, termed as probability of

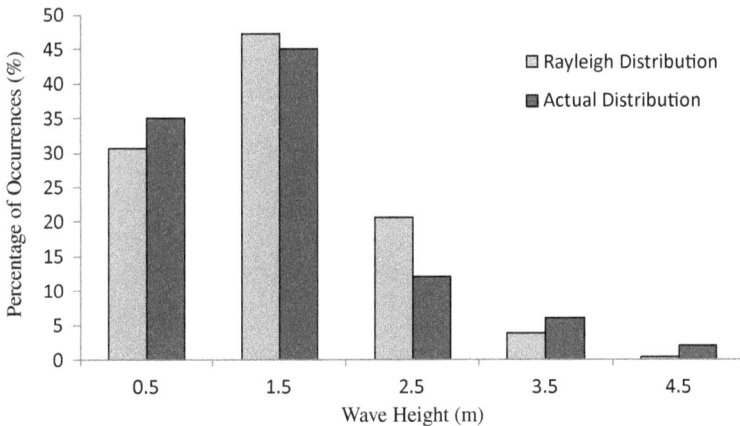

Fig. 6.16 Wave height histogram along with the theoretical Rayleigh distribution.

exceedance, is given by

$$p[H > H_i] = 1 - \int_0^{H_i} \frac{2H_i}{H_{rms}^2} e^{-H_i^2/H_{rms}^2} = e^{-H_i^2/H_{rms}^2}.$$

For example, If $H_I = 3\,\mathrm{m}$ and $H_{rms}^2 = 2.99\,\mathrm{m}^2$, then $p = e^{-3^2/11.96} = 0.05$.

The probability of the wave height being greater than 3m will be 0.05 or 5%. Alternatively, out of number of waves N, $Ne^{-(H_i^2/H_{rms}^2)}$ will be higher than H_i. From such a formulation, one can find out H_{mean}, $H_{1/3}$, $H_{1/10}$, etc.

Problem 6.3.

Compute the wave characteristics and compare with that of prediction formula.

Wave H_t (1)	No. of waves (2)	Cumulative no. of waves (3)	Number × $H_t = $ col (1) × col (2) (4)
0.5	4200	4200	2100
1.5	5400	9600	8100
2.5	1440	11040	3600
3.5	720	11760	2520
4.5	240	12000	1080
	12000		**17400**

$$\bar{H} = \frac{17400}{12000} = 1.45\,\mathrm{m}$$

Solution

Rayleigh distribution $= 0.89 * H_{rms}$,

$\qquad H_{rms}$ for this problem $= 1.73\,\mathrm{m}$,

$\bar{H} = 0.89 \times 1.73 = 1.54\,\mathrm{m}$,

$\bar{H} = \sum H_i f_i / \sum f_i$

$$= \frac{(4200 \times 0.5) + (5400 \times 1.5) + (1440 \times 2.5) + (720 \times 3.5) + (240 \times 4.5)}{12000}$$

$= 1.45\,\mathrm{m}.$

It is seen that the theoretical Rayleigh distribution fits well with the wave data $H_{max} = (1.6\text{–}2.0)\ H_{1/3}$.

In the design of offshore structures, $H_{max} = 2H_{1/3}$ or a higher value is often employed. For the design of a breakwater, Goda (2010) has proposed the use of the relation $H_{max} = 1.8H_{1/3}$.

Chapter 7

Wave-Induced Forces on Structures

7.1 Introduction

Loads on offshore structures can be classified in two main categories mentioned below.

(i) Functional loads are loads due to the existence, use and treatment of the structure under ideal circumstances for each design condition. Such loads may be static, or dynamic loads from drilling and operation of cranes, etc. All external loads which are in fact responses to function loads are to be considered as functional loads, that is, support reactions and buoyancy forces.

(ii) Environmental loads are defined as loads, which are not a necessary consequence of the structure. Such loads are loads due to wind, wave currents, ice, etc. The classification of the loads on offshore structures is also projected in Fig. 7.1.

The precise evaluation of the forces exerted upon a structure by an ocean wave is extremely complicated because of many interacting phenomena. The most important of these factors are (i) the nonlinearity of the water particle displacements and kinematics, (ii) the variability of the wave profiles and forces, (iii) turbulence, (iv) modification of the wave properties by the presence of the structure and (v) the possibility of dynamic effects such as vortex shedding and structure resonance. Wave force equations also represent an unsteady, non-uniform flow condition for which dynamic effects are non-deterministic. The calculation of wave forces first involves the selection of an appropriate wave theory to describe the particle kinematics and displacements for the given design wave condition and second, a choice of semi-empirical coefficient or coefficients to reconcile the assumptions of theory with the findings of laboratory and field investigations. The linear wave theory can be used if $H/gT^2 < 0.001$ in deep and intermediate water conditions. In general, offshore structures are designed for $H/gT^2 > 0.001$ and in such case, the linear theory underestimates the wave force.

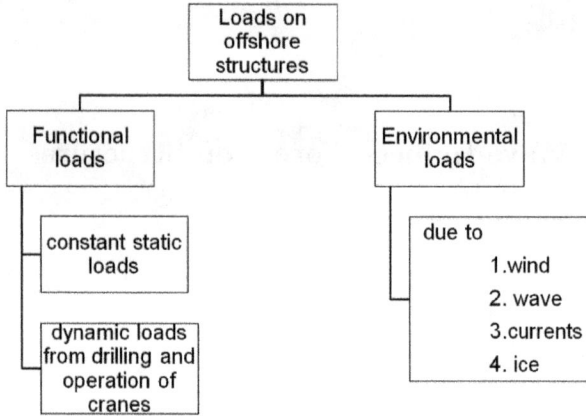

Fig. 7.1   Classification of the loads on offshore structures.

7.2   Various Force Regimes

There are two different approaches to calculate the wave loads. These are the "Design-wave approach" and the "Irregular-wave approach". In the design-wave approach, determining the fluid loading on structures due to waves is still the most common one. The survival condition which may be 50 or 100 years depending on the importance of the structure. The wave force on structures is generally determined from the Morison equation (Morison et al., 1950) discussed below. This equation is applicable for members of offshore structures when $D/L < 0.2$. The total force in such a case is a summation of drag force which is a function of orbital particle velocity and the inertial force which is a function of particle acceleration. However, for large size members more complex theories are necessary to take into account the scattering or radiation of incident wave energy from the member. In this approach, the force will be evaluated for the design wave and period. In the case of irregular-wave approach, the forces for a design wave spectrum are computed either in the frequency domain or presented as probability distribution. The regions of application of the different approaches for the evaluation of wave force based on $KC = u_{max}T/D$ and scattering parameter, $ka = \pi D/L$ are depicted in Fig. 7.2. The sizes of different components of the most common types of offshore structures are projected in Fig. 7.3(a), whereas, the different types of forces acting on different types of marine structures are projected in Fig. 7.3(b).

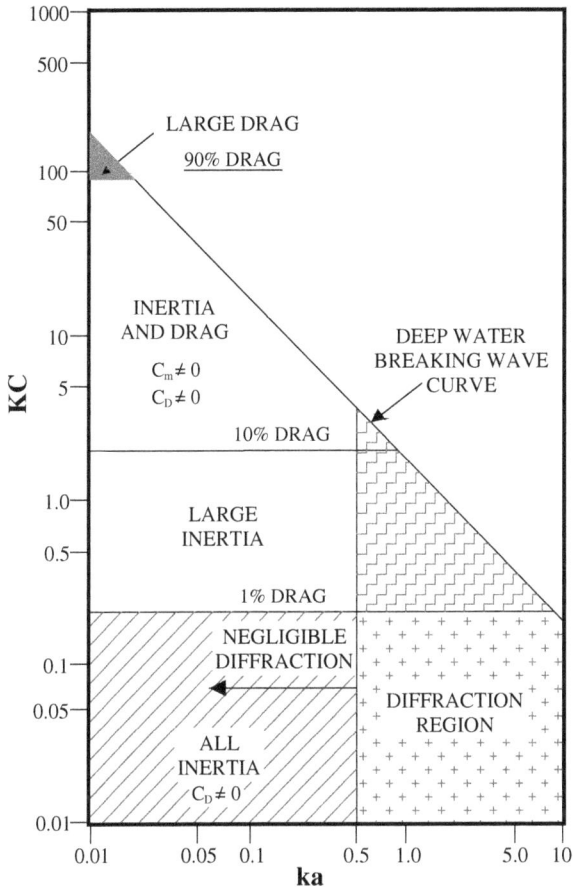

Fig. 7.2   Different wave force regimes (Chakrabarti, 1987).

The different force regimes are as follows:

D/L > 1 Conditions approximate to pure reflection,
D/L > 0.2 Diffraction is increasingly important,
D/L < 0.2 Diffraction is negligible,
$D/L_0 > 0.2$ Inertia force component dominates,
$D/L_0 < 0.2$ Drag force component dominates,

where D = diameter of structure, L = wavelength and L_0 = deepwater wavelength.

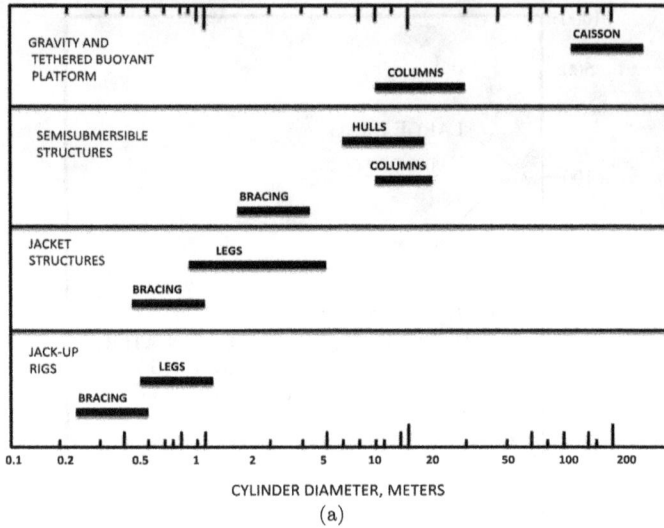

CYLINDER DIAMETER, METERS

(a)

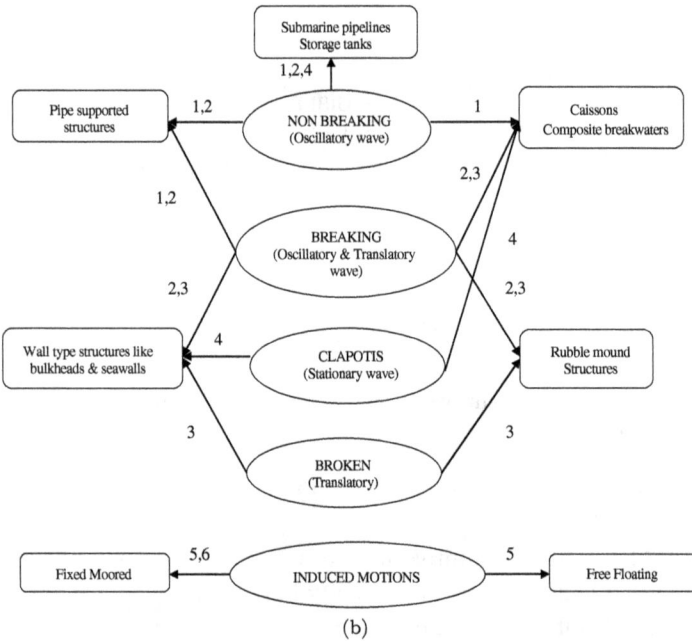

(b)

Fig. 7.3   (a) Perspective on the size of components for typical offshore structures. (b) Types of forces acting on different types of marine structures: 1 — Drag and Lift, 2 — Inertia, 3 — Impact, 4 — Hydrostatic, 5 — acceleration and 6 — Mooring line reaction.

7.3 Wave Forces on Slender Structures

7.3.1 *Fixed cylinder in waves*

When a water particle propagates in a wave, it would carry a momentum leading to the inertia force and while flowing past a cylinder, it undergoes acceleration followed by deceleration. This phenomena necessitates a work to be carried out which is achieved by the application of a force on the cylinder to increase this momentum. The water particle acceleration at the center of the cylinder is proportional to the incremental force on a small segment of the cylinder, dz (in the absence of the cylinder (Fig. 7.4)):

$$dF_I = C_m \cdot \rho \frac{\pi D^2}{4} \dot{u} dz, \qquad (7.1)$$

where dF_I is the inertia force on an incremental element, dz, of a vertical cylinder of diameter D, $\dot{u}$ is the water particle acceleration, C_m is the inertia coefficient (for uniformly accelerated flow, $C_m \cong 2$), and ρ is mass density of fluid.

Due to the presence of the wake region on the downstream side of the cylinder, the drag force component is formed. The wake region on the downstream side possess comparatively low pressure than that on the upstream side, and hence a pressure difference is resulted by the wake between the up and downstream of the cylinder at a given instant of time:

$$dF_D = \frac{1}{2} C_D \rho D |u| u dz, \qquad (7.2)$$

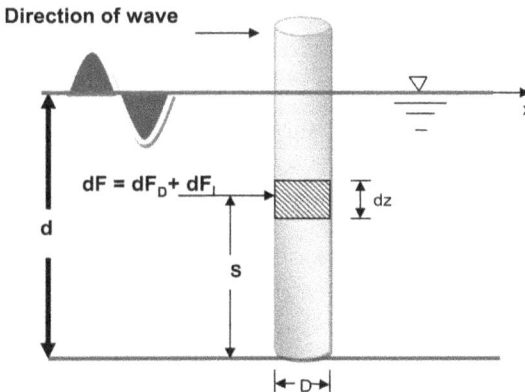

Fig. 7.4   Definition sketch vertical cylinder.

dF_D is the drag force on an incremental segment, dz, of the cylinder, u is instantaneous water particle velocity and C_D is the drag coefficient.

Combining the inertia and drag components of force, the equation of Morison *et al.* (1950) is written as

$$dF_T = \left(C_m \cdot \rho \frac{\pi D^2}{4} \dot{u} + \frac{1}{2} C_D \rho D \, |u| \, u \right) dz. \tag{7.3}$$

If the cylinder extends from the ocean floor to the SWL, the total force on the cylinder is given by the integral,

$$F_T = \int_{-d}^{0} \left[C_m \cdot \rho \frac{\pi D^2}{4} \dot{u} + \frac{1}{2} C_D \rho D |u|u \right] dz, \tag{7.4}$$

where u and $\dot{u}$ are given by

$$u = \frac{\pi H}{T} \frac{\cosh k(z+d)}{\sinh kd} \sin\theta, \tag{7.5a}$$

$$\dot{u} = \frac{-2\pi^2 H}{T^2} \frac{\cosh k(z+d)}{\sinh kd} \cos\theta \tag{7.5b}$$

Substituting Eq. (7.5) in Eq. (7.4), we obtain

$$F_T = C_m \rho \frac{\pi D^2}{4} \int_{-d}^{0} \left(\frac{-2\pi^2 H}{T^2} \frac{\cosh k (z+d)}{\sinh kd} \cos\theta \right) dz$$

$$+ \frac{1}{2} C_D \rho D \int_{-d}^{0} \left(\frac{\pi^2 H^2}{T^2} \frac{\sin\theta |\sin\theta|}{\sinh^2 kd} \cosh^2 k (z+d) \right) dz$$

$$= -C_m \rho \frac{\pi D^2}{4} \frac{2\pi^2 H}{T^2} \frac{\cos\theta}{\sinh kd} \left[\frac{\sinh k (z+d)}{k} \right]_{-d}^{0}$$

$$+ \frac{1}{2} C_D \rho D \frac{\pi^2 H^2}{T^2} \frac{\sin\theta |\sin\theta|}{\sinh^2 kd} \left[\frac{z}{2} + \frac{\sinh 2k (z+d)}{4k} \right]_{-d}^{0}$$

$$= -C_m \rho \frac{\pi D^2}{4} \frac{2\pi^2 H}{T^2} \frac{\cos\theta}{\sinh kd} \times \frac{\sinh kd}{k}$$

$$+ \frac{1}{2} C_D \rho D \frac{\pi^2 H^2}{T^2} \frac{\sin\theta |\sin\theta|}{\sinh^2 kd} \left[\frac{\sinh (2kd)}{4k} + \frac{d}{2} \right],$$

$$\frac{dF_T}{d\theta} = C_m \frac{\rho \pi D^2}{4k} \frac{2\pi^2 H}{T^2} + C_D \rho D \frac{\pi^2 H^2}{T^2} \frac{\cos\theta}{\sinh^2 kd} \frac{1}{4k} (\sinh 2kd + 2kd).$$

$$\tag{7.6}$$

Maximum force is found by substituting the corresponding phase angle, i.e.,

$$\frac{dF_T}{d\theta} = 0 \Rightarrow C_m \frac{\rho\pi D^2}{4k} \frac{2\pi^2 H}{T^2}$$

$$+ C_D \rho D \frac{\pi^2 H^2}{T^2} \frac{\cos\theta}{\sinh^2 kd} \frac{1}{4k} (\sinh 2kd + 2kd) = 0,$$

$$\theta_{max} = \cos^{-1}\left[-\frac{\pi D}{H} \frac{C_m}{C_D} \frac{2\sinh^2 kd}{(\sinh 2kd + 2kd)} \right]. \tag{7.7}$$

Substituting θ_{max} in (7.6), we obtain the maximum total wave force.

The overturning moment about the SWL is given by

$$M = \int_{-d}^{0} dF_T \, z$$

$$= -C_m \frac{\rho\pi D^2}{4} \int_{-d}^{0} \frac{-2\pi^2 H}{T^2} z \frac{\cosh k \, (z+d)}{\sinh kd} \cos\theta \, dz$$

$$+ \frac{1}{2} C_D \rho D \frac{\pi^2 H^2}{T^2} \frac{\sin\theta \, | \sin\theta}{\sinh^2 kd} \int_{-d}^{0} z \cosh^2 k \, (z+d) \, dz$$

$$= -C_m \frac{\rho\pi D^2}{4} \cdot \frac{2\pi^2 H}{T^2} \cdot \frac{\cos\theta}{\sinh kd} \left[\frac{z \sinh k(z+d)}{k} - \frac{\cosh k(z+d)}{k^2} \right]_{-d}^{0}$$

$$+ \frac{1}{2} C_D \rho D \frac{\pi^2 H^2}{T^2} \frac{\sin\theta \, | \sin\theta}{\sinh^2 kd}$$

$$\times \left[\frac{z^2}{4} + \frac{z \sinh 2k(z+d)}{4k} - \frac{\cosh 2k(z+d)}{8k^2} \right]_{-d}^{0}$$

$$= -C_m \frac{\rho\pi D^2}{4} \cdot \frac{2\pi^2 H}{T^2} \cdot \frac{\cos\theta}{\sinh kd} \left[\frac{\cosh k(z+d)}{k^2} + \frac{1}{k^2} \right]$$

$$+ \frac{1}{2} C_D \rho D \frac{\pi^2 H \sin\theta | \sin\theta|}{T^2 \sinh^2 kd} \left[\frac{-\cosh 2kd}{8k^2} - \frac{d^2}{4} + \frac{1}{8k^2} \right]$$

$$= \rho D \frac{\pi^2 H}{2T^2} \frac{1}{k^2 \sinh kd} [-C_m \pi D \cos\theta \, (1 - \cosh kd)$$

$$+ C_D \frac{\sin\theta | \sin\theta|}{8 \sinh kd} H \, (1 - \cosh 2kd - 2k^2 d^2) \bigg]. \tag{7.8}$$

Keulegan and Carpenter (1958) studied the forces of a sinusoidal oscillating two-dimensional flow and found that the non-dimensional forces were found to be related to the parameter $\frac{u_{max}T}{D}$. It was reported that C_D and C_m were smoothly varying functions of the Keulegan–Carpenter number,

$KC = (u_{max} + U)T/D$ and Reynold's number $R_e = (u_{max} + U)D/\nu$. Here, ν is Kinematic viscosity of fluid and U is the current velocity. Iwagaki *et al.* (1983) after a detailed experimental investigation have proposed the following expression for KC number, with $\cos \beta = -U/u_{max}$:

$$KC = \begin{cases} \frac{u_{max}T}{D} \left[\sin \beta + (\pi - \beta)\cos \beta\right] & \text{for } |U| \leq u_{max} \\ \frac{\pi U T}{D} & \text{for } |U| > u_{max}. \end{cases} \tag{7.9}$$

The Morison formulation for a fixed rigid cylinder in the presence of waves and current U can be written as

$$F_T = C_m\, \rho\, \frac{\pi D^2}{4}\, \dot{u} + \frac{1}{2}\, C_D\, \rho D\, |u \pm U|\, (u \pm U). \tag{7.10a}$$

An alternate form of Morison equation has been suggested by Chakrabarti, as shown below.

$$dF_T = C_m\rho\, \frac{\pi D^2}{4}\dot{u} + \frac{1}{2}C_D\, \rho D\, |u|\, u + \frac{1}{2}C_D\rho D\, |U|\, U. \tag{7.10b}$$

Substitute u and $\dot{u}$ in (7.10b) and integrating from sea bottom to SWL, we get

$$\begin{aligned} F_T = {} & -C_m\, \rho\, \frac{\pi D^2}{4}\, \frac{2\pi^2\, H}{T^2}\, \frac{\cos \theta}{k} \\ & + \frac{1}{2}\, C_D\, \rho D\, \frac{\pi^2\, H^2}{T^2}\, \frac{\sin \theta\, |\sin \theta|}{\sinh^2 kd} \left[\frac{\sinh(2kd)}{4k} + \frac{d}{2}\right] \\ & + \frac{1}{2}\, C_D\rho\, Dd\, |U|\, U. \end{aligned} \tag{7.10c}$$

The total overturning moment about the SWL for the combined waves and currents is given by

$$\begin{aligned} M = {} & \int_{-d}^{0} dF_T\, z \\ = {} & \rho D\frac{\pi^2 H}{2T^2}\, \frac{1}{k^2 \sinh kd} \\ & \times \left[-C_m\, \pi D \cos\, \theta\, (1 - \cosh kd)\right. \\ & \left. + C_D\, \frac{\sin \theta|\sin \theta|}{8 \sinh kd}\, H\, (1 - \cosh 2kd - 2k^2\, d^2)\right] \\ & + \frac{1}{2}\, C_D\rho\, D\, |U|\, U\frac{1}{2}d^2. \end{aligned} \tag{7.10d}$$

The coefficients of drag and inertia for a fixed cylinder in waves in the absence and presence of currents as a function of KC number are shown in Figs. 7.5 and 7.6, respectively.

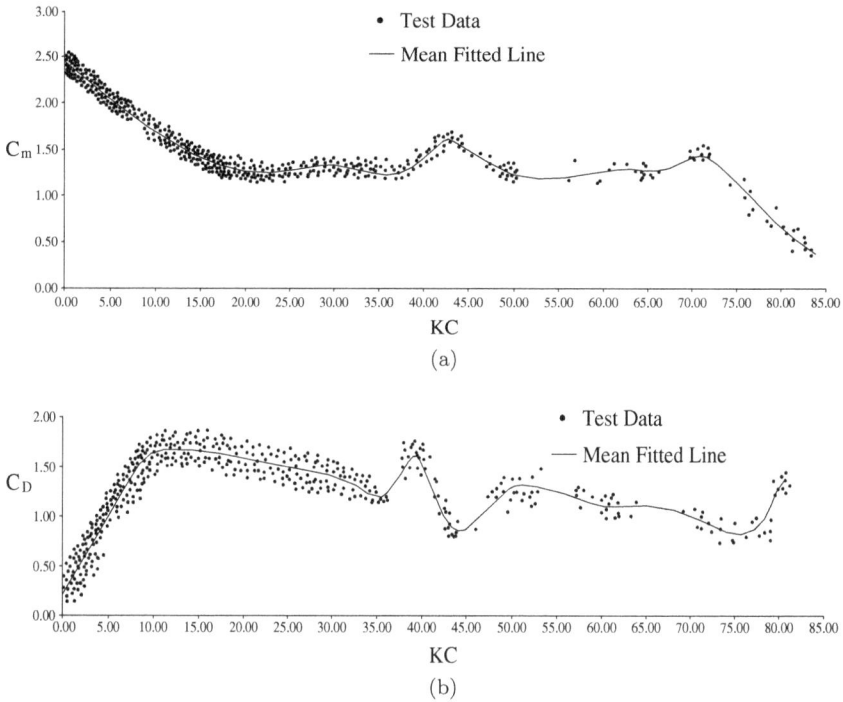

Fig. 7.5 Variation of (a) C_m and (b) C_D with KC (Chakrabarti, 1987).

The C_D and C_m are obtained through measurements, the adopted methodology of which is projected in Fig. 7.7. The measured η, u and the F_T on the cylinder both in phase are simultaneously acquired. The u could also be derived from the measured η. As the measured parameters are in phase, at a phase angle $\theta = 0°$, $F_D = 0$ and $F_T = F_I$, from which C_m can be evaluated. At $\theta = 90°$, $F_I = 0$ and $F_T = F_D$ from which C_D can be evaluated. A more superior method is the least square method that minimizes the error between the measured and computed forces, the detail of which has been discussed in Chakrabarti (1987).

7.3.1.1 *Current design criteria*

The most important single factor in designing of structures for current loads is the selection of the maximum design velocity. The selection of the maximum current velocity has to be based on actual measurements when possible or on other sources of reliable current table data that would

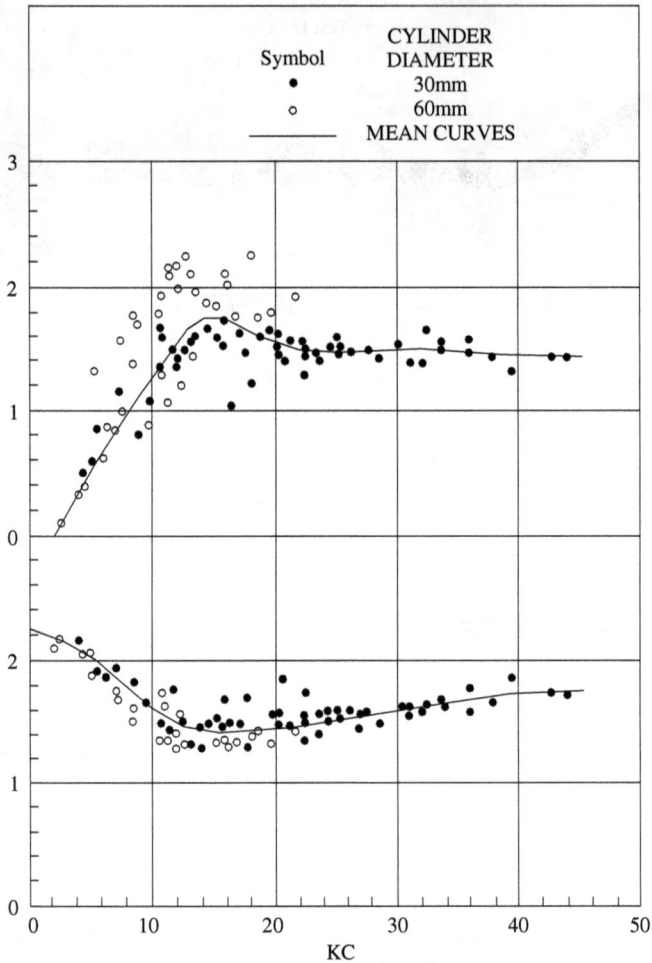

Fig. 7.6   Variation of C_D and C_m with KC for waves and currents (Iwagaki *et al.*, 1983).

consider seasonal variations as well as that during extreme events like storms, cyclones. In general, wherein, measurements are not available or lacking, the vertical distribution of tidal current velocity can be assumed to be distributed as per the one-seventh power law which provides the tidal current velocity at any particular depth, (U_{Tz}) as $U_{Tz} = U_{Ts} (z/d)^{1/7}$, where U_{Ts} is the surface current, and z is the distance above the bottom measured positive upward.

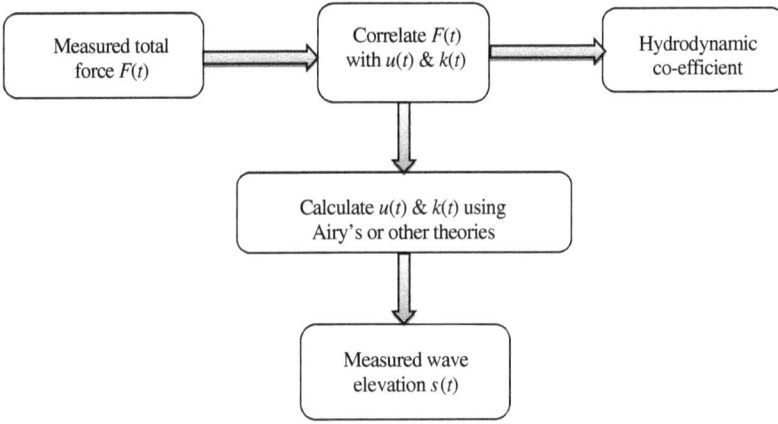

Fig. 7.7   Methodology for evaluating C_D and C_m.

7.3.1.2   *Ratio between F_D and F_I*

Using Airy's linear wave theory,

$$KC(\text{at SWL}) = \frac{\pi H}{T} \frac{L_o}{L},$$

KC basically represents the ratio between drag and inertia forces, which can be derived as below.

$$F_{D\,max} = \frac{1}{2} C_D \rho D u^2_{max}, \quad F_{I\,max} = C_m \rho \frac{\pi D^2}{4} \dot{u}_{max},$$

$$\alpha = \frac{F_{D\,max}}{F_{I\,max}} = \frac{2}{\pi} \frac{C_D}{C_m} \frac{u^2_{max}}{\dot{u}_{max}} \frac{1}{D}. \qquad (7.10e)$$

For a sinusoidal flow, $\dot{u} = \dot{u}_{max} \cos \theta$ and $\dot{u}_{max} = \frac{2\pi u_m}{T}$.
  Hence, $\alpha = \frac{C_D}{C_m} \frac{1}{\pi^2} \frac{u_{max} T}{D} = \frac{C_D}{C_m} \cdot \frac{KC}{\pi^2}$.
  At still water level, for deep waters, $u_{max} = \pi H / T$, and hence

$$\alpha = \frac{C_D}{C_m} \frac{H}{D\pi}. \qquad (7.10f)$$

When H/D is < 1.0 and D/L < 0.1, i.e., KC is less than about 3.0 the flow past a cylinder does not separate and hence the total force is given by the inertia force, the drag force being negligible. For KC greater than 25, particle movement is much greater than D, whereas, for KC less than 5, particle movement is much lesser than D.

7.3.2　Flexible cylinders in waves

For a flexible cylinder in waves, the wave force per unit length is obtained by modifying the Morison's equation by replacing u with relative velocity $(u - v)$ and $\dot{u}$ with relative acceleration, $(\dot{u} - \dot{v})$. The above equation is known as relative velocity model of the formulation of Morison *et al.* (1950).

$$dF_T = C_{m1} \frac{\pi}{4} \rho D^2 \dot{u}\, dz + C_{m2} \frac{\pi}{4} \rho D^2 (\dot{u} - \dot{v})\, dz$$

$$+ \frac{1}{2} C_D \rho D (u - v) |(u - v)|\, dz. \tag{7.11}$$

The above equation is known as relative velocity model obtained by modifying the Morison equation by replacing u with relative velocity $(u - v)$ and $\dot{u}$ with $(\dot{u} - \dot{v})$.

Here v and $\dot{v}$ are velocity and acceleration of the cylinder

$$C_{m1} = 1.0 - 0.12(\pi D/L) \text{ and}$$

$$C_{m2} = 1.0 \quad \text{for } \pi D/L < 0.5$$

$$= 1.54 - 1.08(\pi D/L) \quad \text{for } (\pi D/L) > 0.5, \tag{7.12}$$

wherein C_{m1} and C_{m2} are the inertia coefficients.

7.3.3　Wave forces on an inclined cylinder

Consider a cylinder inclined at an angle α to the vertical, lying in the X–Z plane as shown in Fig. 7.8.

The total force acting on the inclined cylinder is given by the expression

$$F_T = \frac{1}{2} \rho C_D D \int_{-d}^{0} V_n |V_n|\, dz + C_M \rho \frac{\pi D^2}{4} \int_{-d}^{0} \dot{V}_n dz \tag{7.13}$$

where, $V_n = u \cos \alpha + w \sin \alpha$ and $\dot{V}_n = \dot{u} \cos \alpha + \dot{w} \sin \alpha$,
where V_n = component of the instantaneous water particle velocity normal to the cylinder axis, and $\dot{V}_n$ = component of the instantaneous water particle acceleration normal to cylinder axis.

Sundar *et al.* (1998) carried out tests on the measurements of pressures around a cylinder and through its integration computed the C_D and C_m for a cylinder inclined along and against the wave direction. The results were compared with that of Sarapkaya *et al.* (1982) and Chakrabarti (1980). The trend in their variations with KC are shown in Fig. 7.9.

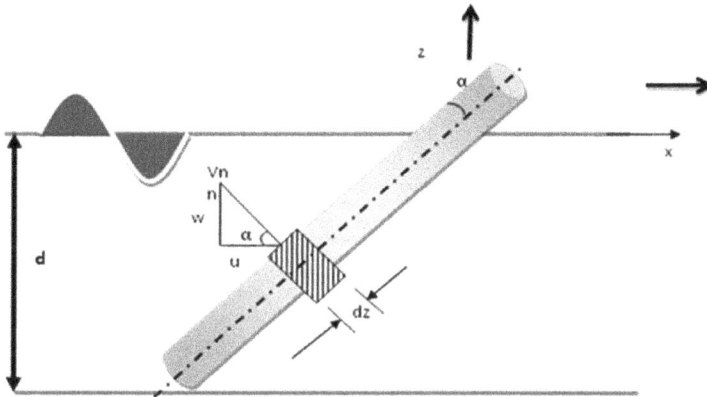

Fig. 7.8   Inclined cylinder — definition sketch.

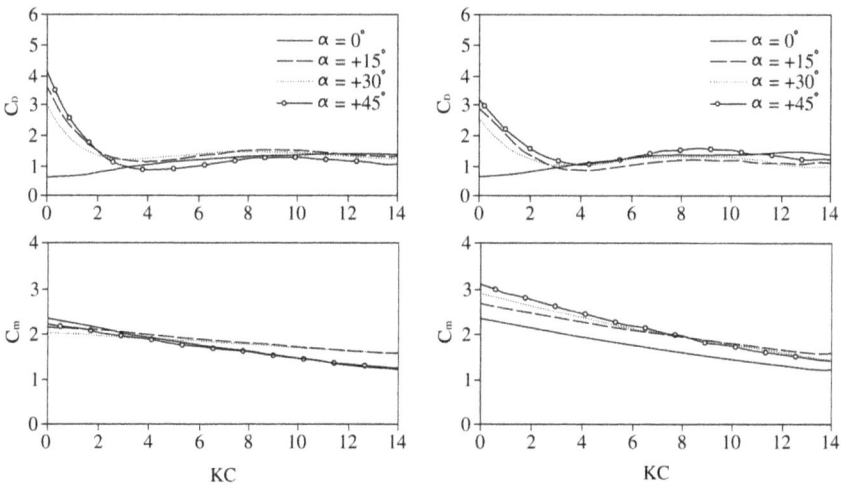

Fig. 7.9   Variation of C_D and C_m for inclined cylinders (Sundar *et al.*, 1998).

7.3.4   *Submerged cylinders*

The Morison equation was originally formulated to calculate the wave force on a vertical cylinder, which was later justified by Parker (1977) that the same equation holds good to evaluate the wave force on submarine pipelines provided the water particle kinematics are calculated at the centerline of the pipeline. In addition to the horizontal wave force, pipelines experience a vertical force, which would act upward from the sea floor being a maximum, when the pipeline is resting on the floor. However, if there is a gap between the pipeline and the sea floor a negative lift force acting downward will tend to decrease the total vertical force (Fig. 7.10(a)). The horizontal force per unit length is given by

$$f_h = \frac{1}{2}C_D\rho D\, u\, |u| + C_{m\rho}\frac{\pi D^2}{4}\dot{u}. \tag{7.14}$$

The vertical force per unit length acting on a pipeline is

$$f_v = C_L\frac{\rho D}{2}\, u^2 + C_{mv}\frac{\rho\pi D^2}{4}\dot{w} \tag{7.15}$$

where C_L and C_{mv} are the vertical coefficients, respectively. In the case of large diameter pipelines resting on the sea floor the horizontal wave force per unit length, f_h and vertical wave force per unit length, f_v are given by

$$f_h = \left[C_m\rho\pi D^2\dot{u}\right]/4, \tag{7.16}$$

$$f_v = \left[C_L\rho Du^2\right]/2. \tag{7.17}$$

For such a case according to potential flow theory

$$C_L = 4.49 \quad \text{and} \quad C_m = 3.29.$$

Based on preliminary laboratory tests, Priest (1971) obtained the following empirical relationship for the forces as a function of H and D.

$$\frac{f_h}{\gamma dD} = 0.18\left(\frac{H}{D}\right)^{1.63},$$

$$\frac{f_v}{\gamma dD} = 0.16\left(\frac{H}{D}\right)^{1.56}. \tag{7.18}$$

For a pipeline resting on a sloping seafloor (Fig. 7.10(b)), the different force components due to waves and currents acting, as well as the requirement of the its stability are provided below.

$$F_D + F_I - F_f + w\sin\beta = 0 \ \text{(horizontal)}, \tag{7.19a}$$

$$F_N + F_L - w\cos\beta = 0 \ \text{(vertical)}, \tag{7.19b}$$

$$F_f = \mu F_N \ \text{(for stability)}. \tag{7.19c}$$

The minimum submerged weight of the pipe to remain at the sea bed is

$$w = \frac{F_d + F_i + \mu F_l}{\mu\cos\beta - \sin\beta}, \tag{7.19d}$$

where μ is the Coefficient of friction, F_D the Drag force; F_I the inertia force, F_N the normal force; F_L the lifting force F_f the friction force and w the submerged weight.

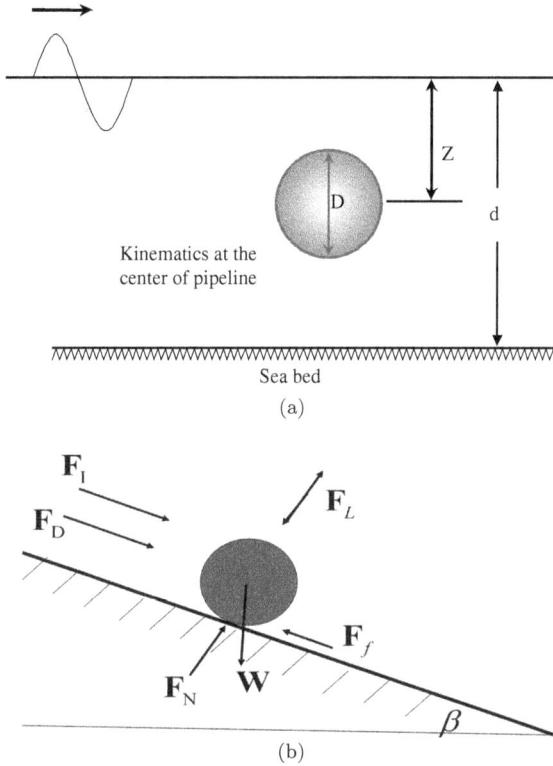

Fig. 7.10   (a) Submerged cylinder — definition sketch. (b) Free-body diagram of a pipe over a sloping bed exposed to waves and currents.

7.4   Froude–Krylov Forces

Froude–Krylov pressure forces are derived from ideal, hydrodynamic flow, and linear wave theory. The diffraction forces are usually calculated on Froude–Krylov pressure forces. Experimental flow coefficients aid to modify these forces. It is assumed that the wave pressure field is completely undisturbed by the presence of the structure.

Here, as an example, the total forces and moment on a single, submerged solid column or cylinder shown in Fig. 7.11 are as follows:

$$F_h(t) = C_h F_{fx}, \tag{7.20a}$$

$$F_v(t) = C_v F_{fz}, \tag{7.20b}$$

$$M_o(t) = C_o \left(\overline{z} F_{fx} + \overline{x} F_{fz} \right) = C_o M_f, \tag{7.20c}$$

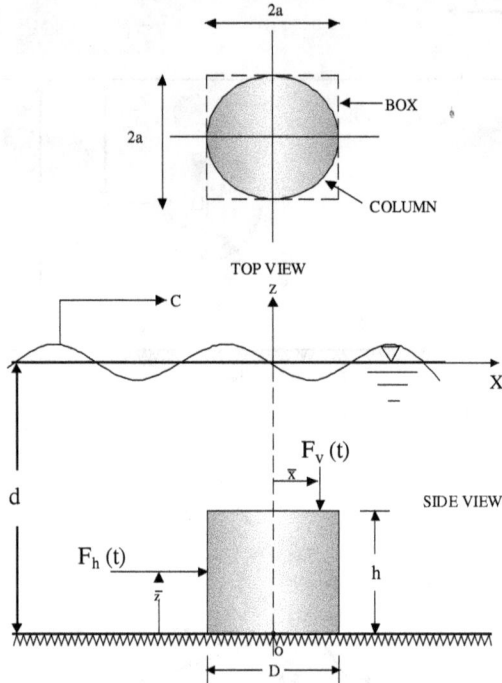

Fig. 7.11　Froude–Krylov force — definition sketch.

where C_h, C_v and C_o are the respective flow coefficients. The terms F_{fx} and F_{fz} produce the overturning moment, due to the net pressure-induced forces on the vertical sides acting at $\bar{z}$ and on the top horizontal surface acting at $\bar{x}$. The F_{fx} and M_o for the structure shown in the above figure is obtained as discussed below.

Assuming linear theory, the dynamic pressure due to propagating wave is given as

$$p(x, z, t) = \frac{\gamma H}{2} \cdot \frac{\cosh k(d + z)}{\cosh kd} \cdot \cos(kx - \sigma t). \qquad (7.21)$$

Integration of the pressure difference over the vertical sides of the box normal to a right traveling wave will give the parameter F_{fx},

$$F_{\text{fx}} = 2a \int_{-d}^{h-d} [p(a, z, t) - p(-a, z, t)] \, dz. \qquad (7.22a)$$

Substituting Eq. (7.21) in Eq. (7.22), integrating and using trigonometric identities, we get

$$F_{fx} = 2\gamma a H \frac{\sinh kh}{k \cosh kd} \sin ka \sin \sigma t \qquad (7.22b)$$

Integration of the pressure difference over the horizontal sides of the box will give F_{fz}:

$$F_{fz} = 2a \int_{-a}^{a} [p(x, -d, t) - p(x, h - d, t)] \, dx, \qquad (7.23a)$$

$$F_{fz} = 2\gamma a H \frac{\cosh(kh) - 1}{k \cosh kd} \sin ka \cos \sigma t. \qquad (7.23b)$$

M_f is sum of two integrals. The first is the moment about "0" due to the side pressure forces, (M_h), and the second is the moment about "0" due to the pressure forces on the top, (M_v). These moment components are as follows:

$$M_h = 2a \int_{-d}^{(h-d)} [p(a, z, t) - p(-a, z, t)] \, (d - (-z)) \, dz \qquad (7.24a)$$

$$M_v = 2a \int_{-a}^{a} [p(x, -d, t) - p(x, h - d, t)] \, x \, dx \qquad (7.24b)$$

and

$$M_f = M_h + M_v, \qquad (7.24c)$$

$$M_h = \frac{2a\gamma H}{k^2} \frac{\sin(ka)}{\cosh(kd)} [1 + kh \sinh(kh) - \cosh(kh)] \sin(\sigma t),$$

$$M_v = \frac{2a\gamma H}{k^2} \cdot \frac{\cosh(kh) - 1}{\cosh(kd)} [\sin(ka) - ka \cos(ka)] \sin(\sigma t).$$

Approximate formulas for C_h, C_v and C_0 are given by

$$C_h = 1 + 0.75 \left(\frac{h}{D}\right)^{1/3} \left[1 - 2.96 \left(\frac{D}{L}\right)^2\right],$$

$$C_v = \begin{cases} 1 + 7.3 \left(\frac{D}{L}\right)^2 \left(\frac{h}{D}\right) & \text{for } \frac{\pi h}{L} < 1, \\ 1 + 1.57 \left(\frac{D}{L}\right) & \text{for } \frac{\pi h}{L} > 1, \end{cases}$$

$$C_0 = 1.9 - 1.1 \left(\frac{D}{L}\right). \qquad (7.25)$$

The above equations are restricted to the following ranges:

$$\frac{h}{d} < 0.6 \quad \text{for } C_h, C_v \text{ and } C_0,$$

$$0.3 < \frac{h}{D} < 2.3 \quad \text{for } C_h \text{ and } C_v \text{ only,}$$

$$0.6 < \frac{h}{D} < 2.3 \quad \text{for } C_o \text{ only.} \tag{7.26}$$

7.5 Linear Diffraction Problem

When the relative characteristic dimension D/L is greater than 0.2, the incident wave gets scattered. Application of Morison equation to this regime is invalid. However, it may be used, provided the force is predominantly inertial which means the drag term be dropped in the Morison equation. The "KC" at the SWL on applying linear wave theory may be written as

$$KC = \frac{\pi H/L}{(D/L)\tanh kd}, \tag{7.27}$$

where the maximum H/L is given approximately 0.14 tanh (kd). Then, the value of KC is KC < 0.44 / (D/L). That is, when D/L > 0.2 (for wave diffraction increasingly important), KC will be less than 2.2 and more specifically it will be less than 1. In any type of wave-structure interaction problem, for example, wave loads on structures, the number of variables like H, L, d, D, etc. to deal with is more and hence it becomes essential to present the results in a non-dimensional form and more so, this facilitates understanding clearly the effect a particular parameter like wave steepness, H/L on the force on a pile in different relative water depths, d/L in which case the force also should be in a non-dimensional form. The non-dimensional wave force on (say) a fixed bottom mounted and surface piercing large cylinder in the diffraction regime may be written as

$$\frac{F}{\rho g H D^2} = f\left(d/L, H/L, D/L, Re\right). \tag{7.28}$$

Since diffraction is important, the Re can be omitted. As we deal with linear waves, H/L is small and hence can be neglected. This allows us to deal with just two parameters instead four:

$$\frac{F}{\rho g H D^2} = f\left(d/L, D/L\right). \tag{7.29}$$

Through the above equation, it is now possible to present the results through experiments or numericals the variation of the non-dimensional force as function of D/L for different d/L.

The linear diffraction problem involves the determination of the velocity potential that satisfies the Laplace equation within the fluid region.

$$\nabla^2 \varphi = 0. \tag{7.30}$$

The above equation is subjected to the following boundary conditions as well as the Sommerfeld radiation condition:

$$\frac{\partial^2 \phi}{\partial t^2} + g \frac{\partial \phi}{\partial z} = 0 \quad at\ z = 0, \tag{7.31}$$

$$\eta = \frac{-1}{g} \left(\frac{\partial \phi}{\partial t} \right) |_{z=0}, \tag{7.32}$$

$$\frac{\partial \phi}{\partial z} = 0 \quad at\ z = -d, \tag{7.33}$$

$$\frac{\partial \phi}{\partial n} = 0 \quad \text{at the body surface.} \tag{7.34}$$

Here, "n" denotes distance in a direction normal to the body surface. The summation of incident wave potential ϕ_i and scattered wave potential ϕ_s gives the total velocity potential:

$$\phi = \phi_i + \phi_s. \tag{7.35}$$

The Sommerfeld radiation condition, i.e., at large distances from the structure, ϕ_s, corresponds to an outgoing wave:

$$\lim_{r \to \infty} r^{1/2} \left[\frac{\partial \phi_s}{\partial r} - i k \phi_s \right] = 0, \tag{7.36}$$

where "r" is the radial ordinate. The body surface boundary condition expressed as

$$\frac{\partial \phi_s}{\partial n} = \frac{-\partial \phi_i}{\partial n} \quad \text{at the body surface} \tag{7.37}$$

shows the dependence of ϕ_s and ϕ_I. Hence, Eqs. (7.30)–(7.34) applied to ϕ_s along with Eqs. (7.36) and (7.37) define the problem in terms of ϕ_s with which ϕ can be determined. Then, the pressure throughout the fluid can be evaluated using the linearized Bernoulli equation

$$p = -\rho g z - \rho \frac{\partial \phi}{\partial t}. \tag{7.38}$$

Appropriate integration of the pressure acting on the body surface may then be carried out to obtain sectional forces and overturning moments as required. Although there are several available theories, the experimental results of Mogridge and Jamieson (1976) and Hogben and Standing (1974)

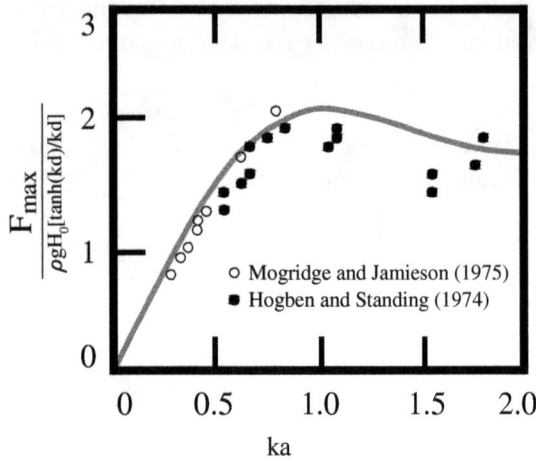

Fig. 7.12   Force coefficient for large vertical cylinder.

superposed over the closed form solution is presented in Fig. 7.12 as this is quite straightforward for the evaluation of wave force. The readers are suggested to refer to Sarpkaya and Isaacson (1981) or Chakrabarti (1987) for the details of other numerical methods. In this chapter, only the linear diffraction theory of MacCamy and Fuchs (1954) is discussed.

7.5.1 Linear diffraction theory for wave forces

7.5.1.1 For circular cylinder

One of the first applications of diffraction theory to ocean engineering structures was carried out by MacCamy and Fuchs (1954) for determining the wave forces on a large vertical circular cylinder. In the formulation of the problem, the coordinate axes were taken to originate at the still water level with x-axis in the direction of wave travel and the z-axis originating upwards from SWL. The incident velocity potential is written as

$$\phi_i = \frac{Hg}{2\sigma} \frac{\cosh k\,(d+z)}{\cosh kd} \cos(kx - \sigma t). \qquad (7.39)$$

Bernoulli's Eq. (7.38) provides the pressure. The horizontal force acting on the pile per unit length is computed as

$$f_h\,(t) = \frac{2\rho g H}{k} \frac{\cosh k\,(d+z)}{\cosh kd} A\,(ka) \cos\,(\sigma t - \alpha), \qquad (7.40)$$

where

$$A\left(ka\right) = \frac{1}{\sqrt{\left(J_1'\left(ka\right)\right)^2 + \left(Y_1'\left(ka\right)\right)^2}},$$

$J_1'(ka)$ is the differential value of Bessel J function of first kind and first order whose argument is ka, $Y_1'(ka)$ is the differential value of Bessel Y function of the second kind and first order whose argument is ka and orientation angle, $\alpha = \tan^{-1}\left[J_1'\left(ka\right)/Y_1'\left(ka\right)\right]$.

This theory is confined to large vertical circular cylinders resting on the ocean bed and piercing the free surface and waves of small steepness which poses a serious practical limitation.

7.5.1.2 *For non-circular caissons: Square caisson*

It was assumed by Mogridge and Jamieson (1976) that the wave force on a large square cylinder of size "b" can be expressed as an inertial force if the coefficient of mass used includes the effects of wave reflection and diffraction. Thus the inline force per unit length may be expressed as

$$f_h = C_{ms}\rho b^2 \dot{u}, \tag{7.41}$$

where C_{ms} is the coefficient of mass for square section.

Substituting for $\dot{u}$ in Eq. (7.41), we obtain

$$f_h = C_{ms}\rho b^2 \frac{2\pi^2 H}{T^2} \frac{\cosh k(d+z)}{\sinh kd} \cos(kx - \sigma t). \tag{7.42}$$

According to the assumption this f_h will be equal to inertial force acting on a circular cylinder of equivalent diameter D_e,

$$C_{ms}\rho b^2 \dot{u} = C_m \frac{\rho \pi D_e^2}{4} \dot{u}, \tag{7.43}$$

$$D_e = 2b\left(\frac{C_{ms}}{\pi C_m}\right)^{1/2}. \tag{7.44}$$

It was further assumed $C_{ms} = C_m$.

Hence,

$$D_e = \frac{2b}{\sqrt{\pi}}. \tag{7.45}$$

According to MacCamy and Fuch's linear diffraction theory, the f_h acting on the circular caisson of diameter, D_e, is

$$f_h = \frac{2\rho g\, H}{k} \frac{\cosh k\left(d+z\right)}{\cosh kd} A\left(ka_e\right)\cos\left(\sigma t - \alpha\right), \tag{7.46}$$

where

$$A\left(Ka_e\right) = \frac{1}{\sqrt{J_1'^2\left(ka_e\right) + Y_1'^2\left(ka\right)}} \quad \text{and} \quad \alpha = \tan^{-1}\left(\frac{J_1'\left(ka_e\right)}{Y_1'\left(ka_e\right)}\right)$$

with $a_e = \frac{D_e}{2}$, $\alpha = $ Orientation angle.

Equating (7.42) and (7.46), we have

$$C_{ms} = \frac{L^2}{\pi^2 b^2} A(ka_e) \frac{\cos\left(\sigma t - \alpha\right)}{\cos \sigma t}, \tag{7.47}$$

$$C_{ms}^* = \frac{L^2}{\pi^2 b^2} A(ka_e), \tag{7.48}$$

so that

$$C_{ms} = C_{ms}^* \frac{\cos\left(\sigma t - \alpha\right)}{\cos \sigma t}. \tag{7.49}$$

Here C_{ms} is equal to C_{ms}^* when a is approximately zero, that is, for b/L approaching zero or equal to 0.52.

The total force in the x direction is obtained by integrating the force per unit length as given by Eq. (7.42) for $x = 0$, through the depth of water from the bottom to the SWL:

i.e.,

$$F_h(t) = C_{ms}^* \frac{\rho \pi b^2 H L}{T} \cos \sigma t. \tag{7.50}$$

Substituting the expression for C_{ms} from Eq. (7.49) into Eq. (7.50) and when $\sigma t = \alpha$, $F_h(t)$ is a maximum given as

$$F_{max} = C_{ms}^* \frac{\rho \pi b H^2 L}{T^2}. \tag{7.51}$$

The overturning moment about the base of the cylinder will be

$$M(t) = \int_{-d}^{0} f_x(z + d)\, dz. \tag{7.52}$$

With $\sigma t = \alpha$, the maximum overturning moment about the base is given by

$$M_{max} = C_{ms}^* \frac{\rho g b^2 H L}{4\pi} [kd \tan hkd + \sec hkd - 1]. \tag{7.53}$$

Mogridge and Jamieson (1976) have presented the design chart for C_{ms}^* as a function of b/L for two angles of orientation of the square cylinder with wave direction, $\beta = 0°$ and $45°$ as shown in Fig. 7.13. The variation of M_{max} as a function of relative water depth, d/L projected in Fig. 7.14 permits us to obtain the maximum moment.

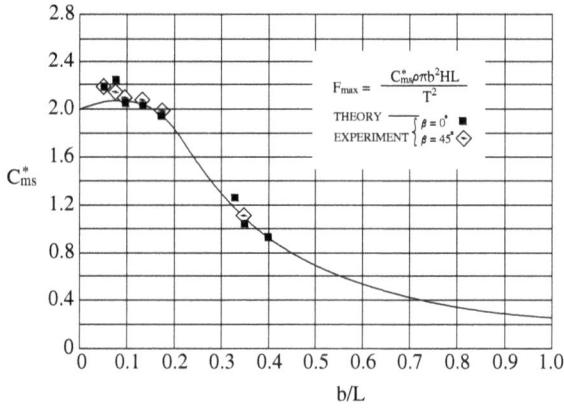

Fig. 7.13   Variation of C_{ms}^{*} with b/L.

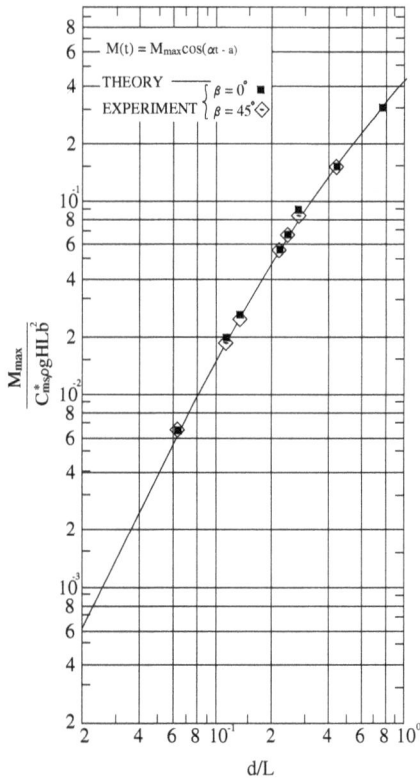

Fig. 7.14   Non-dimensional moment versus d/L.

7.6   Wave Slamming

Wave slam results from the sudden immersion of a structure in water during the passage of a wave. The forces on the member will increase due to buoyancy, drag, and virtual inertia. In addition to these forces predicted by Morison's equation there is found to be a transient peak load on immersion. Apart from early interest in marine craft, recent research in slam forces relates to circular cylinders. Cylindrical elements are common in offshore platforms and it is difficult to omit horizontal tubes from the entire area swept by the design waves. It is considered most appropriate to relate the slam force to a velocity squared term analogous to the viscous drag force. A slam force coefficient is thus defined which is a maximum at the instant of immersion of the cylinder. The resultant slam force on the member is a maximum when the axis of the cylinder is parallel to the water surface. Therefore, slam forces are alleviated on diagonal tubes and absent on vertical tubes. However, the presence of nodes or appendages may reduce slamming loads on horizontal members. Slam forces become important in relation to the other loads when the Froude number of the impact velocity to member diameter exceeds 0.6.

Wave climate in the offshore is not sufficiently coherent to cause a slam load which acts instantaneously along the length of a member. This condition is most likely to occur to members in conditions such as the barge launch of a steel jacket structure. In waves, the implications of fatigue damage to a lightly damped member may be of greater concern to the designer than the maximum static slam force. Horizontal members in the splash zone will be subjected to wave slamming forces given by

$$F_s = 0.5 \, \rho C_s \, D \, u^2, \qquad (7.54)$$

where F_s is the wave slamming force per unit length and C_s the slamming coefficient $\geq 3.0 \cong 3.6$ for tubular members.

If the slamming force is impulsive, dynamic amplification should be considered. DeT Norske Veritas (DNV) recommends that for a horizontal member, the factors of 1.5 and 2.0 should be used for the end moment and the mid-span moment, respectively.

7.7   Wave Forces on Walls and Rubble Mound Structures

Most offshore structures are pile supported and hence the foregoing discussion was on the evaluation of wave forces on cylindrical or tubular members.

In addition, rubble mound and vertical types are the common coastal structures. The evaluation of wave forces on such structures is also important.

7.7.1   Rubble mound structures

In the case of a rubble mound breakwater, like groins or breakwaters, the size of the individual stone/armor or artificial block forming the primary or the cover layer should be designed such that it is stable when exposed to the action of high waves even during severe sea state. The forces acting on a rubble stone on a slope, α is schematically represented in Fig. 7.15. Equating the different forces, we get

$$mg \sin \alpha + \text{(force due to waves)} = \mu mg \cos \alpha, \tag{7.55}$$

where "m" is mass of stone or the armor block and μ is a friction coefficient. The diameter of stone is D_g.

The wave force on the armor, F_a, can be expressed as

$$F_a = 0.5 f_w \rho_w u^2 . D_g^2 . \text{constant}. \tag{7.56}$$

Here, f_w is the friction factor and on assuming a long wave acts on the armor, and $u = (gH_b)^{1/2}$. Hence, the design mass of stone M can be calculated as

$$M = \frac{\rho_s H_b^3}{K_D (\rho_s/\rho_w - 1)^3 (\mu \cos \alpha - \sin \alpha)^3}. \tag{7.57}$$

The breaking wave height H_b can be assumed to be the design wave height, K_D as a constant that takes into account the characteristics of the armor layer and ρ_s and ρ_w are the densities of armor layer, e.g., rock and sea water, respectively. However, the formula of Hudson (1959) is being

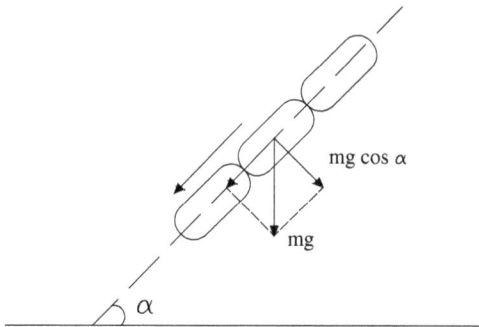

Fig. 7.15   Forces on an armor unit in waves — definition sketch.

widely adopted for obtaining the weight of the armor block which is given by

$$W = \frac{\gamma_s H_D^3}{K_D (S_r - 1)^3 \cot\theta},$$ (7.58)

where H_D is the design wave height in "m", γ_s is the unit weight of the armor unit, θ is the angle of revetment with the horizontal, submerged weight of the armor, $S_r = (\rho_s/\rho_w - 1)$. K_D is a dimensionless stability coefficient, determined through physical modeling. This value for K_D varies for different kinds of armor blocks, type of placing the armor units, number of layers, slope of the surface, and nature of waves whether breaking or non-breaking. K_D is around 2–4 for natural quarry rock, the lower value for breaking waves, whereas, for tetrapods, that is being widely used it is 7 and 8 for breaking and non-breaking waves, respectively. For other artificial interlocking concrete blocks, like CoreLocs, accropods, it is about 10 or even more. Refer USACE for more details.

7.7.2   Vertical walls

The wave forces on wall-type structures particularly in the nearshore may experience non-breaking, breaking, or broken waves. The non-breaking force is primarily hydrostatic. The total hydrodynamic pressure distribution on a wall consists of two components, namely, hydrostatic due to instantaneous pressure component and the dynamic component due to acceleration of the wave orbital acceleration. The common wall-type structures that experience non-breaking wave force can be classified as non-overtopping vertical wall, overtopping vertical wall, vertical wall with rubble foundation, non-overtopping vertical wall with same water depth on both sides, and non-overtopping vertical wall with different water depth on both sides. The total pressure on a vertical wall that was initially deduced by Sainflou (1928) was later modified by Miche (1944) and Rundgren (1958). The total pressure on the wall when the crest or the trough is on wall (positive for crest on the wall, negative for trough on the wall) as shown in Fig. 7.16 is given by

Total pressure $= \gamma d \pm p_1$, where

$$p1 = \frac{1 + K_r}{2} \left(\frac{\gamma H_i}{\cosh 2Kd} \right).$$ (7.59)

The force due to breaking is many times higher. Hence, utmost care should be given in the evaluation of forces in the design of coastal structures

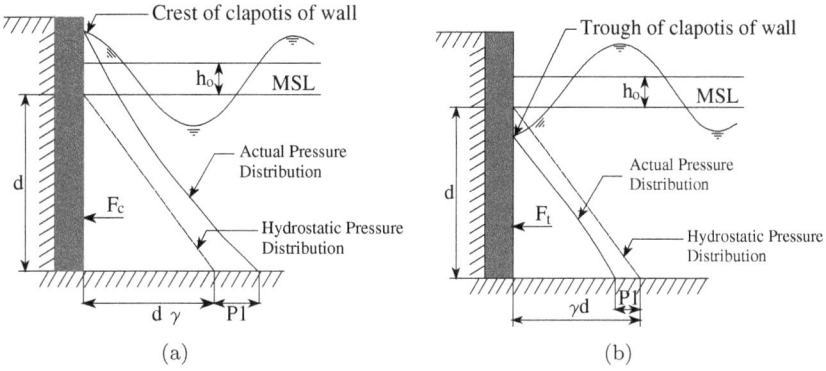

Fig. 7.16 Pressure distribution due to wave (a) crest and (b) trough on a vertical wall.

located in a possible breaker zone. There are basically three practical formulas to evaluate/calculate the breaking wave pressure force. Initially it was Hiroi (1919) who provided the formula for the breaking wave force, F_b given as 1.5 γ_wgH. It is believed that this formula has been used in Japan for more than 70 years with the maximum height for the breaking wave force considered as 1.25 H from the mean water level.

According to the formula of Minikin (1963), the maximum force p_m at mean water level is given by

$$p_m = 101\gamma\frac{H_b}{L_l}\frac{d_s}{d_l}\,(d_l + d_s),$$
(7.60)

p_m is the maximum dynamic pressure, d_s is the depth at the toe of the wall, d_l is the depth of one wavelength in front of the wall and L_l is one wavelength in water of depth, d_l.

The pressure distribution is shown in Fig. 7.17. The area under the dynamic pressure distribution is the force given by

$$R_m = \frac{p_m H_b}{3}.$$
(7.61)

The dynamic pressure distribution leads to the force and overturning moment about the toe as

$$M_m = R_m d_s = \frac{p_m d_s H_b}{3}.$$
(7.62)

The hydrostatic contribution to force and moment, which is not included, should be added.

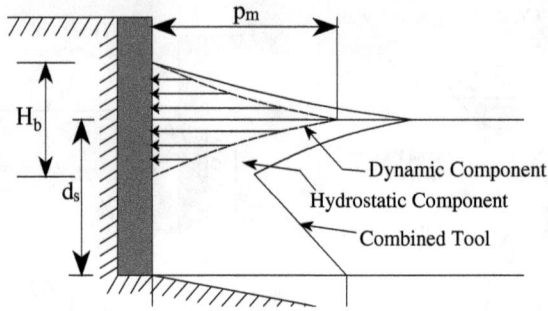

Fig. 7.17   Pressure distribution due to breaking wave on a vertical wall.

Minikin's method was originally derived for composite Breakwaters. Herein, d_s is the depth at the toe of the vertical wall:

$$D = d_s + L_d m, \tag{7.63}$$

where L_d is the wavelength in a depth equal to d_s, and m is the nearshore slope.

The above are the force and moment due to the dynamic component, to which the ones corresponding to hydrostatic pressure should be added to obtain the total force and moment. Refer to Fig. 7.17 for the pressure distribution. The total breaking wave force on the wall per unit length is/ total breaking wave force per unit length of the wall is given by

$$R_t = R_m + \frac{\gamma \left(d_s + \frac{H_b}{2}\right)^2}{2} = R_m + R_s, \tag{7.64}$$

where R_s is the hydrostatic component of breaking wave on a wall, and the total moment about the toe is

$$M_t = M_m + \frac{\gamma \left(d_s + \frac{H_b}{2}\right)^3}{6} = M_m + M_s, \tag{7.65}$$

where M_s is hydrostatic moment.

7.8   Random Wave Forces

7.8.1   *General*

The estimation of wave forces on structures due to random waves can be carried out through a frequency-domain analysis or probabilistic approaches, which are briefly discussed below. In drag/inertia regime, wave loads on slender piles can well be described by the equation of Morison *et al.* (1950).

This would require the statistical description of the fluid velocities and accelerations. Further, it is assumed that $\eta(t)$ is a Gaussian process and linear wave theory is applicable.

Borgman (1967) had shown that $u(t)$ and $\dot{u}(t)$ are statistically independent and that each follows a Gaussian probability distribution:

$$u = 2\pi f \frac{\cosh k\,(d+z)}{\sinh kd} \left(\frac{H}{2}\sin\theta\right), \tag{7.66a}$$

$$\dot{u} = -4\pi^2 f^2 \frac{\cosh k\,(d+z)}{\sinh kd} \left(\frac{H}{2}\cos\theta\right), \tag{7.66b}$$

$$S_u(f) = |R_u(f)|^2 S_\eta(f), \tag{7.67a}$$

$$S_{\dot{u}}(f) = |R_{\dot{u}}(f)|^2 S_\eta(f), \tag{7.67b}$$

where

$$R_u(f) = 2\pi f \frac{\cosh k(d+z)}{\sinh kd} \quad \text{and} \quad R_{\dot{u}}(f) = 4\pi^2 f^2 \frac{\cosh k(d+z)}{\sinh kd}.$$

The details of establishing the above relationships are discussed in Section 6.8.2.

Once the probabilistic and spectral characteristics u and $\dot{u}$ are known as stated above, the same for the forces due to random waves on a pile can be derived through the Morison formulation.

The total force/unit length F may be

$$f(t) = k_d\,u(t)\,|u(t)| + k_i\dot{u}(t), \tag{7.68}$$

where $k_d = \frac{1}{2}\rho D C_D$ and $k_i = \frac{\rho\pi D^2}{4} C_m$.

7.8.2   Probabilistic approach

The probability density function of the sectional force $f(t)$ on a vertical pile as derived by Pierson and Holmes (1965) may be written as

$$p(F) = (2\pi k_i \sigma_u \sigma_{\dot{u}})^{-1} \int_{-\infty}^{\infty} \exp\left[\frac{-1}{2}\left(\frac{u^2}{\sigma_u^2} + \frac{\dot{u}^2}{\sigma_{\dot{u}}^2}\right)\right] du. \tag{7.69}$$

Here σ_u, $\sigma_{\dot{u}}$ are deviations of u and $\dot{u}$ and may be obtained once $S_{\dot{u}}(f)$ and $S_u(f)$ are calculated. Borgman (1967) expressed this distribution in an alternate form

$$p(\varsigma) = \left(\frac{\alpha}{8\pi}\right)^{1/2} \exp\left(\frac{-\varsigma^2}{2}\right)$$

$$\times \left\{\exp\left[(\alpha+\varsigma)^2/4\right] U(0,\alpha+\varsigma) + \exp\left[(\alpha-\varsigma)^2/4\right] U(0,\alpha-\varsigma)\right\},$$

$$(7.70a)$$

where $\varsigma = F/k_i\sigma_{\dot{u}}$ is a dimensionless representation of F, U(0, x) is the parabolic cylinder function (Abramowitz and Stegun, 1965) which may be expressed in terms of modified Bessel function and $\alpha = k_i\sigma_a / 2k_d\sigma_u^2$, "$\alpha$" being ratio of Inertia/Drag.

The above distribution is symmetric with zero mean and standard deviation:

$$\sigma_\varsigma = (1 + 3/4\sigma^2)^{1/2}.$$

7.8.3 Spectral density of force

Borgman (1967) showed that

$$R_F(z) = k_d^2\sigma_u^4 G[R_u(z)/\sigma_u^2] + k_i^2 R_{\dot{u}}(z), \qquad (7.70b)$$

where

$$G(r) = [(4r^2 + 2)\sin^{-1}(r) + 6r(1-r)^{1/2}]/\pi.$$

The above function may be expressed as a power series in r

$$G(r) = \frac{1}{\pi}\left[8r + \frac{4}{3}r^3 + \frac{1}{15}r^5\right]. \qquad (7.71)$$

A linearized approximation leads to (where $r = 1$)

$$R_F(z) = \frac{8}{\pi}k_d^2\sigma_u^2 R_u(z) + k_i^2 R_{\dot{u}}(z). \qquad (7.72)$$

Hence, it follows that

$$S_F(f) = \frac{8}{\pi}k_d^2\sigma_u^2 S_u(z) + k_i^2 S_{\dot{u}}(f), \qquad (7.73)$$

$$S_F(f) = S_\eta(f)\left\{4\pi^2 f^2 \frac{\cosh^2 k(d+z)}{\sinh^2 kd}\left[\frac{8}{\pi}k_d^2\sigma_u^2 + 4\pi^2 f^2 k_i^2\right]\right\}. \qquad (7.74)$$

The above expression is for the evaluation of the sectional force, i.e., force at any elevation, z, which integrated from the seabed to the free surface will yield the total force on the pile. The different steps to be followed to determine the spectral density of force is shown in Fig. 7.18.

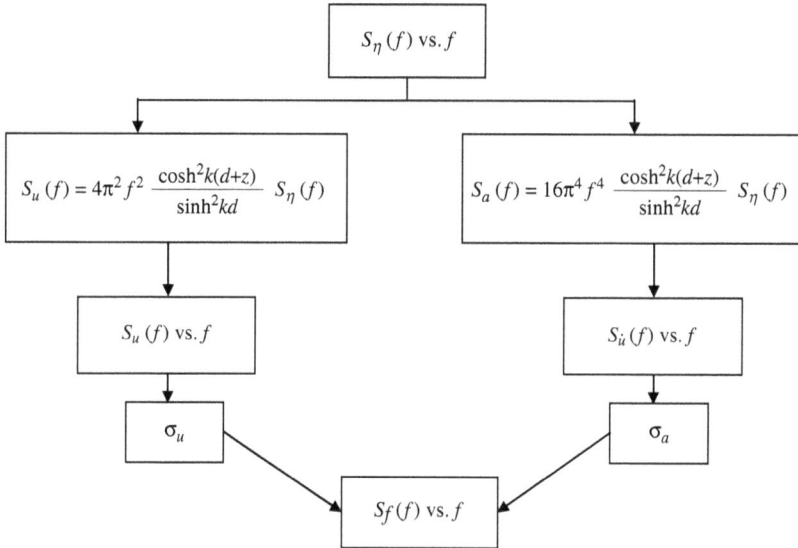

Fig. 7.18 Procedure for evaluation of random wave forces in the frequency domain.

If the force on the pile is expected to be predominantly inertia which depends on the characteristics of the pile and waves (Chapter 6), the drag force could be neglected and the inertia part of equation Morison *et al.* (1950) can be written for a vertical cylinder per unit length as

$$S_{fI}(w) = [RAO]_{fI}^2 \, S_\eta(w), \tag{7.75}$$

where

$$[RAO]_{fI} = 0.25 C_m \pi D^2 gk \, \frac{\cosh k(d+z)}{\cosh kd}. \tag{7.76}$$

Further, in the event the pile is exposed to deep water waves over the entire frequency range of the incident waves, the wave force spectrum on a single cylinder can then be obtained as follows:

$$S_F(f) = \left(\frac{\pi}{4} \rho g D^2 C_m\right)^2 S_\eta(f). \tag{7.77}$$

Under the assumption that $S_\eta(f)$ is narrow banded and force being inertial, it is quite straightforward to show that like the wave heights, the force amplitudes also follow the Rayleigh distribution, in which case, the probability density of the inertial force amplitudes f_i is given by

$$p(f_i) = \frac{f_i}{f_{irms}^2} \exp\left[\frac{-f_i^2}{2f_{irms}^2}\right]. \tag{7.78}$$

If drag component dominates, then the probability density function of the total force does not follow Rayleigh and the statistics of the force becomes more complicated. The proceeding analysis relates to the sectional force F on a pile. The total force and overturning moment spectra are eventually given as

$$S_{FT}(f) = S_\eta(f)\left\{\frac{8}{\pi}\left[\frac{2\pi f k_d}{\sinh kd}\int_{-d}^0 \sigma_u(z)\cosh\left(k(d+z)\right)dz\right]^2\right.$$
$$\left. + \frac{(2\pi f)^4 k_i^2}{k^2}\right\}, \tag{7.79}$$

$$S_m(f) = S_\eta(f)\left(\left\{\frac{8}{\pi}\left[\frac{2\pi f\,k_d}{\sinh kd}\int_{-d}^0 (z+d)\sigma_u(z)\cosh\left(k(d+Z)\right)dz\right]^2\right\}\right.$$
$$\left. + \left\{(2\pi f)^2\frac{k_i\left[(kd)\sinh(kd)+1-\cosh(kd)\right]}{k^2\sinh(kd)}\right\}\right). \tag{7.80a}$$

7.8.4   Wave forces on large bodies

The wave loads on large structures due to the action of random waves are relatively simpler compared to the loads on slender structures. This is because of the absence of the nonlinear drag force component. The spectral density for the wave force can be expressed as (Sarpkaya and Isaacson, 1981)

$$S_F(f) = |H(ka)|^2 S_\eta(f).$$

For a body of known size in given water depth, $|H(ka)|^2$ is a known function of ka (from regular wave concept). If the force amplitude, F_m on the body has been determined in the form of a coefficient C_F defined as

$$C_F = \frac{F_m}{\rho g\,Ha^2}, \tag{7.80b}$$

then

$$|H(ka)|^2 = \left[2\,g\,\rho\,a^2 C_F(ka)\right]^2.$$

7.8.5   Effect of current on S_η

The influence of a steady current U flowing in the direction of a random wave field has been considered by Tung and Huang (1973). The wave spectrum is modified by the presence of a current U to

$$S_{\eta m}(\sigma) = \frac{4}{\alpha(1+\alpha)^2} S_\eta(\sigma), \tag{7.81}$$

where $S_{\eta m}(w)$ is the modified wave spectrum due to the current U:

$$\alpha = 1 + \frac{4U\sigma}{g}. \tag{7.82}$$

This formula indicates that increased values in the modified specific density occurs when U is negative and that this increase is greater at higher frequencies.

It is reported that, the presence of even a moderate current can appreciably affect statistical properties of the wave force.

Worked Out Problems

Problem 7.1

Given D (dia of pile) $=1.0$ m, T (wave period) $=10$ s, H (wave height) $=2$ m, d (depth) $=10$ m, Draw the phase variation of the wave force on a pile along the water depth.

Solution

$H/D = 2$, $L_0 = 1.56 \times 10^2 = 156$ m, $d/L_0 = 0.0641$. From wave tables we get $d/L = 0.1082$ (intermediate waters), wavelength, $L = 92.37$ m, $D/L = 0.01082$.

For the present problem, $H/D > 1$ and $D/L < 0.2$, and hence both inertia and drag force are dominant.

Now, $kd = 2\pi d/L = 0.6802$, $\sinh kd = 0.7339$, $\cosh kd = 1.2404$, $\tanh kd = 0.5916$,

$u_{max} = (\pi H / T \tanh kd) = 1.062$ and $KC = u_{max}T/D = 10.62$.

From Figs. 7.5(a) and 7.5(b), for $KC = 10.62$, one can obtain $C_M = 1.55$, $C_D = 1.65$.

Drag force (per unit length), $F_D = \frac{1}{2}C_D\rho D|u|u$ where $u = \frac{\pi H}{T}\frac{\cosh k(z+d)}{\sinh kd}\sin\theta$.

Inertia force (per unit length), $F_I = 0.25\ C_m\rho\dot{u}$ where, $\dot{u} = \frac{-2\pi^2 H}{T^2}\frac{\cosh k(z+d)}{\sinh kd}\cos\theta$.

Sample calculation at $z = -0.2d$, substituting for the variables, we get $F_i = -769.34 \cos \theta$ N/m, and at $\theta = 180°$, maximum $F_i = 769.34$ N/m, $F_D = 818.25 \sin \theta |\sin \theta|$ N/m and at $\theta = 90°$, maximum $F_D = 818.25$ N/m.

The phase variation of the drag, inertia, and the total forces for different z/d are projected in Fig. 7.19.

Problem 7.2

Draw the total maximum force variation along the depth for the previous question, i.e., given $D = 1.0$ m, $T = 10$ s, $H = 2$ m and $d = 10$ m.

Solution

From problem 7.1, $C_m = 1.55$, $C_D = 1.65$.

To determine the maximum total wave force,

$$\theta_{max} = \cos^{-1}\left[-\frac{\pi D}{H}\frac{C_m}{C_D}\frac{2\sinh^2 kd}{(\sinh 2kd + 2kd)}\right] = \cos^{-1}[-0.49954] = 119.97°.$$

Substituting θ_{max}, C_m, C_D and for other variables in Eq. (7.10c), with $z = 0.0$ and $U = 0$, we get $\mathbf{F_{Tmax}} = \mathbf{1126.11\ N/m}$.

Repeating the calculations for other z/d, the variation of maximum total force along the depth can be obtained as shown in Fig. 7.20.

Problem 7.3

A pile of diameter 1.5 m rests on the seabed in a water depth of 15 m and subjected to the action of waves of height 5 m and period 10 s. Determine the phase variation of wave force.

Solution

Data given: $D = 1.5$ m, $T = 10$ s, $H = 5$ m, $d = 15$ m.

$L_0 = 1.56\ T^2 = 156$ m, $d/L_0 = 0.0961$, from the wave table, $d/L = 0.1376$, $kd = 0.864$, $\tanh kd = 0.6983$, $\sinh kd = 0.9756$, $\cosh kd = 1.3971$, $k = 0.0576$, $\sinh 2kd = 2.7259$, $\cosh 2kd = 2.9035$.

Using the above data, u_{max} is estimated from, $u_{max} = \frac{\pi H}{T}\frac{1}{\tanh kd}$.

Substituting for the variables,

$u_{max} = 2.249$ and $\cos\beta = -U/u_{max} = 0$ (since current velocity $= 0$) and $\beta = \pi/2$,

thus, for

$$|U| \leq u_{max} \quad KC = \frac{u_{max}\,T}{D}\,[\sin\beta + (\pi - \beta)\cos\beta].$$

For the above data, $KC = 14.996$, for which from Fig. 7.5(c), $C_m = 1.55$, $C_D = 1.5$.

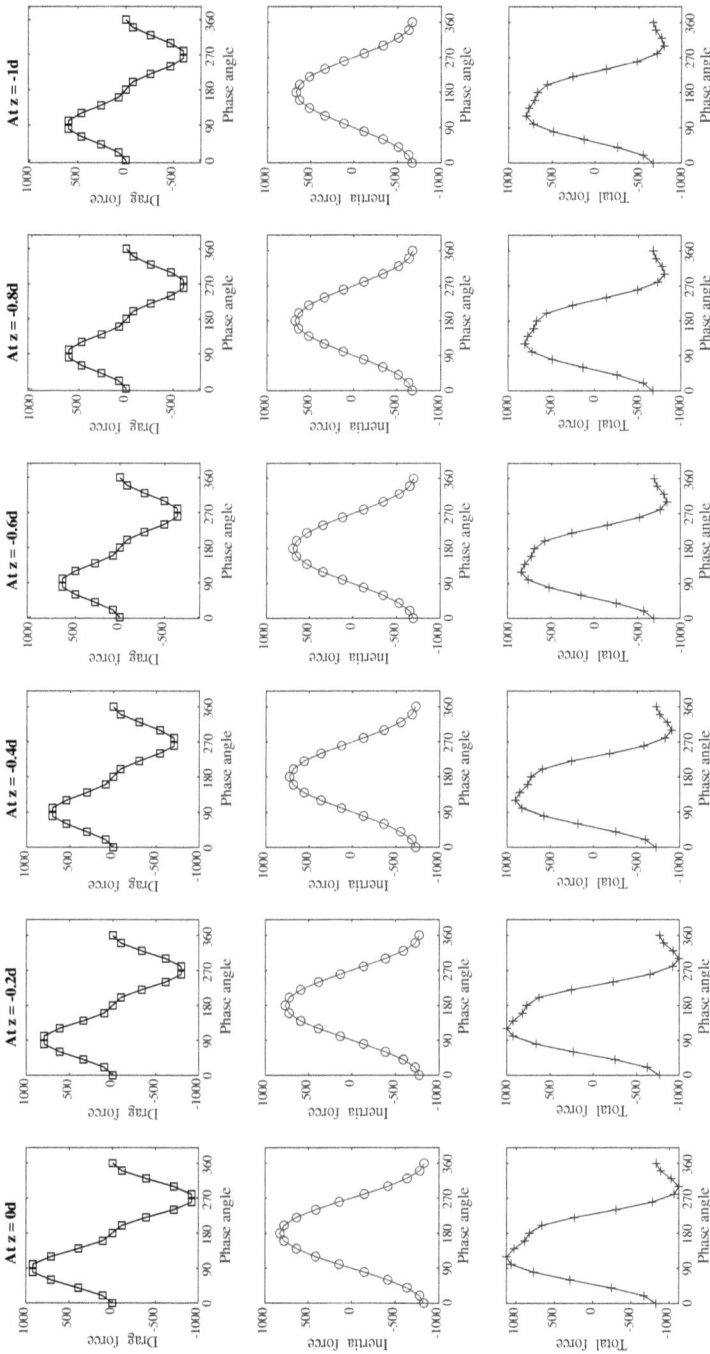

Fig. 7.19   Force (per unit length) variation at various points along the depth.

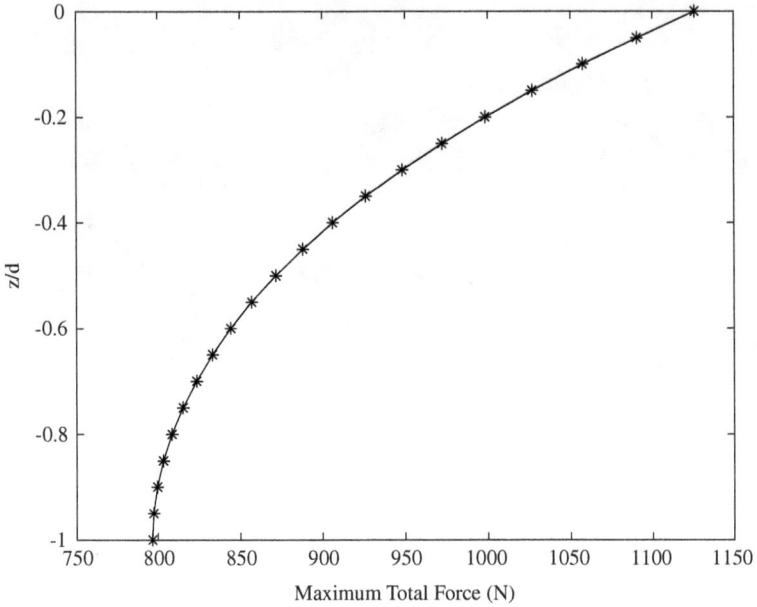

Fig. 7.20　Maximum total force variation along the depth.

The sectional wave force and the total force are given by Eq. (7.10b) and (7.10c), respectively. Substituting for the different variables in Eq. (7.10c), with U = 0.0, we get the phase variation of total wave force as given below.

$$F_T = -47857.3 \cos\theta + 57472.5 \sin\theta \mid \sin\theta \mid \text{N}.$$

The phase variation of drag and inertia and total wave force are shown in Fig. 7.21.

To determine the maximum total wave force,

$$\theta_{\max} = \cos^{-1}\left[-\frac{\pi D}{H}\frac{C_m}{C_D}\frac{2\sinh^2 kd}{(\sinh 2kd + 2kd)}\right] = \cos^{-1}(-0.41457) = 114.6°.$$

Substituting this value for $\theta_{\max}$ and for other variables in the expression for the total force given by Eq. (7.10c), we get $\mathbf{F_{Tmax} = 67431.36}$ **N**.

Similarly, substituting this value for $\theta_{\max}$ and for other variables in the expression for the overturning moment about the SWL given by Eq. (7.8), we get $\mathbf{M_{Tmax} = -397553.6}$ **Nm**. Position of maximum total force from the MSL is given by $\mathbf{M_{Tmax}/F_{Tmax} = -5.895}$ **m**.

Position of maximum total force from the sea bottom is given by $-(-15 - (-5.895)) = \mathbf{9.104}$ **m**.

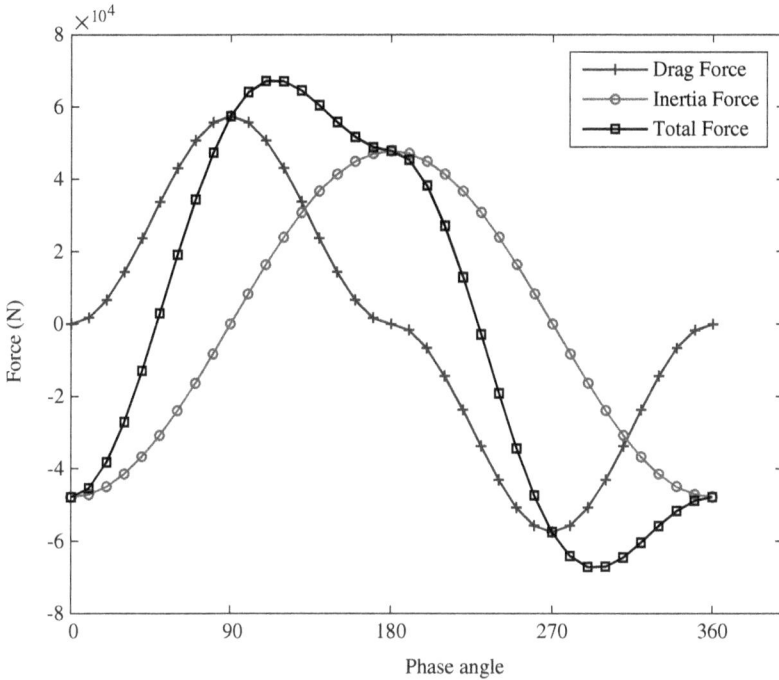

Fig. 7.21   Variation of total wave force with respect to the phase angle.

Problem 7.4

A pile of diameter 1.5 m rests on the seabed in a water depth of 15 m exposed to ocean waves of height 5 m and period 10 s. Determine (i) the phase variation of wave force and (ii) the distribution of total wave force along the depth in the presence of a uniform current of 0.5 m/s in the direction of wave propagation.

Solution

With given H, T, and d, u_{max} can be calculated as 2.249 m/s and given U = 0.5 m/s

$\cos \beta = -U/u_{max} = -0.2223$ and $\beta = 102.85°$, KC = $(2.249 \times 10/1.5))$ [$\sin \beta + (\pi - \beta) \cos \beta$]

With $\beta = 102.85°$, KC = 10.13.

From Fig. 7.6, for a KC value of 11.08, $C_m = 1.6, C_D = 1.4$. Substituting the values for the parameters in Eq. (7.10c),

we get,

$F_T = -49401.1 \cos \theta + 43236.9 \sin \theta | \sin \theta | + 4016.3$ N.

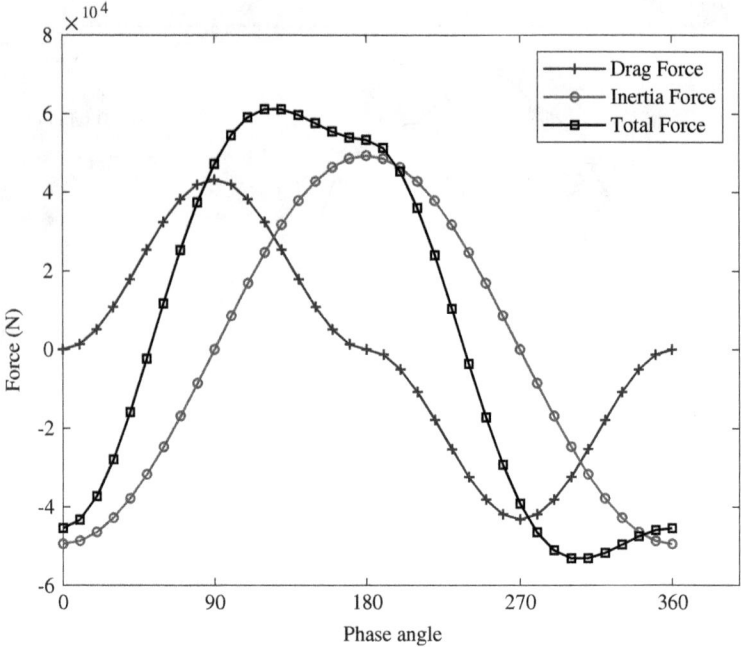

Fig. 7.22   Variation of total force due to waves and following current with respect to the phase angle.

The last term in the above equation is the contribution from the uniform current.

The phase variation of drag and inertia and total wave force with following current and the total maximum force along the depth are shown in Figs. 7.22 and 7.23, respectively.

The phase angle of the maximum total wave force from Eq. (7.7) can be obtained as

$$\theta = \cos^{-1}(-0.4605) = 117.42°.$$

Substituting this value for θ in the expression obtained for the total force Eq. (7.10c), we get

$F_{Tmax} = 15790.95$ N.

The overturning moment about the SWL is obtained from Eq. (7.10d) as

$M_{Tmax} = -190852.7$ Nm.

Hence the position of maximum total force from the SWL is given by **$M_{Tmax}/F_{Tmax} = -12.086$ m.**

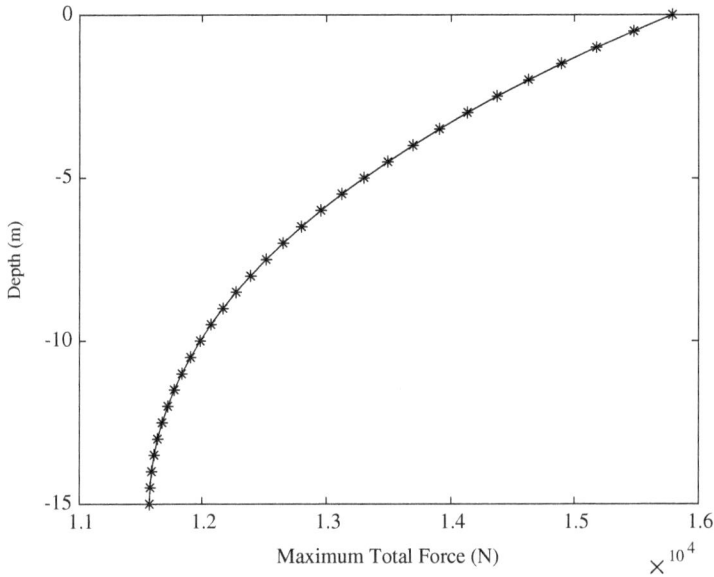

Fig. 7.23   Variation of maximum total force due to waves and following current along the depth.

Position of maximum total force from the sea bottom is given by
$-(-15 - (-12.806)) = \mathbf{2.9138}$ **m**.

Problem 7.5

A pile of diameter 1.5 m rests on the seabed in a water depth of 15 m and subjected to the action of waves of height 5 m and period 10 s. Determine (i) the phase variation of wave force and (ii) the distribution of total wave force along the depth in the presence of a uniform current of 0.5 m/s in the opposite direction of wave propagation. (iii) Compare the variations of the total wave force due to waves, waves with following and opposing currents.

Solution

Assume uniform current over the entire depth.

$U = -0.5$ m/s and u_{max} with the values for H, T, and d can be calculated as 2.249.

$\cos \beta = -U/u_{max} = 0.2223$ and hence $\beta = 77.15°$.

$KC = (2.249 \times 10/1.5) \, [\sin \beta + (\pi - \beta)\cos \beta]$.

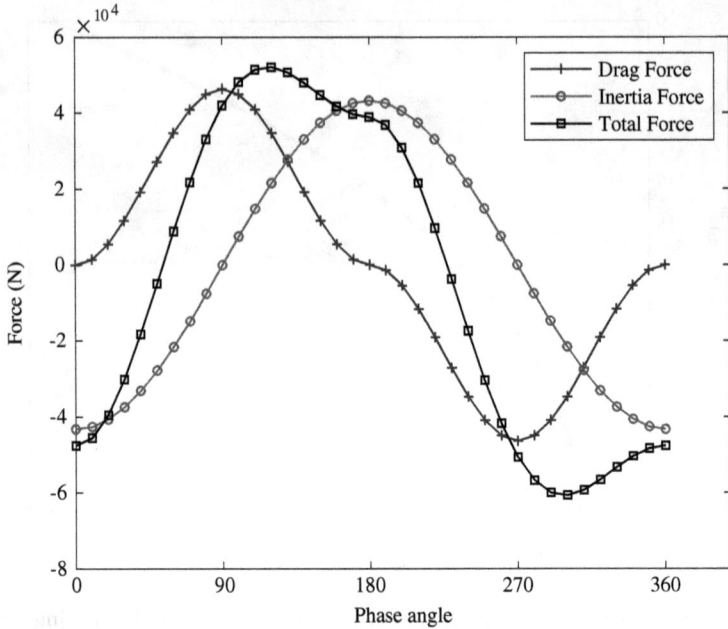

Fig. 7.24   Variation of total force due to waves and opposing current with respect to the phase angle.

Using the above equation, $KC = 20.6$, for which, from Fig. 7.6, $C_m = 1.4$, $C_D = 1.5$.

Substituting for the variables in Eq. (7.10c), the total force, will be $F_T = -43225.9 \cos\theta + 46325.3 \sin\theta \mid \sin\theta \mid + 4303.1$ N.

The last term in the above equation is the contribution from the uniform current. Substituting all the parameters in Eq. (7.7), we get $\theta_{max} = 112.09°$.

Substituting this value for $\theta_{max} = 112.09°$ in Eq. (7.10c), the total force will be $\mathbf{F_{Tmax} = 10627.25}$ **N**.

Substituting this value for $\theta_{max} = 112.09°$ in Eq. (7.10d), the total moment about the SWL will be $\mathbf{M_{Tmax} = -147130.12}$ **N-m**.

Position of maximum total force from the SWL is given by $\mathbf{M_{Tmax}/F_{Tmax} = -13.845}$ **m**.

Position of maximum total force from the bottom is given by $-(-15 - (-13.845)) = 1.155$ m.

The phase variation of drag, inertia, and the total wave force with an opposing current and the total maximum force along the depth are projected in Figs. 7.24 and 7.25, respectively.

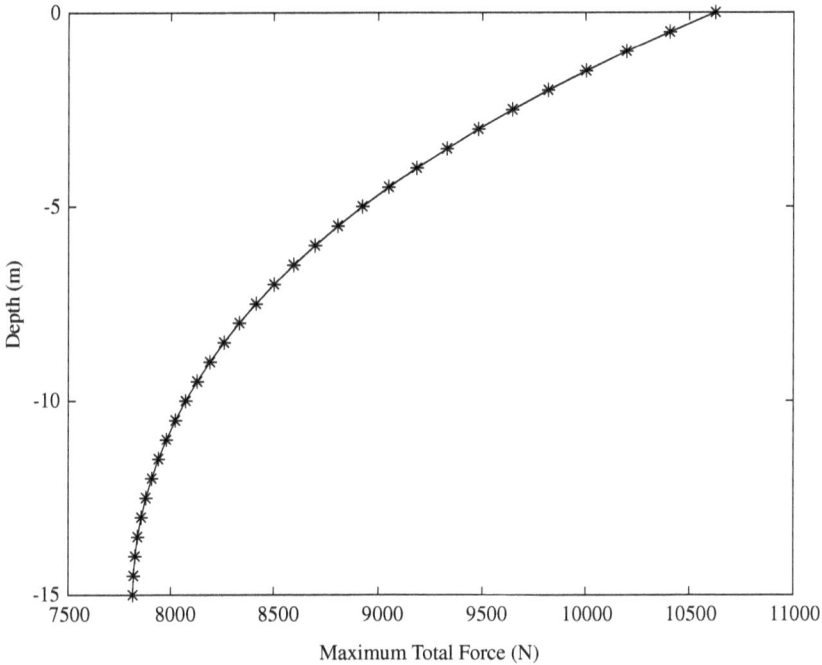

Fig. 7.25 Variation of maximum total force due to waves and opposing current along the depth.

Problem 7.6

Determine the variation of wave force on a pile in two water depths of 10 m and 100 m. For each of the water depths, consider different elevations, $z = -0.25d$, $-0.5d$, $-0.75d$ and $-d$.

Solution

Find out "KC" number at SWL. Find C_d and C_m using the following chart, for the corresponding KC number and follow the procedure explained in the earlier problems.

Substitute these values in the equation of force (drag, inertia, and total force).

The phase variations of the drag, inertia, and the total forces on the pile in $d = 10$ m and 100 m superposed in Fig. 7.26. Show that the force exerted on a pile by the same wave characteristics in shallower water depth is larger.

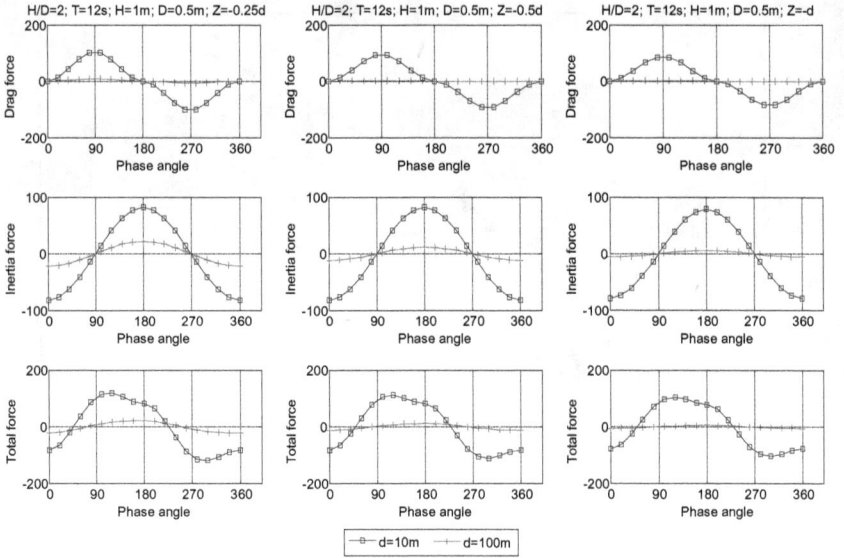

Fig. 7.26   Variation of force (per unit length) in water depth = 10 m and 100 m.

Problem 7.7

Examine the effect of wave height on the force at $z = -0.25d$, $-0.5d$, $-0.75d$ and $-d$ by considering waves of period = 12 s and height = 2 m and 4 m incident on a pile of diameter, 0.5 m in a water depth of 10 m.

Solution

Find out "KC" number at SWL. Find C_d and C_m using the following chart, for the corresponding KC number and follow the procedure explained in the earlier problems.

Substitute these values in the equation of force (drag, inertia, and total force).

The phase variations of the drag, inertia, and the total forces on the pile in H = 2m and 4m superposed in Fig. 7.27. Show that the force exerted on a pile in a constant water of 10 m by higher wave height is larger.

Problem 7.8

Examine the effect of wave period on the force at $z = -0.25d$, $-0.5d$, $-0.75d$ and $-d$ by considering a wave of height = 2 m but with period, T = 6 s, 8 s and 12 s incident on a pile of diameter, 0.5 m in a water depth of 10 m.

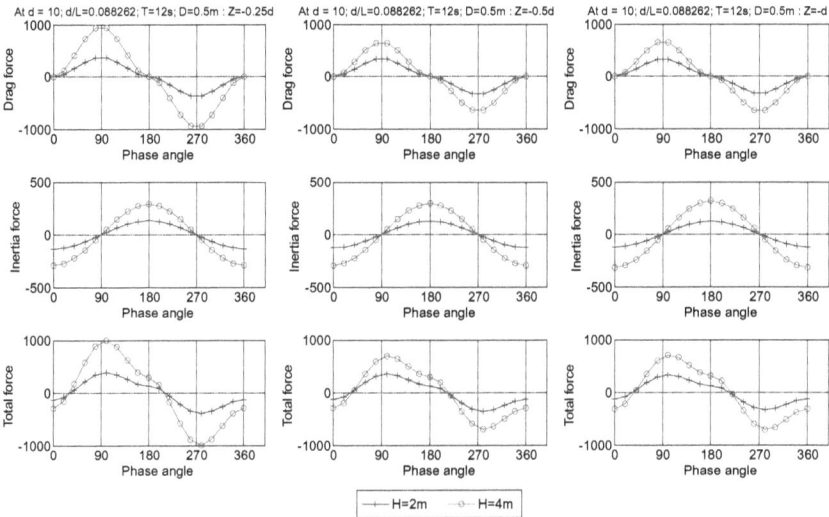

Fig. 7.27   Effect of wave height on the force.

Solution

Find out "KC" number at SWL. Find C_d and C_m using the following chart, for the corresponding KC number and follow the procedure explained in the earlier problems.

Substitute these values in the equation of force (drag, inertia, and total force).

The phase variations of the drag, inertia, and the total forces on the pile in T = 6 s, 8 s and 12 s superposed in Fig. 7.28. Show that for higher value of T, KC will be higher resulting in the domination of viscous effects resulting in higher drag component compared to inertia. The total force for long period waves will be higher as can be seen in the results.

Problem 7.9

A pile of diameter 1.25 m in a water depth of 14 m is exposed to the action waves and currents. Bring out the effect of opposing and following current of 0.5 m/s over waves as well as in the absence of current on the force exerted on the pile at MSL. The wave period may be assumed to be a constant of 10 s, while wave height may be assumed in order to achieve a constant KC number for the three different flow condition as stated.

Solution

Wave period (T) = 10 s, wave height (H) = 3 to 5 m.

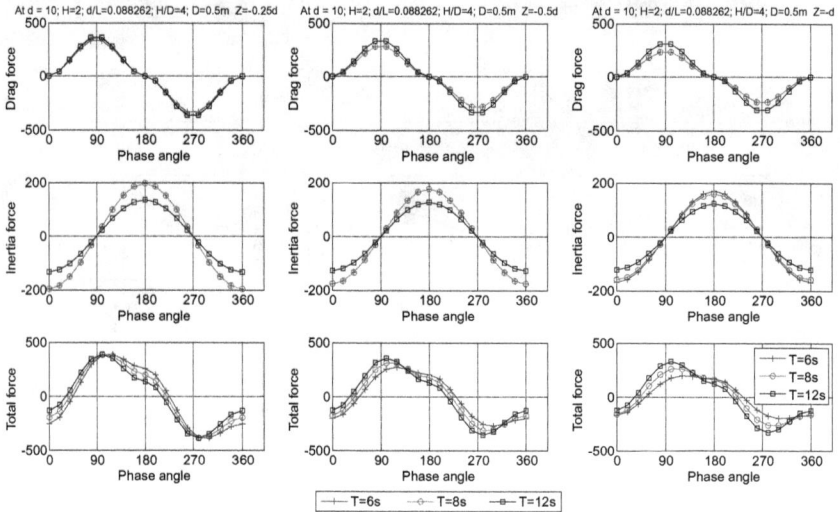

Fig. 7.28　Effect of wave period on the force.

For the present problem, $L_0 = 1.56 \, T^2 = 156$ m, $d/L_o = 0.09$.

From the wave table, for $d/L = 0.1319$, $kd = 0.82876$, $\tanh kd = 0.67981$, $\sinh kd = 0.92694$, $\cosh kd = 1.3635$, $k = 0.82876/14 = 0.05919$, $\sinh 2kd = 2.5278$, $\cosh 2kd = 2.71844$, we know

$$KC = \frac{u_{max}T}{D} \left[\sin \beta + (\pi - \beta) \cos \beta \right].$$

For a $KC \approx 12$, H for waves alone, waves plus following current, and waves plus opposing current will be 3.25 m, 4.85 m, and 3.3 m, respectively. The C_D and C_m for waves alone are obtained from Fig. 7.5, whereas, for waves plus current, they are obtained from Fig. 7.6.

The wave force at MSL ($z = 0.0$) is determined from Eq. (7.3) for waves alone and Eq. (7.10b) for waves superposed on current.

The phase variation of total wave force for without, following, and opposing current are shown in Fig. 7.29.

The force on the pile is found to be higher when it is exposed to the action of opposing current superposed waves, whereas a reverse trend in its variation is seen for following current superposed over waves.

Problem 7.10

For the pile mention in Problem 7.9, draw the variation of the maximum total force at MSL for a range of KC by changing wave heights.

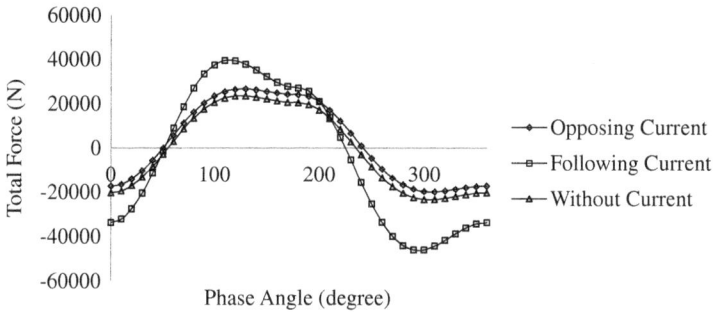

Fig. 7.29   Phase variation of total force at MSL for waves and waves superposed on currents.

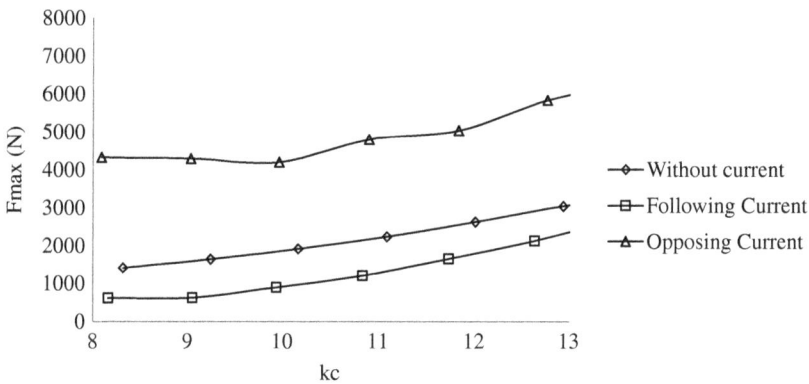

Fig. 7.30   Total force variation with KC for waves and waves superposed on currents.

Solution

For the present problem, $L_0 = 1.56\ T^2 = 156$ m, $d/L_0 = 0.09$.

From the wave table, for $d/L = 0.1319$, $kd = 0.82876$, $\tanh kd = 0.67981$, $\sinh kd = 0.92694$, $\cosh kd = 1.3635$, $k = 0.82876/14 = 0.05919$, $\sinh 2kd = 2.5278$, $\cosh 2kd = 2.71844$.

Using the above data, the KC value for the three flow conditions is calculated and the hydrodynamic coefficients are selected as explained in problem 7.9 and the total maximum forces are evaluated.

The variation of maximum total wave force for the three flow conditions are shown in Fig. 7.30.

The maximum total force on the pile is found to be higher when it is exposed to the action of opposing current superposed waves, whereas a reverse trend in its variation is seen for following current superposed over waves.

Problem 7.11

Find the total horizontal force, vertical force, and overturning moment on the single, submerged solid column with height 4 m and width 6 m for a wave of H = 0.75 m and T = 8 s. Calculate the force for T = 12 s retaining the other parameters same.

Solution

Total horizontal force: $F_h(t) = C_H F_{fx}$.

Total vertical force: $F_v(t) = C_V F_{fz}$.

Overturning moment: $M_0(t) = C_0(M_h + M_v)$.

We first compute the flow coefficient, by using (7.25).

h = 4 m; d = 10 m; D = 6 m, a = 6/2 = 3 m; h/d = 0.4, < 0.6, h/D = 0.67; $L_0 = 1.56\ T^2 = 1.56 * 64 = 99.84$ m, $d/L_0 = 0.1002$, d/L = 0.1409 (from wave table).

Therefore, L = 70.93 m, $k = 2\pi/L = 0.089$, $kh = (2\pi/L)\ h = 0.354$, $\sinh(kh) = 0.36$, $\cosh(kd) = 1.419$, $\sin(ka) = 0.26$, $\cosh(kh) = 1.06$, $\cos(ka) = 0.96$. D/L = 6/70.93 = 0.0846.

Using Eq. (7.26), one can obtain the coefficient,

$C_H = 1.641$, $C_V = 1.034$ (*Note*: $\frac{\pi h}{L} = 0.1 < 1$), $C_0 = 1.806$.

With the above parameters, we get the required series

$$F_{\text{fx}} = 2\gamma aH. \frac{\sinh(kh)}{k\cosh(kd)} \sin(ka)\sin(\sigma t),$$

$$F_{\text{fx}} = 2 * 1020 * 9.81 * 3 * 0.75 * \frac{0.36}{(0.089 * 1.419)} * 0.26 * \sin\sigma t$$

$$= 34057.8 * \sin(\sigma t)N,$$

$$F_{\text{fz}} = 2\gamma aH \frac{\cosh(kh) - 1}{k\cosh(kd)} \sin(ka)\cos(\sigma t),$$

$$F_{\text{fz}} = 2 * 1020 * 9.81 * 3 * 0.75 * \frac{1 - 1.06}{(0.089 * 1.419)} * 0.26 * \cos(\sigma t)$$

$$= 5977.2 \cos(\sigma t)N.$$

Therefore, $F_h = 55898.4$ N and $F_v = 6185.7$ N.

$$M_h = \frac{2a\gamma H}{k^2} \frac{\sin(ka)}{\cosh(kd)}(1 + hk\sinh(kh) - \cosh(kh))\sin(\sigma t),$$

$$M_h = \frac{2 * 3 * 1020 * 9.81 * 0.75}{0.089^2} \frac{0.26}{1.419}(1 + 4 * 0.089 * 0.36 - 1.06)\sin\sigma t$$

$$= 68820.7 * \sin\sigma t \text{ Nm},$$

$$M_v = \frac{2a\gamma H}{k^2} \frac{\cosh(kh) - 1}{\cosh(kd)}(\sin ka - ka\cos ka)\sin\sigma t,$$

$$M_v = \frac{2 * 3 * 1020 * 9.81 * 0.75}{0.089^2} * \frac{(1 - 1.064)}{1.419}$$

$$* (0.26 - (0.089 * 3 * 0.96)) * \sin\sigma t$$

$$= 1597.5 \sin\sigma t \text{ Nm}.$$

The phase variations of the wave forces and the moment on a submerged rectangular tank are projected in Fig. 7.31.

For $T = 12$s,

$F_h = 42866$ N, $F_v = 2911$N, $M_h = 52179$Nm, $M_v = 479$ Nm.

Problem 7.12

A single point-mooring buoy 3 m in diameter is anchored in 100 m water depth. The SPM has a draft of 15 m. If the significant wave height is 2 m and the period is 10 s, calculate the maximum wave force acting on the buoy. Specific weight of sea water is 1020 N/m³

Solution

$L_0 = (gT^2/2\pi) = 156.13$ m, $d/L_0 = 100/156.13 = 0.64$, and hence deepwater condition

$k_0 = 2\pi/L_0 = 0.0402.$

The force/length on the buoy from Morison equation is

$$f_h = \frac{1}{2}C_D\rho Du|u| + C_m\rho\frac{\pi D^2}{4}\dot{u}.$$

From linear theory,

$$u = \frac{\pi H}{T}\frac{\cosh k(d+z)}{\sinh kd}\sin\theta.$$

For deep waters,

$\cosh k(d + z) = e^{k(d+z)}$ and $\sinh kd = e^{kd}$.

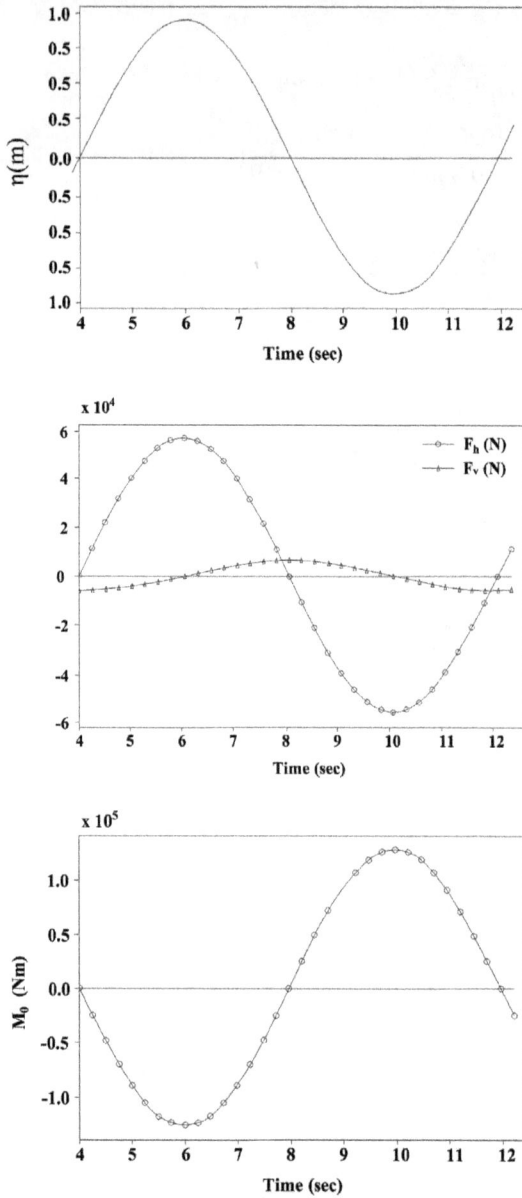

Fig. 7.31   Phase variation of wave forces and moment on a submerged rectangular tank.

Hence,

$$u = \frac{\pi H}{T} \frac{e^{k(d+z)}}{e^{kd}},$$

$$u = \frac{\pi H}{T} e^{kz} \sin \theta.$$

Similarly, $\dot{u} = \frac{-2\pi^2 H}{T^2} e^{kz} \cos \theta.$

Drag Force $= \frac{1}{2}\rho C_D D \left\{\frac{\pi H}{T} e^{kz}\right\}^2 \sin \theta |\sin \theta| = \frac{\pi^2 H^2}{2T^2} \rho C_D D e^{kz} \sin \theta |\sin \theta|$

Total drag force $F_D = \frac{\pi^2 H^2}{2T^2} \rho C_D D |\sin \theta| \sin \theta \int_{-15}^{0} e^{2kz}.$

With $C_D = 0.6$, $C_M = 1.0$, depth of the buoy $= -15$,

$$\rho = 1020 \text{ kg/m}^3.$$

$$F_{Dmax} = \frac{\pi^2 (2)^2}{2 \times (10)^2} \times \frac{1020}{9.81} \times 0.6 \times 3 \int_{-15}^{0} e^{2kz} = \frac{72482.32}{1962.2} \int_{-15}^{0}$$

$$e^{2kz} = (36.94/2 \times 0.0402)[1 - 0.299] = 322.07 \text{ kgf}$$

$$= 9.81 \times 406.8 = \mathbf{3159.57\,N}.$$

Maximum total inertial force

$$= C_M \frac{\rho \pi D^2}{4} \left(\frac{-2\pi^2 H}{T^2} \cos \theta\right) \int_{-15}^{0} e^{kz} dz$$

$$= -1.0 \times \frac{1020}{9.81} \times \pi \times \frac{(3.0)^2}{4} \times \frac{2 \times \pi^2 \times 2}{(10)^2} \int_{-15}^{0} e^{kz} = \frac{290.15}{k}[1 - 0.603]$$

$$= \frac{115.19}{0.028} = 2865.4\text{kgf} = 2865.4 \times 9.81 = 28109.69\text{N}.$$

Since $F_1 \gg F_D$, $H/D = 0.67 < 1$, drag force can be neglected.

Problem 7.13

For a pile of diameter 8 m, find the maximum wave force acting on the structure for the following wave characteristics by using closed-form solution. Wave height $(H) = 2$ m, wave period $(T) = 10$ s, water depth $(d) = 10$ m, mass density of the fluid $(\rho) = 1020$ kg/m^3.

Solution

$H = 2$ m, $T = 10$ s, $d = 10$ m, $D = 8$ m, $\rho = 1020$ kg/m^3, deep water wavelength $= 1.56T^2 = 1.56 \times 10 \times 10 = 156$ m,

$d/L_0 = 10/156 = 0.064$, and corresponding $d/L = 0.1082$,

$L = 92.42$ m, $k = 2\pi/L = 0.0679$, $ka = 0.0679 \times 4 = 0.272$.

From Fig. 7.12, for ka $= 0.272$, the corresponding $\frac{F_{max}}{\rho g Had[\tanh(kd)/kd]} =$ 0.8.

$$F_{max} = 0.8 * 1020 * 9.81 * 2 * 4 * 10[\tanh(0.0679 * 10)/(0.0679 * 10)]$$
$$= 557.3 * 10^3 N \approx 557 kN.$$

Problem 7.14

For a square caisson of size 8 m $\times$ 8 m, find the maximum wave force and moment acting on the structure for the following wave characteristics. Wave height (H) $= 2$ m, wave period (T) $=10$ s, water depth (d) $=10$ m, mass density of the fluid (ρ) $= 1020$ kg/m^3.

Solution

H $= 2$ m, T $= 10$ sec, d $= 10$ m, b $= 8$ m, $\rho = 1020$ kg/m^3, deep water wavelength $= 1.56T^2 = 1.56 * 10 * 10 = 156$ m, d/L$_0$ $= 10/156 = 0.064$, and corresponding d/L $= 0.1082$, L $= 92.42$ m, k $= 2\pi/L = 0.0679$, b/L $=$ 0.0865.

From Fig. 7.13, for b/L $= 0.0865$, C$^*_{ms}$ $= 2.05$.

$$F_{max} = C^*_{ms}\rho\pi b^2 HL/T^2$$
$$= 2.05 \times 1020 \times 3.14 \times 8^2 \times 2 \times 92.42/10^2$$
$$= 776.71 \times 10^3 N = 776.71 kN.$$
$$M_{max} = C^*_{ms}\rho g b^2 HL/4\pi(kd\tanh kd + \text{sech}\, kd - 1)$$
$$= 2.05 \times 1020 \times 9.81 \times 8^2 \times 2 \times 92.42/4\pi[(0.0679 * 10)$$
$$\times \tanh(0.0679 * 10) + \text{sech}(0.0679 * 10) - 1]$$
$$= 7387.26 \times 10^3 Nm$$
$$= 7387\ kNm.$$

Problem 7.15

At vertical wall 4.3 m high is sited in seawater with d$_s$ $= 2.5$ m. The wall is built on a bottom slope of 1:20 (m $= 0.05$). Reasonable wave periods range from T $= 6$ s. The breaking wave height for 6 s wave is 2.8 m. Find the maximum pressure, horizontal force, and overtopping moment about the toe of the wall for the given slope.

Solution

For T = 6 s, d_s = 2.5 m,

$$\frac{d_s}{L_0} = \frac{2.5}{56.2} = 0.04448,$$

$$\frac{d_s}{L_d} = 0.08826, \quad L_d = 28.3 \text{ m},$$

$$d_l = d_s + L_d m = 2.5 + 28.3 * (0.05) = 3.9 \text{ m}.$$

For this local depth, d_l = 3.9 m,

$$\frac{d_l}{L_o} = 0.0694, \quad \frac{d_l}{L_l} = 0.1134, \quad L_l = \frac{3.9}{0.1134} = 34.4 \text{ m},$$

i.e., $L_l \cong 35$ m

From Eq. (7.60),

$$p_m = 101 * 10 * \frac{2.8}{35} * \frac{2.5}{3.9} (3.9 + 2.5),$$

$$= 331 \text{ kN/m}^2,$$

$$R_m = \frac{p_m H_b}{3} = \frac{331 \times 2.8}{3} = 309 \text{ kN/m}.$$

$$R_T = R_m + [\gamma(d_s + (H_b/2))^2/2]$$

$$= 309 + [9.81(2.5 + (2.8/2))^2/2] = 328.13 \text{ kN/m}.$$

$$M_m = R_m d_s = 309 \times 2.5 = 772.5 \text{ kNm/m}.$$

$$M_T = M_m + [\gamma(d_s + H_b/2)^3/6]$$

$$= 772.5 + [9.81(2.5 + 2.8/2)^3/6] = 869.5 \text{ kNm/m}.$$

(*Note*: For further details, refer to **USACE**).

Chapter 8

Ocean Wave Energy

8.1 General

The effect that greenhouse gas production and global warming having on our earth is well known and has become a topic of great interest. With the great advancement in technology for the past few decades, several countries are on the lookout for alternative energy sources of energy to meet the ever-increasing demand. The most popular sources of alternate sources of energy are solar, wind, nuclear, hydroelectric, geothermal, ocean, and biomass. Although there has been a steady progress in the technology development of each of the above forms of energy, the energy in the ocean in abundance as well as clean needs a lot more research and development of concepts for its exploitation. Considering the long-term requirement for energy resources, ocean energy turns out to be the most promising. The untapped energy that is being held by the world oceans is tremendous, wherein untiring efforts toward perfecting the technology toward an increase in the efficiency of the system will certainly be rewarding.

8.2 Ocean Energy

The different sources of energy from the ocean are from waves, tides, currents and wind. The main two forms of energy in which significant progress has been made are from waves and tides. The recent interest is on extracting power from offshore wind, which although, the technology is proven, installing it in the marine environment still poses challenges. Apart from the above, the temperature gradient in the ocean can effectively be utilized for the generation of electricity. This is termed as ocean thermal energy conversion referred to as OTEC which in fact is fairly constant for a site. Where a large temperature gradient exists within a shorter distance from the shore, this technology can be economical. However, this is not the case with waves and tides where the source of energy is intermittent. Ocean energy has an ample resource of renewable energy which is environmentally friendly, clean and easy to harness. The ocean energy is a good replacement for conventional energy resources like fossil fuels which are on the verge of diminishing and not environmentally friendly. The salient details of wave energy alone are highlighted in this chapter.

8.3   Wave Energy Potential

There is approximately 8000–80,000 TW/yr of wave energy in the entire ocean and on an average about 1–10 TW with each wave crest transmitting 10–50 kW per meter. Cornett (2008) derived from analysis of wave climate predictions generated by the WAVEWATCH-III (NWW3) wind-wave model of Tolman (2002) spanning over 10-year period from 1997 to 2006, the global distribution of annual mean wave power thus derived is presented in Fig. 8.1.

The wave power per meter wave front in deep water for a regular sine wave is given by

$$P = \frac{\rho T (gH)^2}{64\pi} = \alpha H^2 T. \tag{8.1}$$

In the above equation, the coefficient $\alpha = 0.98$ kW/m^2.

For random waves, power per meter wave front in deep water is given by

$$P = \alpha_J H_{mo}^2 T_e, \tag{8.2}$$

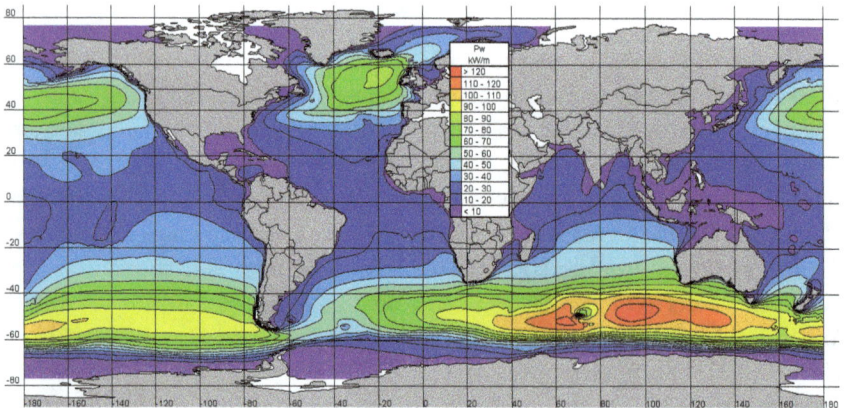

Fig. 8.1   Global annual mean wave energy levels predicted with WAVE WATCH III (Cornett, 2008).

where $\alpha_J = \alpha/2 = 0.49$ kW/m^2 in deep water, H_{mo} is the significant wave height, and, T_e is the energy transport period which is approximately 20% longer than the zero-up crossing period T_z.

8.4   Wave Energy Map for India

One of the earliest works reported on the distribution of wave power potential along the Indian coast was due to Narasimha Rao and Sundar (1982). The data from National Institute of Oceanography (NIO) were employed for this exercise. The data were collected from ships and Indian daily weather reports covering the period 1968 to 1973. For understanding the wave statistics, this amount of data would suffice. The season-wise distribution of mean wave height, wave period, and the power potential along the Indian coast are reported in Table 8.1.

Further, to explore the wave energy potential along our Indian Coast in detail, 10-years simulation wave data have been utilized. The third-generation wind-wave model (WAM) has been employed to generate 10 years wave data from 1993 to 2002 in Indian Ocean (Sannasiraj, 2007). The hindcast wind of QUICKSCAT with a resolution of $0.25° \times 0.25°$ has been utilized.

The distribution of wave power in a finite range and the length of coastline over which it is spread are grouped and presented in Table 8.2. The wave energy concentration along the coastal stretch of different maritime states is projected in Table 8.3. It is to be noted that the revised estimate shows the total power available along the coast line is about 50 GW.

8.5   Wave Energy Conversion

8.5.1   *Levels of harnessing of wave energy*

Primary Conversion: The oscillatory motion of ocean waves can be stored in the form of potential and kinetic energy in an energy convertor which could be effectively trapped through a floating or an oscillating element within the convertor.

Secondary Conversion: This is the second level trapped energy being converted into a productive and beneficial form through the utilization of drives and control systems like devices for level control and power take off. The conversion of the kinetic/potential energy to rotational energy by means of a turbine thereby results in the rotation of a shaft.

Tertiary Conversion: This is the last and final level that the energy from rotatory motion is transmitted to electric generators which converts the harness power into electricity.

Table 8.1   Wave power potential along the Indian Coastline (Narasimha Rao and Sundar, 1982).

Location	Southwest monsoon			Northeast monsoon			Non-monsoon		
	Mean wave height (m)	Mean wave period (s)	Wave power (kW/m)	Mean wave height (m)	Mean wave period (s)	Wave power (kW/m)	Mean wave height (m)	Mean wave period (s)	Wave power (kW/m)
10−15°N and coast-85° E (of Madras)	1.71	5.80	16.62	1.53	5.86	13.44	1.14	5.5	7.00
15−20°N and coast-85° E (of Visakha-patnam)	2.04	8.25	33.65	1.60	6.28	15.75	1.24	7.10	10.70
20−25°N and coast-85−95° E (of Kolkata)	1.96	7.66	28.84	1.33	8.01	13.88	1.72	6.51	18.87
5−10° N and coast-75−80° E (of Cape Comorin)	1.78	6.29	19.52	1.22	5.35	7.80	1.29	5.46	8.90
10−15°N and coast-70° E (of Cochin)	2.03	6.77	27.34	1.03	5.05	5.25	1.01	5.38	5.37
15−25°N and coast-70° E (of Bombay)	2.63	6.93	46.98	1.00	5.00	4.90	1.01	5.25	5.24

Table 8.2   Wave power contour.

Contour power level (kW/m)	Contour length (km)	Total power flux crossing contour (GW)
0–5	1530	3.825
5–10	822	6.165
10–15	1634	20.425
15–20	665	11.64
20–25	400	9

Table 8.3   Wave power contour along different maritime states.

State	Contour length (km)	Total power flux crossing contour (GW)	Contour power level (kW/m)
Gujarat	0–5	465	1.2
	5–10	110	0.8
	10–15	325	4.1
Maharashtra	0–5	90	0.2
	5–10	115	0.9
	10–15	120	1.5
	15–20	210	3.7
	20–25	130	2.9
Karnataka	0–5	130	0.3
	15–20	215	3.8
	20–25	95	2.2
Kerala	0–5	130	0.3
	5–10	60	0.5
	10–15	65	0.8
	15–20	125	2.2
	20–25	85	1.9
Tamil Nadu	0–5	75	0.1
	5–10	110	0.8
	10–15	529	6.6
	15–20	115	2.0
	20–25	90	2
Andhra Pradesh	0–5	150	0.4
	5–10	85	0.6
	10–15	550	6.9
Orissa	0–5	235	0.6
	5–10	160	1.2
	10–15	45	0.6
West Bengal	0–5	255	0.6
	5–10	182	1.4

8.5.2   *Classification of wave energy converters*

As per their horizontal size and orientation

Point absorbers are devices that are very small compared to wavelength. This system has a typical power rating of a few hundred kilowatts. Hence, a large power plant would consist of hundreds or perhaps thousands of such units, which needs to dispersed in a very long and relatively narrow array along the coast.

Line absorbers that are counterpart of point absorbers are elongated floating structures. The length of these systems are comparable to or greater than wavelength. Line absorbers can be further classified as terminators and attenuators; the former is aligned along the wave direction, whereas the latter is aligned normal to the wave direction.

As per different location with respect to the coastline

Wave energy convertors are normally placed along the coastline, over a cliff, while the nearshore wave energy devices are located in the shallow waters, whereas offshore convertors are of floating types.

As per their locations with respect to the mean water level

Wave energy converters like oscillating water column device are usually fixed gravity type, however, they can also be considered as floating units. Here, the front lip wall of the devices will be submerged below the mean sea level. Several types of floating devices have been proved to be effective in extracting power from the ocean waves. A few devices are also of semi-submerged type.

8.5.3   *Conversion process*

One of the earliest floating device is the backward bent duct buoy (BBDB) which utilizes a long horizontal water-filled duct held up by a float on the water surface with opening of the duct facing away from the incident waves. The duct is connected to a vertical chamber like OWC device and the oscillation of the air–water interface drives an air turbine. Lighter buoys operating in shallower waters adopt the BBDB design. The details of the characteristics of BBDB have been discussed by Seymour (1992).

The different process by which energy in waves can be converted to electrical energy was discussed by Hagerman (1995) and later modified by Brooke (2003). The variation from one concept to another depends on the mode of oscillation of the device for the effective energy absorption, and

Wave Energy Conversion Process	Concept
oscillating water column	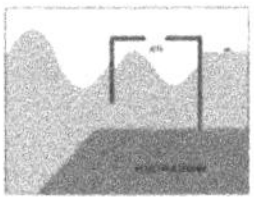
Reservoir filled by wave surge	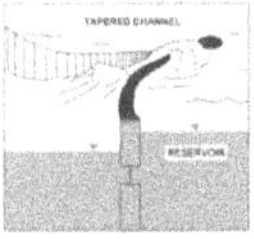
Pivoting flaps	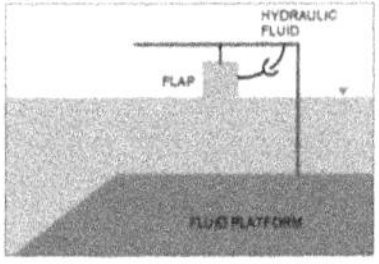

Fig. 8.2 Classifications of wave energy devices (onshore).

type of reaction point. These oscillations like heave, surge, pitch, and yaw or combinations of these motions are utilized for the absorption of the energy as shown in Figs. 8.2 (on shore), 8.3 (near shore), and 8.4 (off shore) devices, respectively. Detailed information are provided by Seymour (1992).

8.6 Wave Energy Devices

8.6.1 Types

Depending on the type and their location with respect the coast and offshore, a number of devices have been developed to extract the wave energy for conversion into electricity. The devices are briefly discussed as follows.

The Oscillating water column (OWC) device acts as an interface in which the entry and exit of the wave within a chamber leads to oscillation of the air column present inside it that can be utilized for driving a turbine.

The hinged contour device consists of a series of floating sections held in position, the movement of which is through hinges. These floating sections exposed to waves, move in accordance to the wave motion linked to hydraulic rams that drive the hydraulic motors which in turn drive the alternators. The entire system is moored to the sea floor. The Pelamis that falls under this category is an offshore WEC for power generation from the waves developed by Pelamis Wave Power Ltd of Scotland. The

Freely floating oscillating water column (**Backward bent duct buoy, BBDB**)	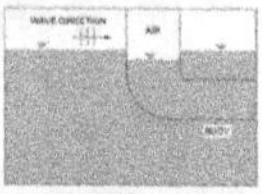
Moored floating oscillating water column (**Backward bent duct buoy, BBDB**)	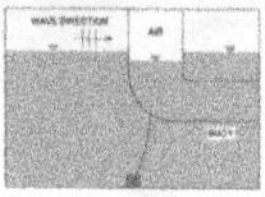
Bottom mounted oscillating water column	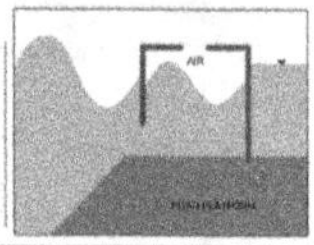
Reservoir filled by direct wave action	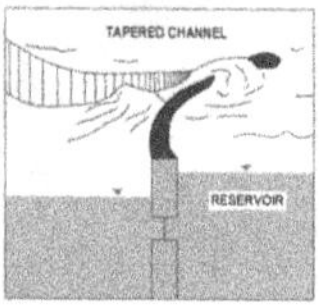
Flexible pressure device	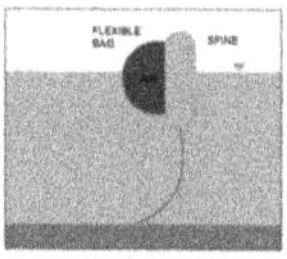
Submerged buoyant absorber with sea floor reaction	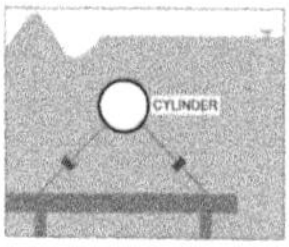
Heaving float in bottom mounted or moored floating caisson	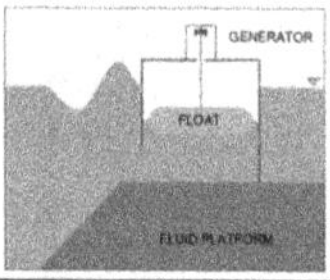

Fig. 8.3   Classifications of wave energy devices (nearshore).

device consists of cylindrical sections linked by hinge joints. The device is laid perpendicular to the wave crest and is kept in position by compliant mooring system. Being compliant, it can get over the adverse effects of severe environmental conditions for its survivability. On wave activation, the relative motion between adjacent units is captured by hydraulic ram to generate electricity. The first prototype tested in Scotland between 2004 and 2007 had a length of 120 m and the diameter was 3.5 m.

Freely heaving float with sea floor reaction point	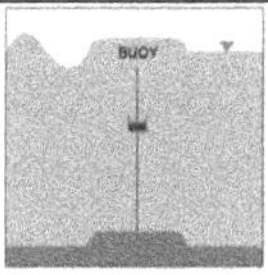
Freely heaving float with mutual force reaction	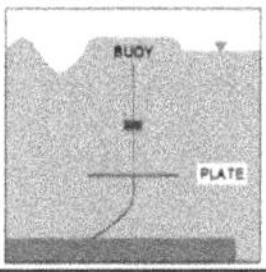
Countering float with mutual force reaction	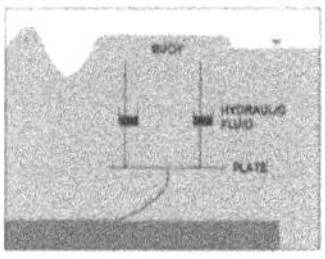
Countering float with sea floor reaction point	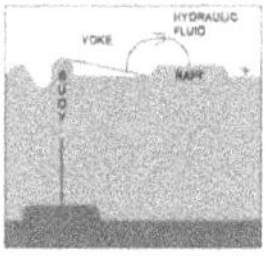
Pitching float with mutual force reaction	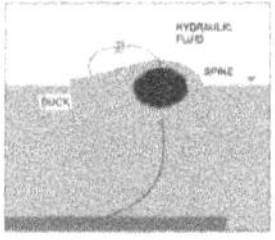
Flexible bag with spine reaction point	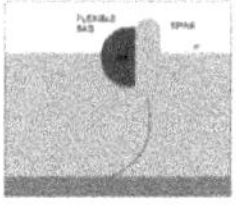
Reservoir filled by direct wave action	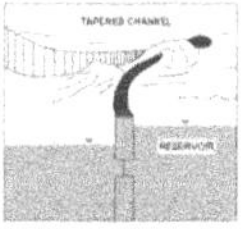

Fig. 8.4   Classifications of wave energy devices (offshore).

The buoyant moored device basically is floating type with the main structure that is responsible for the conversion of energy in ocean waves to electrical will undergo motion as per the wave motion. The device is anchored to the sea floor. Among the various mechanisms exercised for electricity generation via turbines, the basic mechanism involves the application of hydraulic pumps or pumps supplying seawater under pressure in order to drive the turbines.

The overtopping device has a ramp over which the waves propagate and overtop creating a head of water that could probably be used to run a turbine. These devices are usually floating types but can also be of fixed type.

Among all the above types of devices, OWC has taken the lead and is widely preferred. Full-sized OWC prototypes were built in Norway (in Toftestallen, near Bergen, 1985), Japan (Sakata port, 1990) (Ohneda *et al.*, 1991), India (Vizhinjam, near Trivandrum, Kerala state, 1990) (Ravindran and Koola, 1991), Portugal (Falcão, AF de O., 2000), UK (the LIMPET plant in Islay island, Scotland) (Heath *et al.*, 2000). The largest of all (2 MW), a nearshore bottom standing plant (named Osprey) was destroyed by the sea in 1995 shortly after having been towed and sunk into place near the Scottish coast. Smaller shoreline OWC prototypes (also equipped with Wells turbine) were built in Islay, UK (1991) (Whittaker and Stewart, 1993), and more recently in China. The Australian company Energetech developed a technology using a large parabolic-shaped collector to concentrate the incident wave energy (a prototype was tested at Port Kembla, Australia, in 2005) (IEA-OES, 2008).

A summary of the OWC devices in the Ocean are provided in Table 8.4.

Table 8.4   Summary of the OWC devices.

Location	Type	Rated output	Operation period
Shanwei, Guangdong, China	Coastal OWC	100 kW	Since 2005
Dawanshan, China	Coastal OWC	3 kW	Since 1990
Isle of Islay, Scotland	Coastal OWC (Islay 1)	75 kW	1988–1999
Kujukuri, Japan	OWC with pressure storage	30 kW	Since 1987
Niigata, Japan	Breakwater OWC	40 kW	1986–1988
Toftestallen, Norway	Coastline OWC	500 kW	1985–1988
Sanze, Japan	Coastal OWC	40 kW	1983–1984
Vizhinjam, Kerala, India	Nearshore close to a breakwater	50 kW	Since 1996
Pico, Portugal	Coastal	1 MWh	Since 2005
Islay, UK	Coastal	500 kW	Since 2000

8.6.2   *The factors to be considered for wave energy devices*

- The available wave power to be extracted is higher for long waves and hence sites with long waves should be ideal. However, for development of harbors, the criteria is reverse and hence it is more meaningful to combine both structures.
- Wave power is greatest in deep waters progressively reduces towards the shore when depth is less than half of the wavelength. The information on bathymetry in addition to wave characteristics is important.
- Waves are difficult to harness as they vary in direction, wave height, wavelength and are able to withstand impact (durable). Locations with well-defined predominant wave directions (narrow range) could be ideal.
- The system should be designed for both operational wave characteristics for its efficiency as well as for extreme wave characteristics for its stability and survival.
- Distribution of power requires smooth filtering, submarine cables, and grid extension.
- The rules and regulations imposed by the international, national and local bodies/agencies in regard to navigation, fishing, coastal highway, tourism, etc., need to be addressed and considered.
- Locations of huge tidal variations to be avoided as they can affect mooring and the efficiency of the device. Further, if coupled with harbor breakwaters the location may pose problems for the berthing of vessels.

8.7   Oscillating Water Column Linked to Breakwaters

The oscillating water column (OWC) consists of a chamber with a partially submerged front wall that is continuously excited by the ocean waves as shown in Fig. 8.5. Due to wave excitation, during the instant of its crest inside the chamber the air column is compressed, whereas the presence of its trough decompresses the air column. This leads to oscillation of air column and thus, the system acts like a pump which could be easily converted into energy by driving a turbine on board the OWC. The turbine should be capable of rotating in the same direction irrespective of the reversible flow inside the chamber due to the passage of crest and trough. Wells turbine is the most popular one handling this requirement with higher degree of

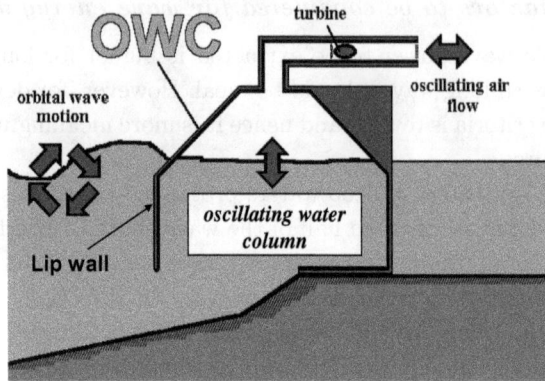

Fig. 8.5   Oscillating water column (OWC) device.

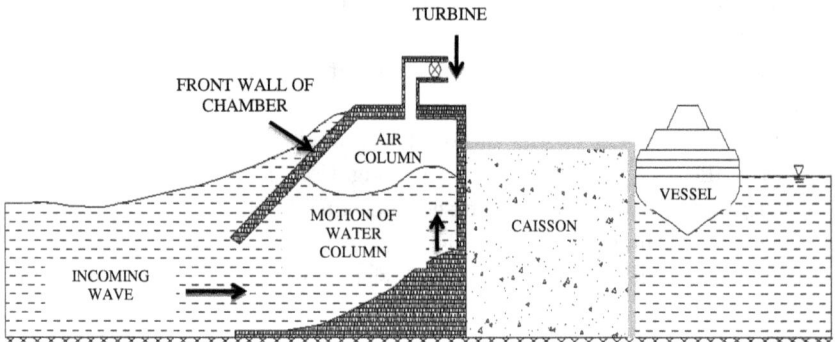

Fig. 8.6   Schematic representation of caisson-type OWC with a sloping bottom with berthing facility on its lee side.

efficiency. The complete concept and details have been discussed by Falcão (2002).

Since several decades, researchers have been involved in the improvement of the performance efficiency of OWC mostly tuning its configuration with respect to the wave climate. Apart from the generation of electricity by OWC wave energy device, these types of devices can also serve as breakwater cum berthing facility, coastal protection structures against erosion and calm water basin for cage culture.

Sundar et al. (2010) presented a comprehensive review on the possible approaches that can make use of the OWC as part of breakwaters and coastal defense systems for the harbor formation. The concept of integration of OWC with breakwaters that can reduce the total cost significantly to bring forth economic security in project planning was highlighted. OWC bottom seated device has been installed in a water depth as low as 3 m. In such situations, the possibility of integrating with breakwaters could be a part of the breakwaters for fishing and small craft harbors, a schematic representation of which is projected in Fig. 8.6. Boccotti (2007) and Boccotti *et al.* (2007) have detailed their experiences with OWC-linked caisson breakwaters in Italy. A new concept of wave energy convertor through breakwater has been reported by Roberts and Shepard (2009). The other possibility could be to consider it as part of detached breakwaters that can in addition to wave energy extraction serve as a coastal protection measure, i.e., a mitigation measure for coastal erosion (Pilarczyk and Zeidler, 1996). Offshore detached breakwaters as a shore defense measure are increasing at a remarkable pace, since they effectively reduce and absorb the incident wave energy. The ends of the breakwaters diffract the waves resulting in a calm region in between them. They are usually of mound type, aligned parallel to the shoreline to serve as a shore-protection measure. These prevent the destructive waves from reaching the shore and protect it against the encroachment of the sea. The shore starts building up towards the breakwater and the formation is called "salient" if the advanced shoreline does not reach the breakwater and if it reaches the breakwater, the formation is known as "tombolo". Offshore breakwaters are expensive and should be used after good model tests as sometimes they bring in precisely the opposite effects as against what is desired. The shoreline variations due to construction of a single offshore breakwater and due to segmented offshore breakwaters are shown in Figs. 8.7(a) and 8.7(b), respectively.

If each of these breakwaters is fixed with one or a series of OWC, the results are expected to be quite fruitful as the costs will be shared and the benefits are more than just coastal protection or energy from ocean waves. A typical concept of incorporating OWC with rubble mound breakwater for this purpose is shown in Figs. 8.8(a)–8.8(c). A typical deepwater berthing facility is shown in Fig. 8.9. This encompasses an approach trestle (an open jetty supported on piles); a breakwater (may be of caisson type or rubble mound type). Any of the foresaid OWC system can easily be integrated into breakwater or even with the approach trestle. This however needs a careful and detailed investigation. Although it is not too ambitious, a proposition that could be achieved in the near future is a harbor (minor

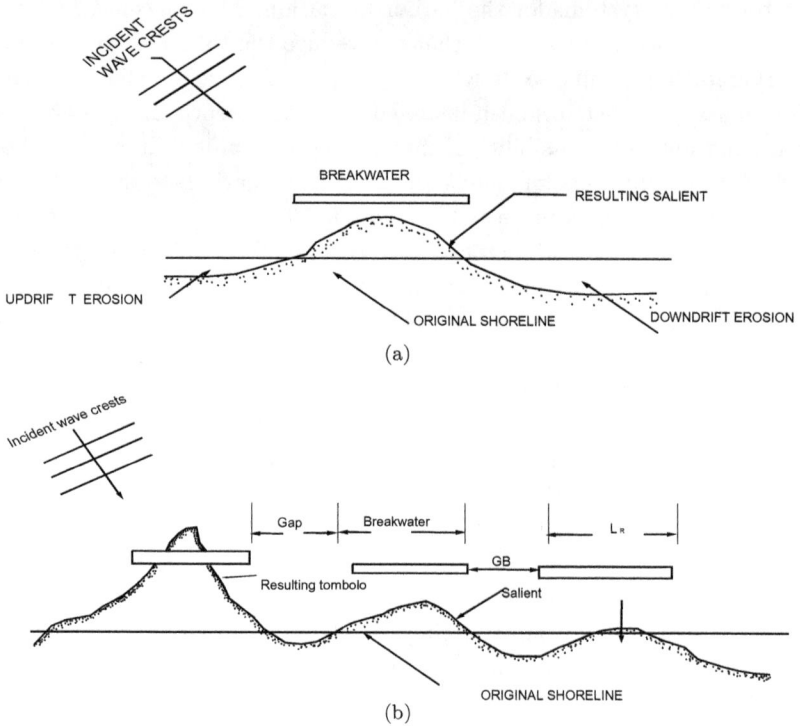

Fig. 8.7   (a) Definition sketch of single detached breakwater. (b) Definition sketch of segmented detached breakwater.

or fishing or commercial) with OWC. Dhinakaran *et al.* (2012) have carried out a comprehensive review of literature on the works carried out on the wave hydrodynamic characteristics of semi-circular breakwaters, that could be considered for the formation of harbors as well as offshore detached breakwaters to serve as coastal protection measure. The concept of integrating OWC with a semi-circular breakwater as projected in Fig. 8.10 could be worth investigating which is expected to have a greater potential for islands.

The motion of waves is a store house/manufacturer of a great amount of power, which does not pollute the environment by emitting greenhouse gases or other harmful waste.

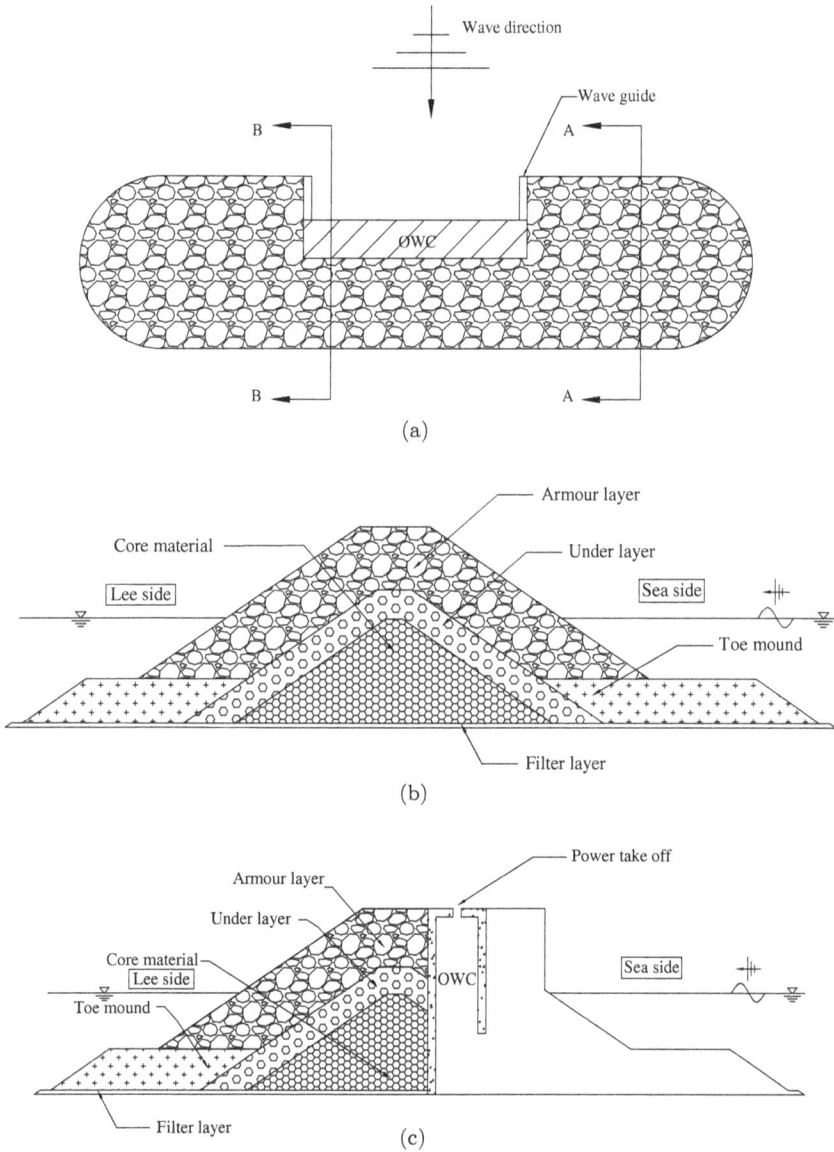

Fig. 8.8   (a) Plan view of OWC integrated with offshore detached breakwater. (b) Cross-section of rubble mound breakwater at A–A. (c) Cross-section of OWC integrated with rubble mound breakwater at B–B.

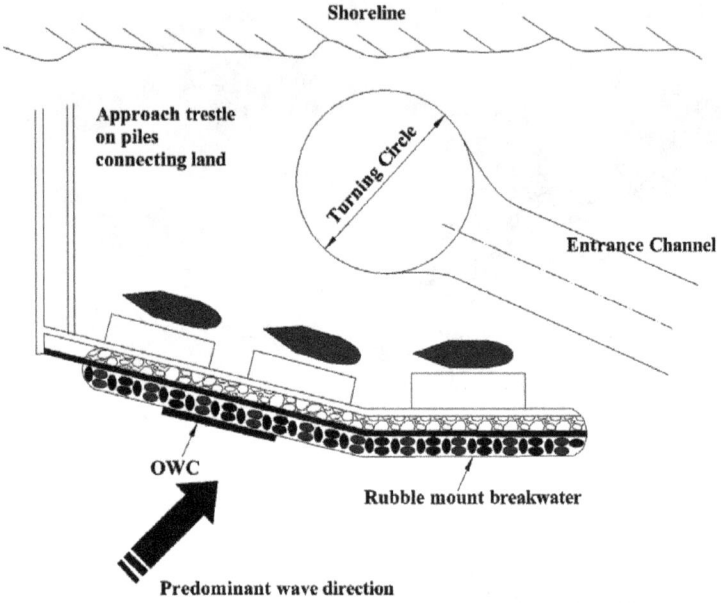

Fig. 8.9   Typical deeper water direct berthing facility formed by breakwater with OWC.

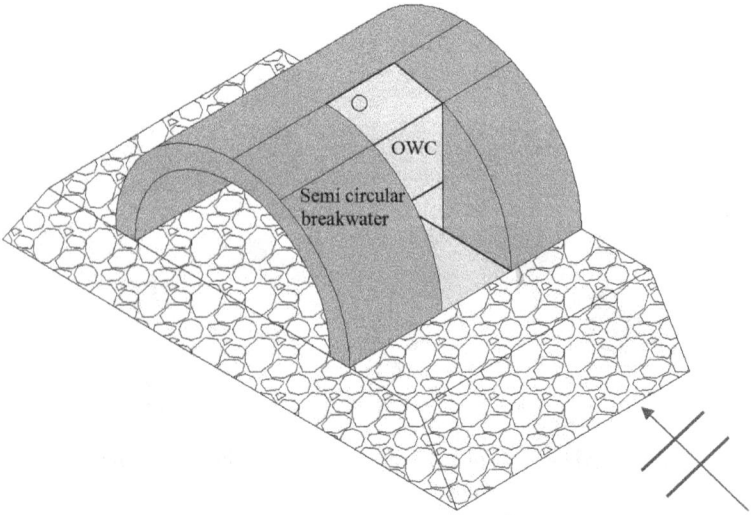

Fig. 8.10   Cross-section of OWC integrated with semi-circular breakwater.

Chapter 9

Physical Modeling

9.1 Introduction

Physical modelling is performed to study the behavior of a prototype, through a model scale. A full-scale model is considered to be ideal, although scaled-down models are preferred since the former is the most expensive one and to reduce costs, one resorts to the latter, which can be handled in a laboratory. The idea is to extrapolate the observations made on the small model to the prototype and to this end dimensional analysis is an important tool.

Prior to the evolution of computers, physical modeling alone was mainly sought to investigate a wide range of problems related to coastal, harbor and offshore engineering. Even today, physical modeling remains the most reliable means of understanding the complex phenomena in the field of applied fluid mechanics, wave hydrodynamics and coastal engineering. The effects of wave structure interaction, nearshore wave characteristics, wave loads on offshore structures, coastal processes, sediment dynamics are some of the widespread problems studied through physical modeling. The random nature of sea state, sediment dynamics, vortex formation, flow through porous media, etc., cannot be precisely modeled using numerical techniques, although it has evolved significantly over the recent years. Though physical modeling is costlier compared to numerical modeling, engineers, researchers, scientists and planners still prefer it over numerical modeling for specific cases to get a real-time picture of the various phenomenon occurring in the vicinity of proposed structures. Dimensional analysis, similitude, scale effects and model laws are the fundamental aspects of physical modeling, which are discussed in the subsequent sections of this chapter.

9.2   Dimensional Analysis

9.2.1   *General*

In simple words, dimension is the measure of any physical quantity such as mass, length, time, temperature, etc. and unit is the means by which a numerical value is assigned to the said dimension such as kg, m, s, Celsius, etc. Dimensional analysis is a technique used to facilitate correlation and interpretation of experimental data. It considers all the variables associated in defining a problem and provides a means of combining the many variables into lesser number of dimensionless groups. This exercise reduces the amount of experimental work needed to determine the effect of any variable on the dependent parameter of the experiment. It is instrumental in the identification of the important independent variables, to arrive at one or more dimensionless parameters.

Dimensional homogeneity: If an equation truly expresses a proper relationship among variables in a physical process, then it will be *dimensionally homogeneous*. The equations are correct for any system of units and consequently each group of terms in the equation must have the same dimensional representation. This is also known as the law of *dimensional homogeneity*. For example, $L_0 = 1.56\ T^2$ is non-homogenous, whereas $L_0 = (g/2\pi)\ T^2$ is dimensionally homogenous.

Dimensional constants: These are normally held constant during a given experimental trial, e.g., acceleration due to gravity, fixed water depth, fixed bed morphology, etc. However, they may vary from one trial case to another. A variable or an equation can be non-dimensional, i.e., without unit (e.g., strain, reflection coefficient, etc.) through suitable substitution of variables involved in the problem. The most commonly non-dimensional parameters in wave hydrodynamics is the relative water depth (d/L) and wave steepness (H/L). The fundamental representation of all the variables reduces to three basic units namely, mass, length and time. Other physical variables defining a problem related to wave mechanics and coastal engineering could be height and period of the wave, sediment size, structural characteristics, etc. Table 9.1 presents the details of the basic unit representation for most of the parameters pertaining to coastal engineering.

9.2.2   *Rayleigh's method*

This method determines a parameter for a dependent variable which depends upon maximum of three to four independent variables. If the

Table 9.1   Unit representation and scale factor for various parameters.

Variable	Unit	Remarks
Geometric		
Length	L	Any characteristic dimension of the object
Area	L^2	Surface area or projected area on a plane
Volume	L^3	For any portion of the object
Moment of inertia area	L^4	
Section modulus	L^3	Area moment of inertia divided by the distance from the neural axis to the extreme fiber
Kinematics and Dynamics		
Moment of inertia mass	ML^2	Taken about a fixed point
Center of gravity	L	Measured from a reference point
Time	T	Same reference point (e.g., starting time) is considered as zero time
Acceleration	LT^{-2}	Rate of change of velocity
Velocity	LT^{-1}	Rate of change of displacement
Displacement	L	Position at rest is considered as zero
Angular acceleration	T^{-2}	Rate of change of angular velocity
Angular velocity	T^{-1}	Rate of change of angular displacement
Angular displacement	None	Zero degree is taken as reference
Spring constant (Linear)	MT^{-2}	Force per unit length of extension
Damping coefficient	MT^{-1}	Resistance (viscous) against oscillation
Damping factor	None	Ratio of damping and critical damping coefficient
Natural period	T	Period at which inertia force = restoring force
Momentum	MLT^{-1}	Mass times linear velocity
Angular momentum	ML^2T^{-1}	Mass moment of inertia times angular velocity
Torque	ML^2T^{-2}	Tangential force times distance
Work	ML^2T^{-2}	Force applied times distance moved
Power	ML^2T^{-3}	Rate of work
Impulse	MLT^{-1}	Constant force times its short duration of time
Force, thrust, Resistance	MLT^{-2}	Action of one body on another to change or tend to change the state of motion of the body acted on
Statics		
Stiffness	ML^3T^{-2}	Modulus of elasticity times the moment of inertia, EI
Stress	$ML^{-1}T^{-2}$	Force on an element per unit area
Moment	ML^2T^{-2}	Applied force times its distance from a fixed point
Shear stress	MLT^{-2}	Tangential force per unit cross sectional area
Hydraulics		
Kinetic energy	ML^2T^{-2}	Capacity of a body for doing work due to its configuration

(Continued)

Table 9.1   (*Continued*)

Variable	Unit	Remarks
Pressure energy	ML^2T^{-2}	Energy due to pressure head
Potential energy	ML^2T^{-2}	Capacity of a body for doing work due to its configuration
Loss to friction	ML^2T^{-2}	Loss of energy or work due to friction
Scour		
Grain size of sediments	L	For same prototype material
Free fall velocity	LT^{-1}	Final velocity of a freely falling particle in a medium
Wave Mechanics		
Wave height	L	Consecutive crest to trough distance
Wave period	T	Time between two successive crests passing a point
Wavelength	L	Distance between two successive crests at a given time
Celerity	LT^{-1}	Velocity of wave (crest, for example)
Particle velocity	LT^{-1}	Rate of change of movement of a water particle
Particle acceleration	LT^{-2}	Rate of change of velocity of a water particle
Particle orbit	L	Path of a water particle (closed or open)
Wave elevation	L	Form of wave (distance from still waterline)
Wave pressure	$ML^{-1}T^{-2}$	Force exerted by a water particle per unit area
Stability		
Displacement (volume)	L^3	Volume of water displaced by a submerged object
Righting and overturning Moment (Hard Volume)	ML^2T^{-2}	Moment about a fixed point of a displaced weight and dead weight, respectively
Natural period	T	Period of free oscillation in still water due to an initial disturbance
Metacenter	L	Instantaneous center of rotation
Center of buoyancy	L	Distance of C.G. of displaced volume from a fixed point
Material Properties		
Density	ML^{-3}	Mass per unit volume
Modulus of elasticity	$ML^{-1}T^{-2}$	Ratio of tensile or compressive stress to strain
Modulus of rigidity	$ML^{-1}T^{-2}$	Ratio of shearing stress to strain

Note: (1) The theory of dimensional analysis is purely algebraic. (2) Problems are expressed by means of dimensionally homogeneous equations. (3) The governing differential equation of a phenomenon or any crude theory can be adopted to find all variables which may influence the same.

independent variables exceed four in number, then obtaining a parameter involving the associated-dependent variables becomes complex. Let X be a dependent variable which is function of say, X_1, X_2, and X_3 independent variables. Then according to Rayleigh's method,

$$X = f(X_1, X_2, X_3). \tag{9.1}$$

It can be rewritten as

$$X = KX_1^a, X_2^b, X_3^c, \tag{9.2}$$

where "K" is a non-dimensional constant and a, b, c are arbitrary powers which are obtained by comparing the powers of fundamental dimensions (dimensional homogeneity).

Example. Find the expression for tidal prism P in an estuary where H is the average tidal range and A is the average surface area of the basin.

Solution

$$P = KA^a H^b, \tag{9.3}$$

where K is a non-dimensional constant. Substituting the dimensions on both sides of Eq. (9.3),

$M^0 L^3 T^0 = K(L^2)^a (L)^b$. Equate powers of M, L, T on both sides.

Power of L: $3 = 2a + b$. By solving for the above equation, we get $a = 1$ and $b = 1$ and substituting values of a and b in Eq. (9.3),

$P = KA^1 H^1 = AH$.

9.2.3   Buckingham's Pi theorem

Buckingham's Pi theorem is another popular method of dimensional analysis to obtain the relationship between the variables involved in each physical problem. It states that, *"If there are "n" variables in a problem and these variables contain "m" primary dimensions (for example, M, L, T) the equation relating all the variables will have (n−m) dimensionless groups".* The dimensionless groups are represented by π and are called as π groups. According to this theorem, the final expression is given by

$$\pi_l = f(\pi_2, \pi_3, \dots, \pi_{n-m}).$$

These π groups formed should be independent of each other and no groups shall be obtained by multiplying powers of other π groups. Compared to the Rayleigh's method of solving simultaneous equations, this method is a lot more advantageous and simpler.

Example. Consider a stationary sphere of diameter D immersed in a fluid flowing past the sphere in a steady flow where velocity of flow, v; and the fluid properties, i.e., density, r, and viscosity, μ. Therefore, a functional relationship is expected between the drag force F_D and these variables,

$$F_D = \phi(D, v, \rho, \mu). \tag{9.4}$$

An exponential form for this relation is

$$F_D = CD^a v^b \rho^c \mu^d, \tag{9.5}$$

where C is an arbitrary dimensionless constant.

Converting this equation to their dimensional equivalent in an MLT system gives

$$\frac{ML}{T^2} = L^a \left(\frac{L}{T}\right)^b \left(\frac{M}{L^3}\right)^c \left(\frac{M}{LT}\right)^d. \tag{9.6}$$

Equating the exponents of each dimension for dimensional homogeneity, we have

For $M : c + d = 1$; for $L : a + b - 3c - d = 1$; for $T - b - d = -2$.

This gives us four unknowns and three equations. Writing the equation in terms of one unknown,

$$F_D = CD^{2-d} v^{2-d} \rho^{1-d} \mu^d, \tag{9.7}$$

where C is a constant.

Rearranging terms,

$$\frac{F_D}{D^2 \rho v^2} = C \left(\frac{vD\rho}{\mu}\right)^{-4}. \tag{9.8}$$

Note that the term within parenthesis is the definition of Reynolds number. Then, the general form of the relationship becomes

$$\overline{F_D} = C\phi(Re), \tag{9.9}$$

where $\overline{F_D}$ is the non-dimensional drag force on the sphere which is represented as a function of Reynolds number (R_e). Note that Eq. (9.4) involved five variables in an MLT system so that only $(5-3=2)$ two non-dimensional quantities are needed (Eq. (9.9)) for a functional relationship. An experiment with a sphere in steady flow produces such a relationship.

Example. Group the variables of a vegetal belt response subjected to both wave and current action. The following variables of vegetal response are of interest for both currents and waves.

$$\text{Vegetal response} = f(\rho, g, h, H, T, B_s, D, D_b, BG, f_1, SP, E, L, \beta, R_u, V), \tag{9.10}$$

where ρ — mass density of water; g — gravitational acceleration; h — depth of water in the flume; H — wave height; T — Wave period; B_s — Width of the structure; D — diameter of the vegetation; D_b — diameter at the root of the vegetation; BG- width of green belt; f_1 — frequency of first mode of the vegetal stem; l — height of the vegetation; SP — center to center spacing between vegetation; E — modulus of plant stiffness; L — Wave length; β — Beach slope; R_u — Run-up; V_f — Flow velocity. The h refers to the following: h_s — depth of water at the toe of the structure; h_{avg} — average depth of water on the up-stream and downstream of flow in open channel.

The variables seen in Eq. (9.10) are grouped as per Buckingham's Pi theorem, in which, $U_{\max}$ is the velocity V_f and L_0 is the deepwater wavelength. The non-dimensional quantities investigated in the study are Darcy's f and Manning's n in steady uniform flow and wave run-up R_u/H, pressures, and forces on structures. They are designated as follows:

$$f = \left(\frac{8H_f gh}{LV_f^2} \right), \tag{9.11}$$

$$n = \left(\frac{R^{(2/3)} * S^{(1/2)}}{V_f} \right), \tag{9.12}$$

$$F* = \left(\frac{F_{\max}}{0.5 * \rho g H^2 B_s} \right), \tag{9.13}$$

$$P* = \left(\frac{P_{\max}}{H} \right). \tag{9.14}$$

9.3   Model Analysis

9.3.1   *General*

Application of general modeling laws to problems in fluid mechanics were started by Reynolds and Froude by conducting a series of experiments to develop a criterion for viscous and inertia effects. From fluid mechanics, this extended to Ocean Engineering modeling considering the different environmental parameters. The terminologies often used in model analysis and their definitions are given below

Prototype is defined as the actual system or a field problem being investigated.

Model is defined as the substitute system that is proportional to the prototype by a scale ratio.

Homologous is defined as the corresponding conditions in the model and its prototype.

9.3.2   *Complete similarity*

In order to obtain accurate results through dimensional analysis, the variables are grouped to form a complete set of dimensionless products such as $\pi_1, \pi_2, \ldots, \pi_n$. If the independent variables $\pi_1, \pi_2, \ldots, \pi_n$ possess the same value for the model and the prototype, then the model is said to be completely similar. Complete similarity is nearly impossible to achieve in our model studies. It is to be noted that the forces such as surface tension, surface roughness, etc., may influence the model but not the prototype and the independent dimensionless variables which are believed to have secondary influences are to be neglected. Scale effects (discussed later in this chapter) of the same magnitude occurs inevitably in all models. To avoid scale effects, larger models are to be employed.

9.3.3   *Applications of model analysis*

Most of the marine structures such as breakwaters, jetties, groins, training walls, seawalls, piers, intake, and outfall structures, etc. can be tested, for their interaction with waves as well as their effect on the coast. The behavior and the stability of coastal structures could be verified in the laboratory by applying a suitable model scale, the results of which could be extended in the field. The wave loads on structures, extent of sedimentation in a harbor approach channel, influences by wave and tides, behavior of estuaries, bottom stability of breakwaters and other gravity structures, behavior of shoreline can be investigated through physical model studies.

9.4   Principles of Similitude

9.4.1   *General*

The behavior of a model when subjected to external forces holds a definitive relation with that of a prototype. The model should be designed and tested

in accordance with certain principles of similarity (or) principles of similitude. The principles not only govern the design, construction and testing,
but are also applicable to the extrapolate model results in order to predict
prototype response.

9.4.2 Similitude in hydrodynamic problems

The principles of similitude adopted for hydrodynamic problems need to
be treated separately from the conventional similitude. The three major
types of similitude, namely, geometric similitude, kinematic similitude and
dynamic similitude are explained in detail as follows. A convention of using
"p" as a subscript for prototype and "m" as a subscript for model is
followed.

9.4.3 Types of similitude

Geometric Similitude: Geometric similarity signifies that the dimensional similarity from the prototype to the model, i.e., similarity of the
physical form. When the horizontal and vertical scales are same, the model
is said to be undistorted and when the horizontal and vertical scales are
different the model is said to be a distorted model. Definitive objectives
can be achieved by departing from geometric similarity. The major problem areas are confining boundaries of laboratory testing facilities, surface
roughness of model, scaling thickness, depth of water, etc. For example, in
case of modeling a stretch of offshore structure deployed at higher depths,
a distorted model is preferred for the study because the horizontal scaling
dimension becomes extremely insignificant.

Kinematic similitude: Kinematic similarity is the similarity of motions
which implies both geometric similarity and similarity of time intervals, in
other words, it requires that the model and prototype have the same length-
scale ratio and the same timescale ratio. Thus, the velocity scale ratio is
the same for both.

Example: Estimate the scaling criterion necessary for kinematically similar
wave motion for gravity waves whose length is given by

$$L = \frac{gT^2}{2\pi} \tanh \frac{2\pi d}{L}.$$

Solution: $2\pi d/L$ is dimensionless, this ratio for model and prototype
should be an invariant for geometrically similar model. If $2\pi d/L$ is the

same, the tangent of the hyperbola will also be the same therefore scaling relation between L and T is arrived from prototype to model ratio of wavelength. Thus, for a kinematic similarity of wave motion,

$$\frac{L_p}{L_m} = \frac{\left[\frac{gT^2}{2\pi} \tanh(2\pi d/L)\right]_p}{\left[\frac{gT^2}{2\pi} \tanh(2\pi d/L)\right]_m}, \quad \frac{L_p}{L_m} = \frac{T_p^2}{T_m^2},$$

$$\lambda_L = (\lambda_T)^2.$$

The kinematic similarity of wavelengths can be achieved by setting $\lambda_T = \sqrt{\lambda_L}$, since "g" is a constant.

Dynamic similitude: Dynamic similarity exists when the model and the prototype have the same length, time, and force scale (mass scale) ratios. Thus, the ratio of any two forces in one system must be the same as the ratio of the corresponding forces in the other system. Kinematic similarity can be achieved without considering any properties of the fluid, which is unlikely for dynamic similarity.

$$\frac{F_m}{F_P} = \frac{M_m a_m}{M_p a_p} = \frac{\rho_m L_m^3}{\rho_p L_p^3} \times \frac{\lambda_L}{\lambda_T^2} = \lambda_p \lambda_L^2 \left(\frac{\lambda_L}{\lambda_T}\right)^2 = \lambda_p \lambda_L^2 \lambda_u^2.$$

The following is a case of pile-supported oscillating water column (PS-OWC) showing the similitude between prototype and model using Froude's scaling law. The PS-OWC is envisaged for use in a water depth d_p, wave period T_p and wave height H_p as indicated in Fig. 9.1(a). The corresponding parameters for the model indicated as d_m, T_m and H_m in Fig. 9.1(b) are arrived at by reducing the prototype dimensions by a scale factor ($\lambda = 25$). Other dimensions in the model are obtained the same way from the prototype dimensions. Whereas wave period T_p in the prototype is reduced by a scale factor $\lambda^{0.5} = 5$ to obtain the wave period T_m for the model. Similarly, to get other hydrodynamic parameters for the model, the Froude scale factors given in Table 9.2 are adopted.

9.5　Scale Effects

As a rule, scale effects can be reduced by applying as large models as possible and by performing model tests at different scales, one may obtain an idea of the degree of the scale effects. A complete similarity between model and prototype is impossible to achieve, since certain quantities such as gravity, cannot be scaled down. The parameters such as gravity, fluid viscosity and density are the same in model as well as prototype. It is impossible to

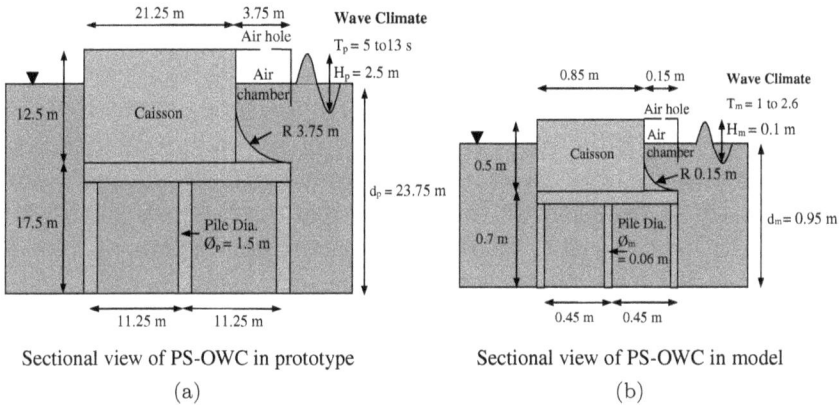

Sectional view of PS-OWC in prototype   Sectional view of PS-OWC in model

(a)   (b)

Fig. 9.1   Geometric, kinematic and dynamic similitude achieved for PS-OWC.

Table 9.2   Froude scale factors for different hydrodynamic parameters.

Hydrodynamic parameters	Dimension	Froude scale
Mass	$[M]$	λ^3
Force	$[MLT^{-2}]$	λ^3
Pressure and stress	$[ML^{-1}T^{-2}]$	λ
Energy and work	$[ML^2T^{-2}]$	λ^4
Power	$[ML^2T^{-3}]$	$\lambda^{3.5}$
Time	T	$\lambda^{0.5}$

generate both gravity driven and fluid viscous related phenomena concurrently. Waves and currents are gravity-driven forces and in model studies it is preferred to simulate gravity properly. This means that viscosity effects will not be properly reproduced. The effects of such non-similarity are called scale effect.

9.6   Model Laws

The laws based on which the models are designed for dynamic similarity are known as model laws or laws of similarity. The model laws are classified based on their application as follows:

- Reynold's model law,
- Froude's model law,
- Euler's model law,

- Weber's model law,
- Mach's model law.

Reynolds model law: Reynolds number, R_e, is the ratio of inertia force and viscous force. If the viscous forces acting on the system are more predominant than the other forces, the models are designed for dynamic similarity based on Reynolds law, which states that the Reynolds number for the model must be equal to the Reynolds number for the prototype. Models based on Reynolds number includes problems based on pipe flow, resistance experienced by sub-marines, airplanes, fully immersed bodies, etc., $R_e = \rho V L/\mu$, where "ρ" is the density, "V" is the velocity, "L" is the characteristic length, and "μ" is the viscosity.

Froude's model law: Froude number, F_r, represents the ratio of inertial forces to gravitational forces and is expressed as $V/(gL)^{0.5}$ where "v" is average velocity of flow. Froude model law is applicable when the gravity force is only predominant force which controls the flow in addition to the force of inertia. Froude number for the model must be equal to the Froude number for the prototype. Froude model law is applicable for free surface flows such as flow over spillways, weirs, sluices, channels, flow of jet from an orifice or nozzle and flow of different densities of fluid over one another.

Euler's model law: Euler's number, E_u, represents the ratio of square roots of inertia force to pressure force and defined as $V/(p/\rho)^{0.5}$ where "p" is the intensity of pressure. This law is applicable when the pressure forces are alone predominant in addition to the inertia force. Dynamic similarity between the model and prototype is achieved when the Euler number for model and prototype are equal. Euler's model law can be applied to fluid flow problems in a closed pipe where turbulence is fully developed so that viscous forces are negligible and gravity force and surface tension force is absent.

Weber's model law: The Weber number, $W_e = \rho V^2 L/\sigma$ is the ratio of the inertial forces to surface tension forces. "σ" denotes surface tension. The Weber number becomes an important parameter when dealing with applications involve two fluid interfaces such as the flow of thin films of liquid and bubble formation. Weber model law is the law in which models are based on Weber's number and dynamic similarity between model and prototype can be achieved when the Weber number for both model and prototype are equal. This law is applicable for capillary rise and fall problems.

Mach's model law: Mach's number $M_a = V/c$ is defined as the square root of the ratio of the inertia force of a flowing fluid to the elastic force. herein, c is the speed of sound (343 m/s at 20°C) and V is the fluid velocity. Mach model law is the law in which models are designed on Mach number, which is the ratio of the square root of inertia force to elastic force of a fluid. Hence where the forces due to elastic compression are prominent in addition to inertia force, the dynamic similarity between the model and its prototype is obtained by equating the Mach number of the model and its prototype.

9.7   Case Studies

9.7.1   *Pile supported breakwater*

Physical model studies are widely used in the field of wave structure inter-action problems and in assessing the stability and of marine structures. Typically to initiate a model study, the scaling law to be followed is first decided. Then based on the prototype details and features of the testing facility scaling law is adopted, through which, the dimensions of the model material, water depth and wave climate are arrived at. The proto-type can either be the one existing at site or that is being planned for. Figure 9.2(a) projects the prototype of Galveston seawall located at Texas, USA (Coastal Engineering Manual, 2006), a sectional view of which supported on piles to act as a Galveston wall shape pile-supported breakwa-ter is shown in Fig. 9.2(b) model to be tested in a flume with a model scale of 1:25.

The hydrodynamic scaling parameters are water depth, wave height and wave period for a two-dimensional study. For a given scale factor λ (25 in this case), the hydrodynamic testing parameters are scaled down as follows:

$$\text{(water depth)}_m = \text{(water depth)}_p/\lambda = 25/25 = 1 \text{ m},$$
$$\text{(wave height)}_m = \text{(wave height)}_p/\lambda = 2.53/25 = 0.1 \text{ m}.$$

Range of $\text{(wave period)}_m = \text{(wave period)}_p/\sqrt{\lambda} = $ from $5/\sqrt{25} = 1$ s to $12/\sqrt{25} = 2.4$ s.

Now, based on the values arrived, a physical model investigation was performed. The hydrodynamic characteristics of the above type pile sup-ported breakwater has been reported by Karthik Ramnarayan *et al.* (2020).

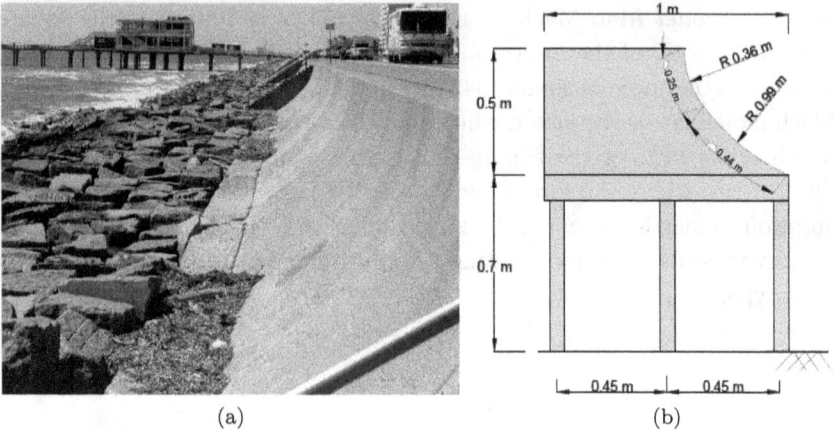

(a)   (b)

Fig. 9.2   Galveston seawall and pile-supported breakwater model.

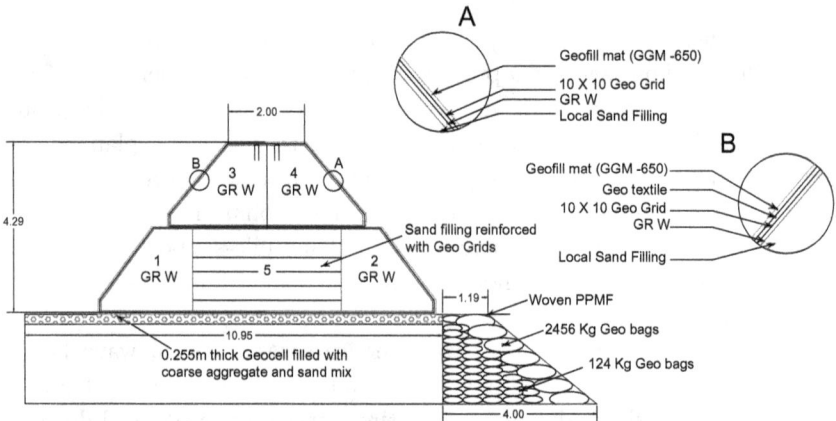

Fig. 9.3   Cross-section of geosynthetic seawall.

9.7.2   Pallana Beach, Kerala (19°17′55.19″ N and 76°23′18.55″ E): Geosynthetic seawall stability

In this example, model scale of a geosynthetic seawall is described. The proposed cross-section comprises of several geosynthetic units such as geosynthetic reinforced soil retaining wall (GRW), geomattress, geogrid, geocell, geobags, woven and non-woven geotextile. The prototype cross-sectional design is shown in Fig. 9.3.

Table 9.3   Testing of seawall section at High tide Level + Storm surge (Water depth at the wave maker = 65 cm, Water depth at the structure = 39 cm).

Sl. No	Prototype			Model (1:8.5)		
	H (m)	T (s)	Duration (h)	H (m)	T (s)	Duration (min)
1	1.22	8.5	3	0.143	2.9	48.33
2	1.6	8.5	3	0.19	2.9	48.33
3	1.94	8.5	3	0.228	2.9	48.33

The model scale adopted is 1:8.5, it was adopted to accommodate the model inside the existing flume facility. The water depths adopted for testing corresponds to the highest low tide level, highest high tide levels and storm surge along the Pallana coast. The components of the sea wall include geosynthetic reinforced soil retaining wall (GRW), geobags (small bags and mega bags), geocells, geogrids, and geotextiles (woven and non-woven). An outer protective layer of geomattress filled with sand slurry was incorporated in the design for protection against vandalism and harmful ultra-violet radiation, the same was not incorporated in the physical model testing because it does not possess any structural stability in the design, also scaled down width for the slurry pockets are not possible to manufacture.

The geotextiles (woven and non-woven) and geogrids of required length are used. One singular GRW unit is 0.25 m high, 0.28 m wide at bottom, 0.12 m wide at top, they are interconnected and stitched using high denure polyester thread. The geobags constituting the excavated toe are scaled down by weight to 0.2 kg bags made from non-woven geotextile and 4 kg geobags made from woven geotextile. The geocell of 0.75 m thickness is adopted to serve as the foundation layer.

The seawall cross-section was tested for its hydrodynamic stability against the predominant wave climate and varying water depths. The results from the 125% of the design wave(overload condition is depicted below. The test conditions are highlighted in Table 9.3. To represent the wave climate for testing, the wave elevation has been prescribed by the random wave spectrum with the given significant wave height and peak period to be generated.

The response of the model structure to the exposed test conditions listed above were visually observed and reported. Based on the observations recorded the results are tabulated in Table 9.4. The categories C1, C2, C3 and C4 denote the percentage of displaced chips above the scour

Table 9.4　Observation of seawall section at High tide Level +Storm surge.

Conditions	Scale	Wave period T (s)		Wave height H (m)		Movement of units		Remarks
		Prototype	Model	Prototype	Model	Category	% of occurrence	
75% of design wave	8.5	8.5	2.9	1.22	0.143	C1	4	1. Chips above the scour apron are slightly displaced
						C2	20	2. Frequent splashing of waves over the crest
						C3	24	3. Average of 5 overtopping occurrence per minute
						C4	2.5	4. Upheaving of the geo grids can be noticed
100% of design wave	8.5	8.5	2.9	1.6	0.19	C1	7.5	1. Chips above the scour apron are slightly displaced
						C2	23	2. Frequent splashing of waves over the crest
						C3	33.6	3. Average of seven overtopping occurrence per minute
						C4	10	4. Visible upheaving of the geo grids

125% of design wave	8.5	8.5	2.9	1.94	0.228	C1	7.5	1. Chips above the scour apron are slightly displaced
						C2	25	2. Frequent splashing of waves over the crest
						C3	38	3. Average of eight overtopping occurrence per minute
						C4	12	4. Visible upheaving of the geo grids

Fig. 9.4   A seawall designed based on physical model study.

apron, percentage of splashing waves over the crest without overtopping, percentage of overtopping and the percentage of upheaving of the geogrid, respectively.

Based on the experimental investigations, it was observed that the model was sturdy to withstand the continued wave action without undergoing any failure and the mean position demarcated in the flume walls for erection of the model remained intact after completion of testing. The results of the tests based on the modeling of wave structure interaction was adopted for construction, the details of which are reported by Sukanya *et al.* (2016). The completed seawall in 2016 is shown in Fig. 9.4.

Chapter 10

Laboratory Generation of Waves

10.1 Introduction

The ocean waves dominate the loading on marine structures and in addition, also dictates the stability of the beaches. The larger the waves, the more are the problems posed to marine trade through shipping and threaten the safety of recreation lovers using the beaches and marinas. For planning and carrying out marine operations related to the exploitation of marine living and non-living sources, the weather window plays a dominant role. Further the wave-induced loading as well as the motion characteristics of floating structures are derived through numerical or physical modeling. There are several wave structures interaction problems that need to be tested in flumes and wave tanks in which waves are generated by external force by oscillating a plunger piston-type wave maker that could be controlled by a computer. Hence, it is important to understand the types of waves that could be generated in the laboratory, the reasons behind testing structures in the different of waves are essential, the details of which are presented in this chapter. The wave tanks constructed across laboratories before the 1980s was two dimensional (2D); capable of generating unidirectional waves. In real time, long period waves may exhibit unidirectional behavior. The history of laboratory wave generation has been initially reviewed by Funke and Mansard (1987) and later detailed by Hughes (1993). Whether it is a narrow tank for 2D wave generation or it is a basin for multidirectional waves, they both should have wave generators and absorbers at either ends, the efficiency of which contribute in no mean measure to the success of the experiments. A variety of waves, namely, regular, random, solitary, Cnoidal, wave groups, freak/Rogue waves, tsunami-like waves, nonlinear waves, etc., are generated in a laboratory wave flume or a basin, the most common of which are projected in Fig. 10.1. Balaji *et al.* (2008) reported the response characteristics of a discus data buoy under the action of nonlinear waves, namely, Cnoidal and stokes waves through an experimental investigation

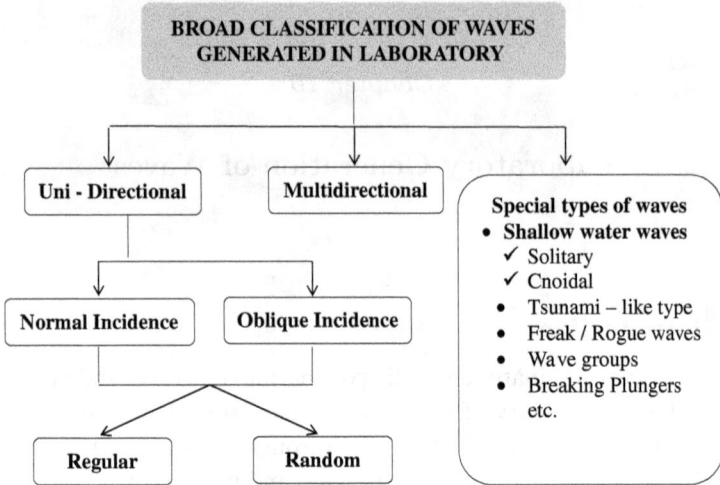

Fig. 10.1   Broad classification of laboratory-generated waves.

carried out in a wave current flume in Department of Ocean Engineering, IITM (DOE-IITM).

Several laboratories worldwide have flumes of varying sizes that are housed with wave makers that can generate waves of varied characteristics. A few flumes are equipped with the facility of generation of currents beneath waves to study the wave-current structure interaction problems. A few have a deeper pit for soil at the test section to facilitate wave-soil structure interaction problems. Rambabu *et al.* (2003) experimentally in a wave current flume in DOE-IITM studied the piles embedded in cohesive soil/sediments to determine an empirical relation to assess the scour depth corresponding to the diameter of obstruction and Froude number. For testing structures under survival conditions, world's largest flumes that can generate maximum wave height from 1.5 m to 3.5 m are shown in Fig. 10.2.

As flumes permit only unidirectional waves, multidirectional waves and oblique waves are generated in laboratory wave basins.

10.2   Testing of Structures to Different Types of Waves

10.2.1   *Regular wave tests*

Testing of coastal and offshore structures in hydrodynamic laboratories with regular sinusoidal waves had remained the normal practice over several decades, for the simple reason, its ease in the design and implementing

Large Wave Flume (FZK) Hannover/Germany:
Hmax = 2.5m

7m

5m | 330m

Electric Power Inst. Tokyo/Japan:
Hmax = 2.0m

6m

3.4m | 180m

Tainan Hydraulics Lab. Tainan, Taiwan:
Hmax = 1.5m

5m

5m | 300m

"Super Tank" Oregon State/USA:
Hmax = 1.3m

4.6m

3.7m | 104m

Delta Flume De Voorst/Netherlands:
Hmax = 2.5m

7m

5m | 233m

Hydraulic Institute St. Petersbrug/Russia:
Hmax = 2.0m

7.5m

4m | 110m

Large Hydro - Geo Flume (LHGF) Japan:
Hmax = 3.5m

12m

3.5m | 185m

Ciem Flume Univ. Catalunya/Spain:
Hmax = 1.6m

5m

3m | 100m

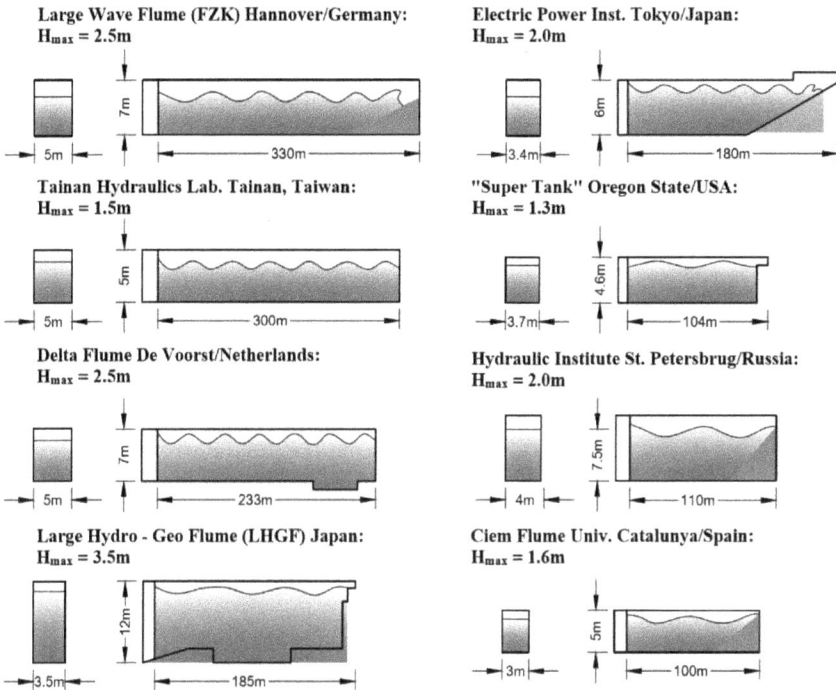

Fig. 10.2   Large flumes in the world (Zhang and Geng, 2015).

the wave maker. In the case of regular wave tests, the design wave is represented by the wave characteristics, namely, its period and height, that may be either representing the structures exposed to operational or survival conditions. The main reason for such tests is that the approach is simple in the design analysis and the determination of the response of structures due to extreme wave conditions (as against a design sea state) is easier. The wave-structure interaction phenomena can be well understood with a regular wave test through flow visualization techniques. One of the main drawback is that if the response of a floating structure over a range of frequencies is to be assessed, the method of testing structures exposed to regular waves alone if resorted to, the tests need to be carried out for several waves each of which with a particular combination of wave height and period. The exercise thus becomes laborious and time consuming as well as the wave maker is likely to be exposed to considerable stress. One of the common tests to check the stability of armor blocks adopted for breakwaters.

10.2.2 *Random wave tests*

Random wave tests are recommended, since such a wave simulates many harmonic components and its effect can be studied over a short duration, thus saving considerable testing time. Combinations of frequencies may give rise to higher order excitations due to coupling mechanisms. In many cases, knowledge about this nonlinearity may be limited and test with random waves is the only way to study them. In the case of random wave test, the design wave environment considers a wave spectrum, i.e., energy spread over a frequency band. A suitable wave spectrum model could be chosen representing an appropriate power spectral density distribution of the wave surface elevation at the site under consideration. A variety of waves like the wave groups, freak waves, breaking waves can be generated basically utilizing the modulation of frequency components which is possible only if the wave maker can generate random waves. Sundar *et al.'s*, (1999) testing-inclined cylinders to random wave in the presence and absence of freak waves have concluded that the dynamic pressures in the former type is greater. The recommended practice for the duration of testing with random waves is to generate at least 100 waves for operational conditions and 1000 waves for survival conditions. These tests were carried in a shallow water wave flume in DOE-IITM.

10.2.3 *Breaking waves*

The principle of constructive interference of different superposed wave components was adopted for the simulation of breaking waves by Chan and Melville (1988). The random waves closer to real ocean waves can be realized by the linear superposition of numerous wave components with varying frequencies and amplitudes. The phase difference between different wave components is chosen as random. To concentrate the wave energy at a predefined location and time, the phases of each wave component can be chosen such that the constructive interference of the prescribed wave components would occur. The details of the generation of breaking waves and its impact on a vertical wall in the wave cum current flume in DOE-IITM are discussed in detail by Rajasekaran *et al.* (2010). Recently, Sruthi and Sriram (2017) investigated the focused wave impacts on the Jacket structure, single leg, and individual sides of the Jacket structure in shallow water wave flume at DOE-IITM.

10.2.4 *Three or multi dimensional waves*

The importance of testing of structures in three-dimensional waves are the forces on long floating cylinder of arbitrary shape in sway, and roll mode is reported to reduce by about 40% due to wave directionality. Measured values of the ratio of the largest force maximum to the standard deviation were found to be 3.98 in long crested seas and 3.31 in short crested seas. The added advantage of multielement wave makers, in addition to the generation of multidirectional sea state, is that it can also generate oblique waves. Generation of oblique waves is of paramount importance in the simulation of different ship headings and to measure the forces due to oblique waves on bottom-mounted model structures, which cannot be rotated. For details, refer to Ploeg and Funke (1985). Sannasiraj *et al.* (2000) through tests in a wave basin in DOE-IITM tested twin floating boxes in directional waves, and reported that responses in the sway, heave, and roll modes for the seaside and leeside structures reduce by about 47%. The experimental investigations of Sundar *et al.* (2002) in a wave basin in DOE-IITM revealed that the measured maximum normal force ratio is about 25–37% less when a vertical cylinder is exposed to directional waves compared to being subjected to long crested waves. Recently, Sithik *et al.* (2020) investigated the crane load impact during the installation of the heavy concrete-type jacket structure under directional wave attack in the wave basin at DOE-IITM. The characteristics of a few wave basins around the world including the one in Department of Ocean Engineering, DOE, IIT Madras, India are shown in Table 10.1. For complete details, refer to ITTC (2017).

10.3 Wave Makers (Types Based on Functionality)

Wave makers are designed to be set in motion within certain frequencies and stroke length such as to closely resemble the actual wave motion mechanism with wave orbital velocity envelopes for different relative water depth conditions as discussed in Chapter 3. In general, the wave makers are of flap, piston, or plunger types as projected in Fig. 10.3.

It is expensive to design a single facility that can cater to the testing of deep, intermediate, and shallow water waves; thus most laboratories have multiple facilities satisfying both the conditions. Wave generators also will be different to achieve the different characteristics. In a natural sea state, most of the energy will be concentrated within the waves possessing a period between 5 s and 25 s. Presumably, the largest scale ratio for model studies

Table 10.1   Characteristic of wave basins around the world.

Name of basin	basin			Segment			
	Size (m)	Depth (m)	No. of segment	Width (m)	Height (m)	Stroke/2 (±)	Wave height (m)
University of Edinburgh	27 × 11	1.2	80	0.30	0.70	15.0°	0.22
HRS (Wallingford)	30 × 48	2.0	90	0.31	–	–	0.50
MARINTEK (Norway)	50 × 72	0–10	144	0.50	1.30	16.5°	0.40
DHI (Denmark)	30 × 20	3.0	60	0.50	1.50	16.7°	0.50
CERC (Washington)	59 × 11	0.76	60	0.46	0.76	0.15 m	0.30
DHL (Delft)	Variable	1.3	80	0.33	1.28	0.40 m	–
NRCC	30 × 50	0.3–2.9	64	0.50	2.00	0.40 M	0.70
DOE–IITM	**30 × 30**	**3.5**	**52**	**0.50**	**2.167**	**15°**	**0.60**
CEDEX — Madrid	26 × 34	1.6	72				0.58
LNHE	50 × 30	0.8	56				0.4
Sogreah Pont de Claix	22.5 × 30	1					0.25
University College CORK	18 × 18	1					0.2
WL, Delft Hydraulics	60 × 26.4	0.7					0.15
ORTC State of Texas	45.7 × 30.5	5.8	48	0.6	2.4		0.9
China Ship Sci Res Center (CSSRC)	69 × 46 × 4	3.5					0.5

in a wave tank is taken as 1:100, the shortest wave period of 5 s will be scaled to 0.5 s (i.e., 2 Hz).

Wave makers can be classified into active and passive type. The active wave makers the motion of which is dictated through a computer have direct contact with water that will be displaced, thus generating user defined waves. By controlling the movement of the device through control system or a personal computer, the desired waveform could be generated in the flume. In the case of passive wave makers, none of the moving parts will be in direct contact with the water column, whereas the air pressure variation triggers the water movement/wave oscillations. The most adopted active wave makers are the plunger type, hinged- flap, piston type, and a combination of flap and piston type. The drives used for wave makers are

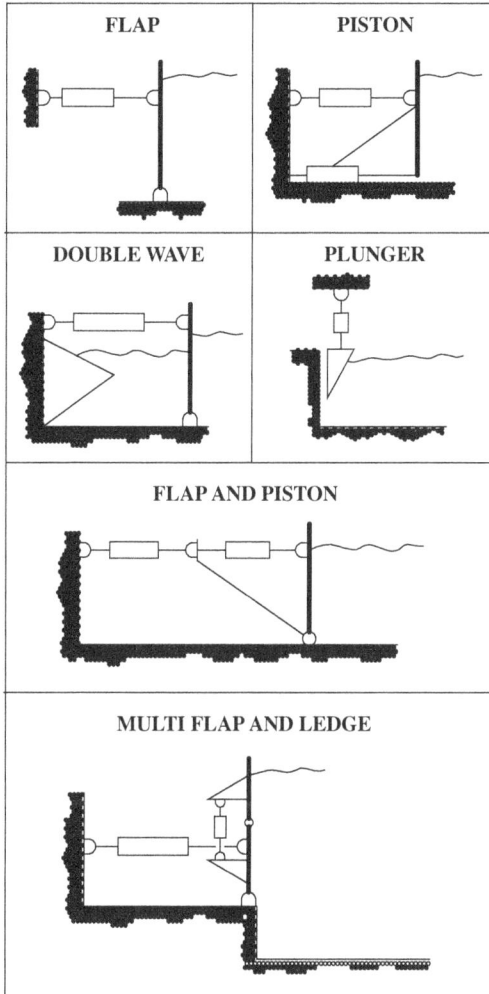

Fig. 10.3   Broad types of wave Makers.

electromechanical, servo electric, eccentric drives, or servo hydraulic. The plungers and flaps are ideal for deep water waves. Piston is best suited to generate shallow water waves. Combinations of flap and piston modes find use in both deep and intermediate water conditions. In any new design water particle kinematics as obtained in actual state should be ascertained before the facility is put into use. A wet back wave maker will have water on either of its side. Dry back is possible by proper sealing. Both systems

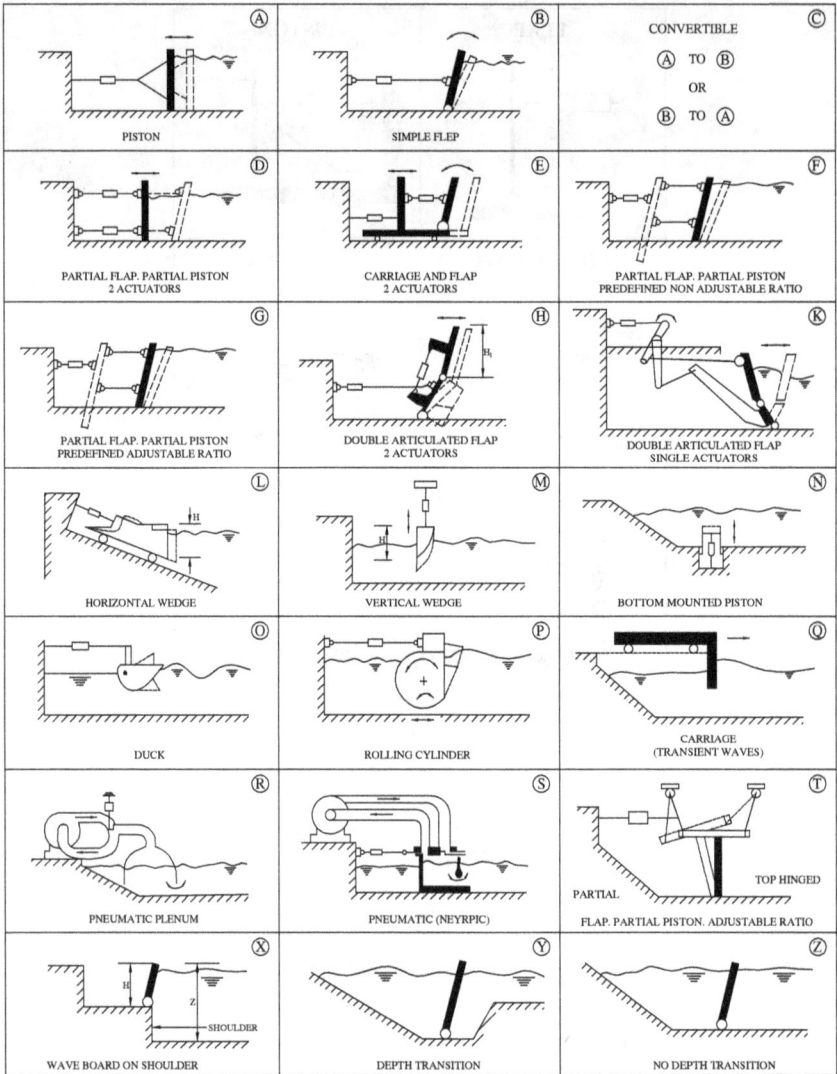

Fig. 10.4   Schematic view of the variety of wave generators (Ploeg and Funke, 1980).

have its own pros and cons. The variety of wave makers as discussed by Ploeg and Funke (1980) are shown in Fig. 10.4.

The *wet back system* does not require the power to balance the static water pressure and the leakage need not be a problem. The drawback of the wet back system is that it draws more power, since it is required to push/displace water on either sides of the wave maker, also a secondary wave absorber needs to be erected for the dissipation of the oscillations on its leeside.

A *dry back* system is commonly preferred over the wet back system in terms of qualitative wave generation and active absorption of incident waves. A rolling seal is used at the sides and at the hinges of the flaps to prevent leaks.

A *pneumatic wave maker* has no moving part in direct contact with the water. The core of the wave generator is a blower with low pressure head which is located at one end of the wave tank connected to a partially immersed chamber with an open bottom. A flapper valve is placed at the bifurcation between the outside vent duct and the chamber and connects alternatively the discharge and intake of the blower to and from the chamber. This introduces pressure differential in the chamber which alternatively draws water up into the chamber and then pushes it down. The cyclic motion of air in the chamber generates waves in tank. A differential pressure is introduced in a chamber by means of flapper valve, which is positioned between outer vent duct and the chamber. This facilitates alternative suction and release of water within the chamber, generates waves in the tank. The position of the flapper valve is controlled by an electric or hydraulic servo drive. The system accepts a flapper position feedback signal from a transducer at the flapper as well as a reference signal (e.g., a signal generated for a particular wave spectrum) and operates an actuator to cause the flapper position to match the reference. The timeseries of the generated waves is a function of the reference signal. A schematic representation of Pneumatic and plunger wave makers is shown in Fig. 10.5 (Hughes, 1993; Chakrabarti, 1987).

A *dual flap wave maker* consists of two hinged flaps and an actuating system driven hydraulically and a control system for the movement of the flaps. The hydraulic systems design is a function work done by the actuator depends on its force and velocity which can be derived from theory of wave makers. A typical design is shown in Fig. 10.6.

A hydraulic system drives the flap wave maker either in a closed or open loop. The displacement of the pumping element/wave board/flap is driven by a variable displacement motor/pump, a servo actuator valve that regulates the shift with the aid of a hydraulic fluid pushed through a pressure pump. The movement of the servo valve is controlled by an electronic control system. The control system for the servo valve consists of a command

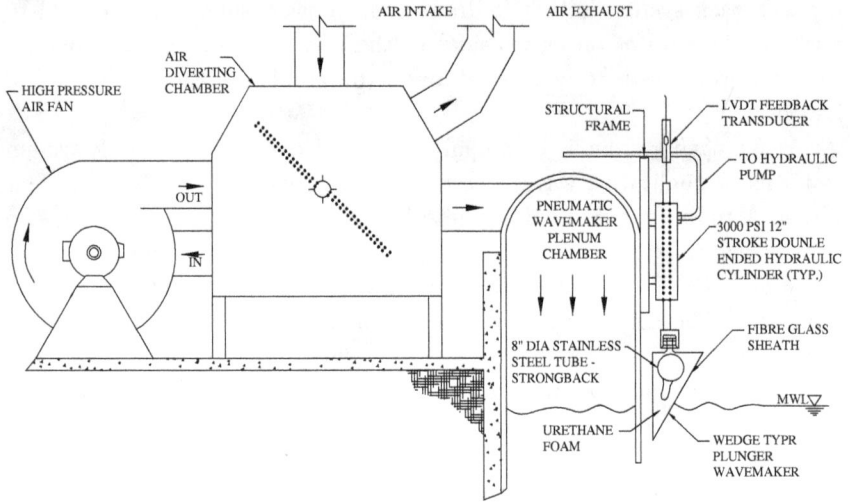

Fig. 10.5   Pneumatic and plunger wave makers.

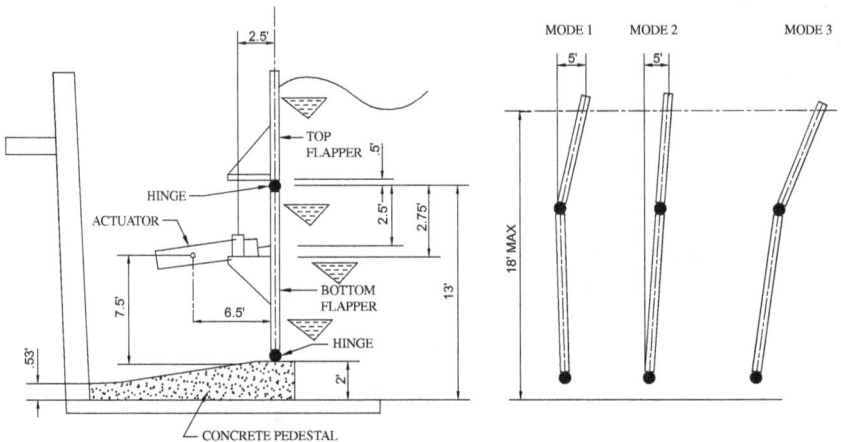

Fig. 10.6   Double-hinged wave generator, configuration as in deep water flume, DOE-IITM.

signal generator and servo amplifier to supply current to the coils of the valve. The command signal generator provides an analog signal representing the desired position of the flap. It can be either a regular or random signal. A computer can be programmed to generate the desired signal.

10.4   General Theory of Mechanical Wave Generation

The general theory of wave making was proposed by Havelock (1929) and later Biesel and Suquet (1951) described theoretical and practical aspects of piston and flap type wave makers. Figure 10.7 is a schematic of a wave maker mounted in a fume. The wave maker moves in a combined rotating and translation modes.

The Laplace equation in terms of velocity potential is given by

$$\nabla^2 \varphi = 0. \tag{10.1}$$

The boundary conditions are, at $z = -d$,

$$\partial \phi / \partial z = 0, \tag{10.2}$$

at $z = n$

$$\frac{\partial \eta}{\partial t} + \frac{\partial \phi \partial \eta}{\partial x \partial t} - \frac{\partial \phi}{\partial z} = 0, \tag{10.3}$$

and

$$\frac{\partial \phi}{\partial t} + \frac{1}{2} \left[\left(\frac{\partial \phi}{\partial x} \right)^2 + \left(\frac{\partial \phi}{\partial z} \right)^2 \right] + g\eta = 0. \tag{10.4}$$

The boundary condition for a wave maker can be formulated by assuming that the board is uniformly flat and impermeable. Considering "s" to be the fluid velocity normal to wave maker, the velocity on the board and

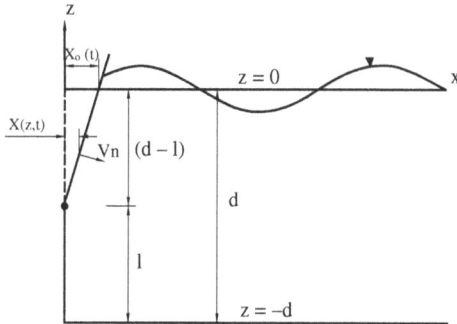

Fig. 10.7   Two-dimensional wave flume definition sketch.

boundary condition is given by

$$-\partial\phi/\partial t = v_n,\tag{10.5}$$

which is specified as, at $x = x(z, t)$,

$$\frac{\partial\phi}{\partial t} = \left(1 + \frac{a}{d+l}\right)\frac{\partial X_0(t)}{\partial t}.\tag{10.6}$$

When $l = 0$, in Eq. (10.6), the boundary condition represents a bottom-hinged flap-type wave maker and when $l = \infty$, it represents the boundary condition of a piston-type wave maker. The wave-maker problem can be solved by standard perturbation techniques. The complete solution is beyond the scope of this chapter. The wave height (H) to stroke S_0 ratio can be obtained for the *piston mode* as

$$\frac{H}{S_0} = \frac{4\sinh^2 kd}{\sinh 2kd + 2kd} = \frac{2(\cosh 2kd - 1)}{\sinh 2kd + 2kd},\tag{10.7}$$

and for *flap-type wave maker* it is

$$\frac{H}{S_0} = \frac{4\sinh^2 kd}{\sinh 2kd + 2kd}\left[\sin kd + \frac{1 - \cosh kd}{kd}\right].\tag{10.8}$$

The wave height stroke ratios for both the modes as a function of relative depth (kd) are superposed in Fig. 10.8. In very shallow water, both the curves in Fig. 10.8 are approximately straight lines that can be obtained by taking the shallow water limits in (10.7) and (10.8).

For piston type,

$$H/S_0 \approx kd,\tag{10.9}$$

For flap type,

$$H/S_0 = kd/2.\tag{10.10}$$

Approximate solution like this can be obtained by equating the volume of water mass contained in a sin wave above the free surface to the volume of water displaced by wave maker at maximum forward stroke. For a variable draft flap-type wave maker defined in Fig. 10.9, the wave height to stroke ratio can be found to be

$$\frac{H}{S_0} = \frac{4\sinh kd}{\sinh 2kd + 2kd}\left[\sinh kd + \frac{(\cosh kl - \cosh kd)}{k(d-1)}\right].\tag{10.11}$$

The plots for various hinge elevation are given in Fig. 10.10. The curve given for $I/h = 0$ is the solution for a flap type board hinged at the bottom.

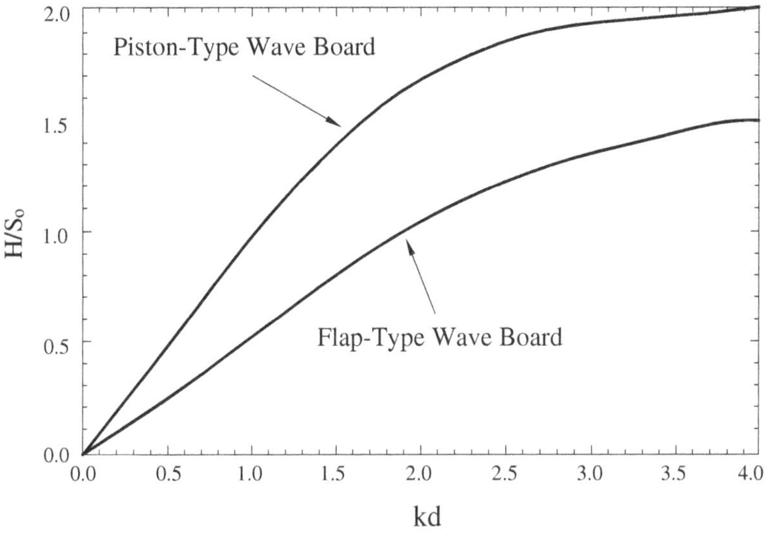

Fig. 10.8   Two-dimensional wave height-to-stroke ratio.

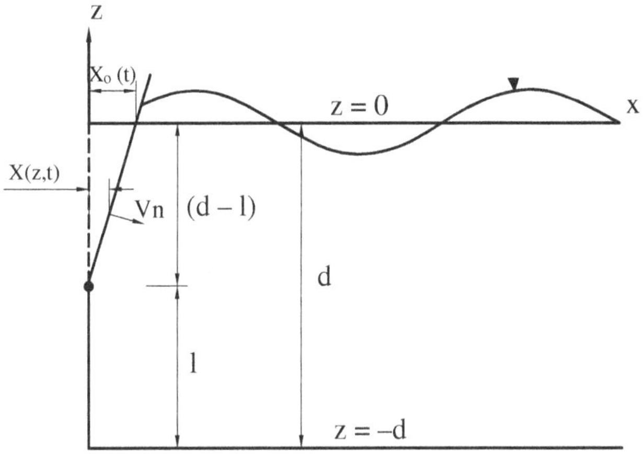

Fig. 10.9   Variable-draft flap-type wave board.

Directional (3D) wave maker: The multidirectional waves discussed in Chapter 6, Section 6.9, can be generated in the laboratory only with multielement wave makers (MEWM). It is also referred to as snake-type wave maker. The MEWM consists of several elementary wave generators driven individually to set its motion with different stroke and frequency,

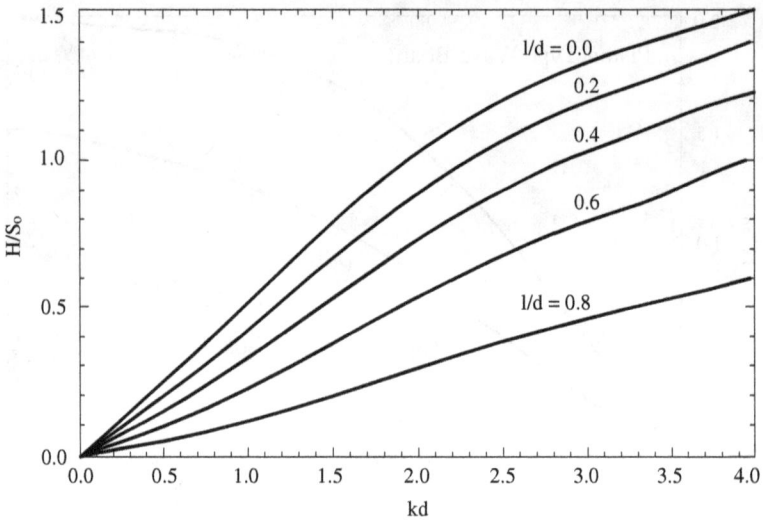

Fig. 10.10   Wave height-to-stroke ratio for variable draft wave maker.

the motions of which are controlled by a personal computer. Directional wave makers are fixed to one of the side walls of a basin which produces waves that propagate at an oblique angle to the axis of the wave maker or it can generate directional waves of desired characteristics which is a value addition for testing of offshore structures and the reasons for the purpose of testing offshore structures in directional have been already emphasized earlier in this chapter.

The basic theory of multielement wave maker was proposed by Biesel (1954). As in the case of 2D wave makers, the theory of 3D wave makers can be established by assuming that the moving surface satisfies certain continuity conditions. However, multielement wave makers do not satisfy these conditions as there are sharp changes in phase and amplitude from one wave board to the other. A schematic representation of the working of MEWM is shown in Fig. 10.11.

10.5   Hydrodynamic Testing Facilities AT DOE-IITM

Since early 1980s, there has been a steady progress in the development of hydrodynamic testing facilities in DOE-IITM. The hydrodynamics testing facilities with the dimensions in DOE-IITM are given in Table 10.2.

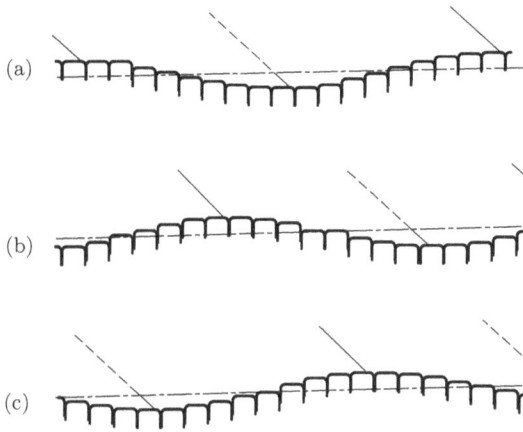

Fig. 10.11   Principle of the MEWM.

Table 10.2   List of hydrodynamic testing facilities in DOE-IITM.

Facility	Dimensions
Wave current flume	30.0 m × 2.0 m × 1.5 m ($d = 1.0$ m)
Shallow water wave flume	72.5 m × 2.0 m × 2.7 m ($d = 0.3$ to 2.0 m)
Wave basin	30 m × 30 m × 3.8 m deep ($d = 3.0$ m)
Deep water flume	90.0 m × 4.0 m × 2.8 m
Glass flume & PIV	20 m × 0.6 m × 1.0 m ($d = 0.6$ m)
Shallow wave basin	19.7 m × 16.5 m × 1.0 m ($d = 0.6$ m)
Towing tank	85 m × 3.2 m × 2.8 m, with selectable carriage speed 5 m/s

10.5.1   *Wave current flume*

The first facility is the wave current flume, the dimensions of which are 30 m length, 2 m width and 3.7 m depth. A false bottom is built 2 m above the original floor, so that the water depth is maintained at 1m with a free board of 0.5 m. A sand pit over the entire width of the flume for a length of 1.5 m and 0.3 m deep facilitates carrying out experiments on wave structure interaction problems. A chamber below the false bottom and the original flume floor serves as a reservoir having inlet and outlet arrangements. The regulated circulation of water from inlet and outlet arrangements of the false bottom provides current velocity at a constant discharge for generating current of up to 0.20 m/s. A plunger type wave generator used to generate only regular waves has been replaced by a piston-type wave maker

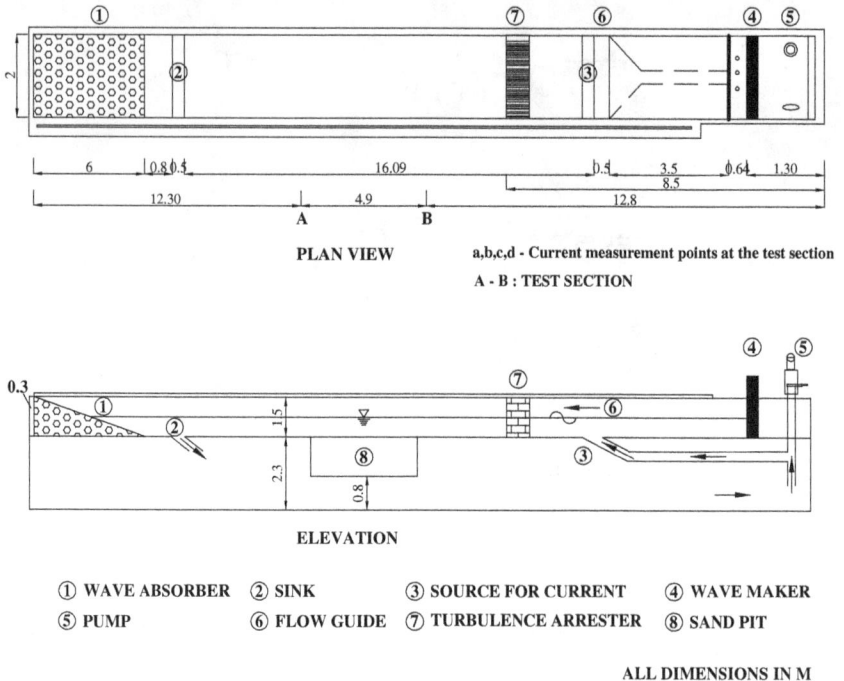

Fig. 10.12   Details of wave current flume.

installed at one end of the flume which can generate random predefined spectral characteristics and user-defined waves controlled by a personal computer. Waves with frequencies ranging between 0.5 Hz and 1.7 Hz with a maximum height of about 0.25 m can be generated in this flume.

The sloping beach end consisting of stone rubbles at 1 in 3.5 slope extending over a length of 6 m was provided at the downstream end to dissipate wave energy and reduce wave reflections. The details of the flume and experimental setup are presented in Fig. 10.12. A view of the wave current flume is shown in Fig. 10.13.

10.5.2   Shallow water wave flume

The wave current flume could facilitate generation of intermediate depth waves and the depth in the flume could be varied marginally. For a wide variation in the water depth, the shallow water wave flume in DOE-IITM was commissioned in mid-1990s. A computer controlled wave maker is installed at one end of the flume and on the other end, a combination of parabolic

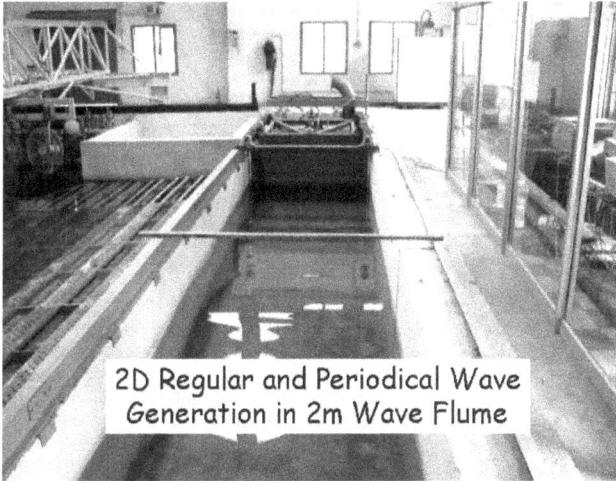

Fig. 10.13   A view of the wave current flume.

absorber and rubble mound wave absorber is placed to minimize the reflection. The water depth in the flume can be varied from 0.3 m to 2.0 m. The wave generating system in this wave flume can generate regular or random waves. The wave maker can be operated either in piston mode, as used in the present study, for the generation of shallow water waves or, in hinged mode primarily for the generation of deep water waves. In the piston mode, the maximum operating water depth is 1.0 m, whereas, in the hinged mode, it is 2.0 m. Perforated plate wave absorber with progressive porosity has been provided on the rear of the wave maker to dissipate the wave energy behind the wave maker as it is of wet back type. A view of this flume is shown in Fig. 10.14. Waves with frequencies ranging between 0.3 Hz and 2 Hz and height closer to the critical wave steepness could be generated in this flume.

The *wave synthesizer*, an application software package, along with analog-to-digital and Input/Output (I/O) modules is installed in a personal computer (PC). The software can control the wave paddles and at the same time acquires the signatures from the instruments used for the tests. This software can also be used for the analysis of acquired data in frequency and time domain. Most of theoretical standard frequency spectra are available in this software for the simulation of the random waves. The schematic sketch of the wave generating system along with main control system is shown in Fig. 10.15.

Fig. 10.14   A view of the shallow water wave flume.

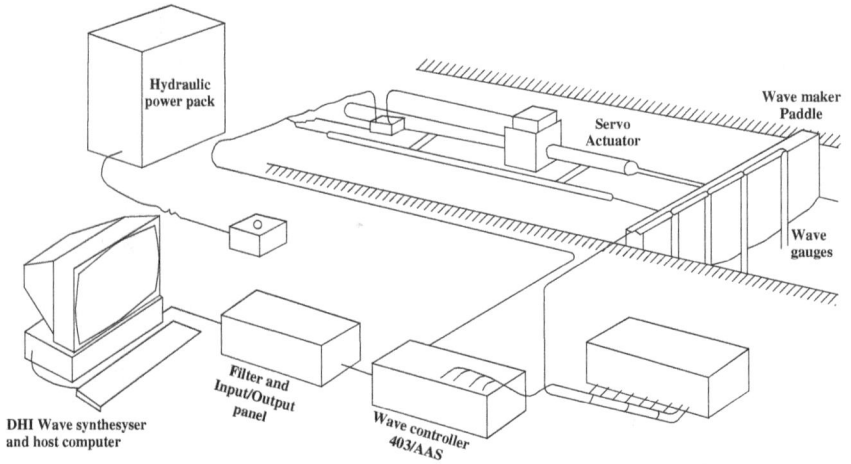

Fig. 10.15   Schematic representation of wave generating system.

10.5.3   *Wave basin*

10.5.3.1   *General*

The wave basin facility in DOE-IITM is equipped with two different kinds of wave makers, namely the long crested wave maker (LCWM) and the multi element wave makers (MEWM). The LCWM can generate long crested unidirectional regular and random waves with frequency ranging between 0.3 Hz and 1.7 Hz with a maximum wave height of 0.25 m. The MEWM is capable of generating multi-directional waves with frequency ranging between 0.2 Hz and 2.0 Hz with a maximum wave height of 0.40 m. They are positioned perpendicular to each other with absorbers housed on the other two sides to dissipate the incident wave energy from either of the wave makers. The plan and sectional view of the wave basin is depicted in Fig. 10.16(a) and Fig. 10.16(b) respectively. A view of the wave basin is shown in Fig. 10.17. Waves with the MEWM can be generated with or without directional spreading of the waves referred to 2D and 3D waves, respectively.

The MEWM consists of 52 paddles, each of 0.5 m width. The paddles are hinged at 1.3 m below the water surface and 1.7 m above the basin floor. Each paddle pivots independently according to the servo actuator motion. The front surface of the 52 paddles is covered by an elastic membrane which ensures that the rear side of the wave maker is dry. The membrane is elastic and tolerates elongations beyond the working range of paddles. The maximum rotation of the paddles about their hinges is $\pm 15°$. MEWM incorporates six main subsystems, such as computer with software, Digital Servo Controllers (DSC), hydraulic power pack and hydraulic servo actuator, the details of which are shown in Fig. 10.18.

10.5.3.2   *Computer and software*

The host computer is a personal computer with several A/D and I/O modules. This hardware is controlled by software. This system can control the 52 paddles of the wave maker and at the same time handle data logging required during the hydraulic model tests. The desired wave maker control signal is synthesized on the PC using the software, WS4. This software is also capable of monitoring other subcomponents of the system like the DSC and the hydraulic power pack.

10.5.3.3   *Digital servo controller*

The digital servo controller (DSC) performs the servo control of the motion of the wave maker. Each of the 52 paddles is controlled by its own DSC.

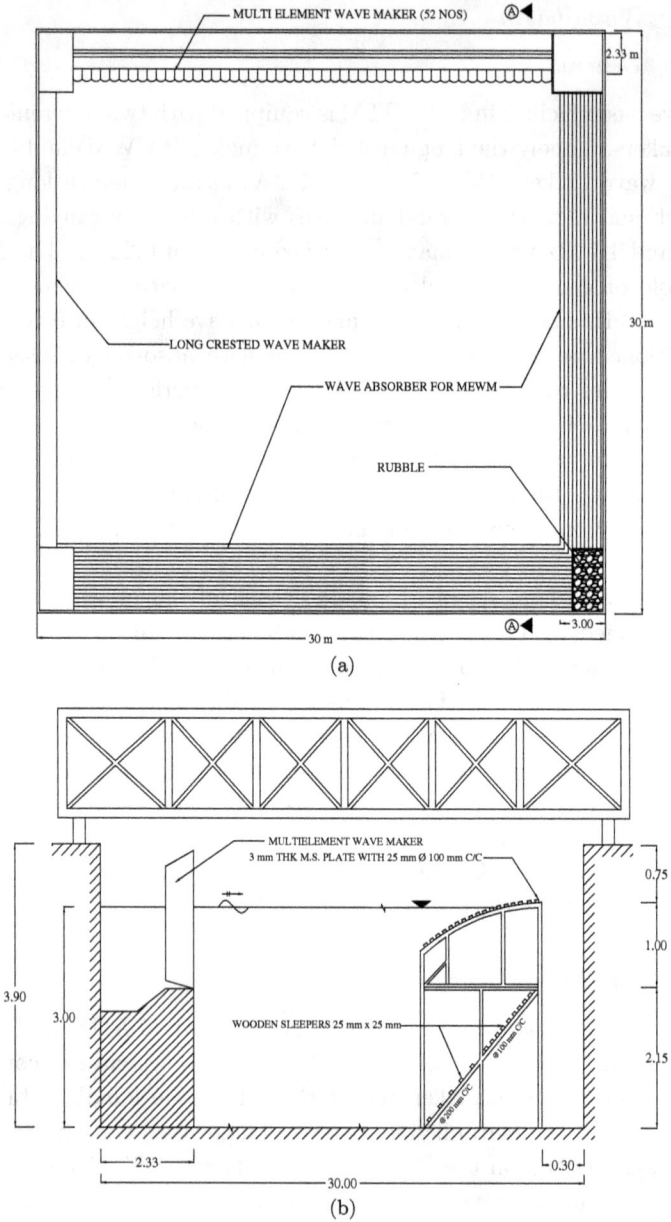

Fig. 10.16   (a) Plan view of the wave basin. (b) Sectional view of the wave basin.

Fig. 10.17   View of the wave basin at the DOE-IITM.

The user can control the DSC only through the host computer. The 52 numbers of DSC are connected in series (daisy chained). The signals from the PC through RS 422 serial communication lines are sent through all the DSC in sequence and returned to the PC in a closed loop. Based on the address, each DSC picks up its relevant command (the servo actuator position signal) and compares it with the actual position of its corresponding paddle. The difference between the two is sent as a correction signal to the servo actuator. The actual position of the paddle is measured using a feedback transducer and is sent back to the DSC for the next command processing. A complete description is highlighted by Krishnakumar (2009).

10.5.3.4   Hydraulic power pack

The pressurized oil flow to the 52 servo actuators is supplied by the hydraulic power pack. This unit is run by three sets of pumps with electric motors. Control and safety features for excess temperature and pressure are also incorporated. Auxiliary components such as oil cooling systems

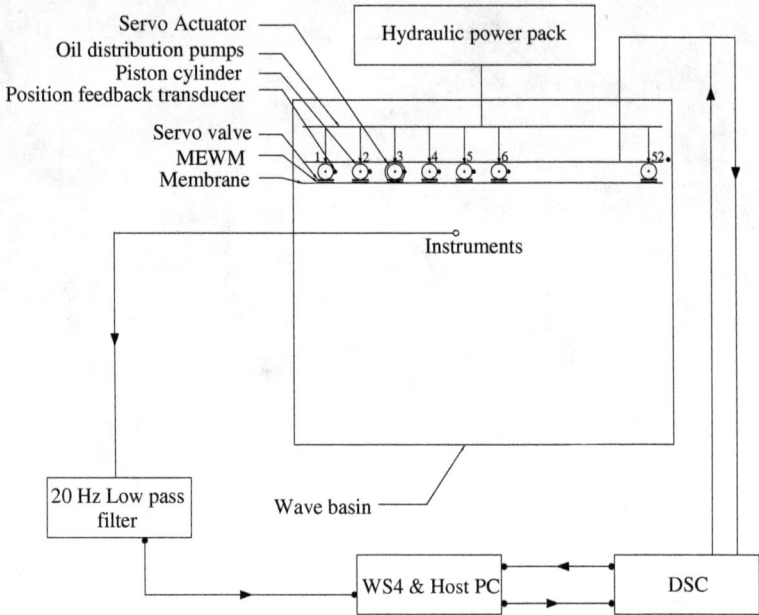

Fig. 10.18   Schematic layout of MEWM control system and data acquisition system.

have also been provided. The power pack is equipped with a programmable logic controller interfaced with the host PC via RS 422 serial communication line. The power pack is controlled for wave generation only through the PC.

10.5.3.5   *Hydraulic servo actuator*

The motion of the wave paddles is governed by the hydraulic servo actuators. An actuator consists of a piston servo valve and a position feedback transducer. According to the control signal received from the DSC, the servo valve controls the instantaneous oil flow to the piston and thereby the motion of the paddle. Since this is a dry back system, the hydrostatic pressure on the paddle and thereby on the piston is balanced by a counterbalancing secondary oil pressure circuit. Hence, the control signal need to only supply the dynamic component of the pressure signal to the piston for paddle motion and thus reducing power consumption.

Fig. 10.19   A view of the deep water wave flume.

10.5.4   *Deep water wave flume*

In order to test large diameter structures, a flume of larger width of 4 m, 2.8 m deep and length 45 m (later extended to 90 m) was commissioned in mid-1990s. One end of the flume is equipped with a twin-flap wave maker, the motion of which controlled by a personal computer can generate both regular and random waves. A wave absorber to absorb the incident wave energy is placed at the other end of the flume. The test section is about 20 m away from the wave maker. A view of this flume is shown in Fig. 10.19.

10.5.5   *Glass flume*

The glass flume can generate steady, unsteady current, and regular waves. The main advantage of this flume is that cross-sectional view of the test model could be observed with ease, due to transparency of the walls. Due to the wall transparency, instruments like particle image velocimetry (PIV) could be employed for capturing images of vortex shedding pattern behind bluff bodies and settlement of sediments on the bed of the flume for studying the influence of nearshore regions. As the flume is relatively smaller, handling and installation of models is easier compared to other facilities.

Fig. 10.20   A view of the shallow wave basin.

10.5.6 *Shallow wave basin*

The flumes and the basin discussed could not cater to the shallow water conditions in particular testing of stability of breakwaters, tranquility inside harbor basin, and similar types of problems. Hence, the three-dimensional shallow wave basin of size 17 m long, 15 m wide, and 1 m deep wave basin commissioned around 2008. Further, testing of oblique waves by conventional method of orienting of the model is much easier as water depth is shallow. The piston-type wave maker is provided at one end of the shallow wave basin for the generation of waves. The wave maker can be operated up to a maximum stroke length of 0.2 m with a possible range of frequency from 0.3 Hz to 4 Hz. An artificial beach (wave absorber) on the other end is provided with a combination of parabolic perforated fiber reinforced polymer (FRP) sheet and rubble mound to absorb the incident waves. The water depth in the shallow wave basin can be varied from 0.3 m to 0.6 m. A pictorial view of a shallow wave basin equipped with wave paddle, absorbing artificial beach, and electric wave actuator is projected in Fig. 10.20. The desired wave maker control signals are generated on a personal computer

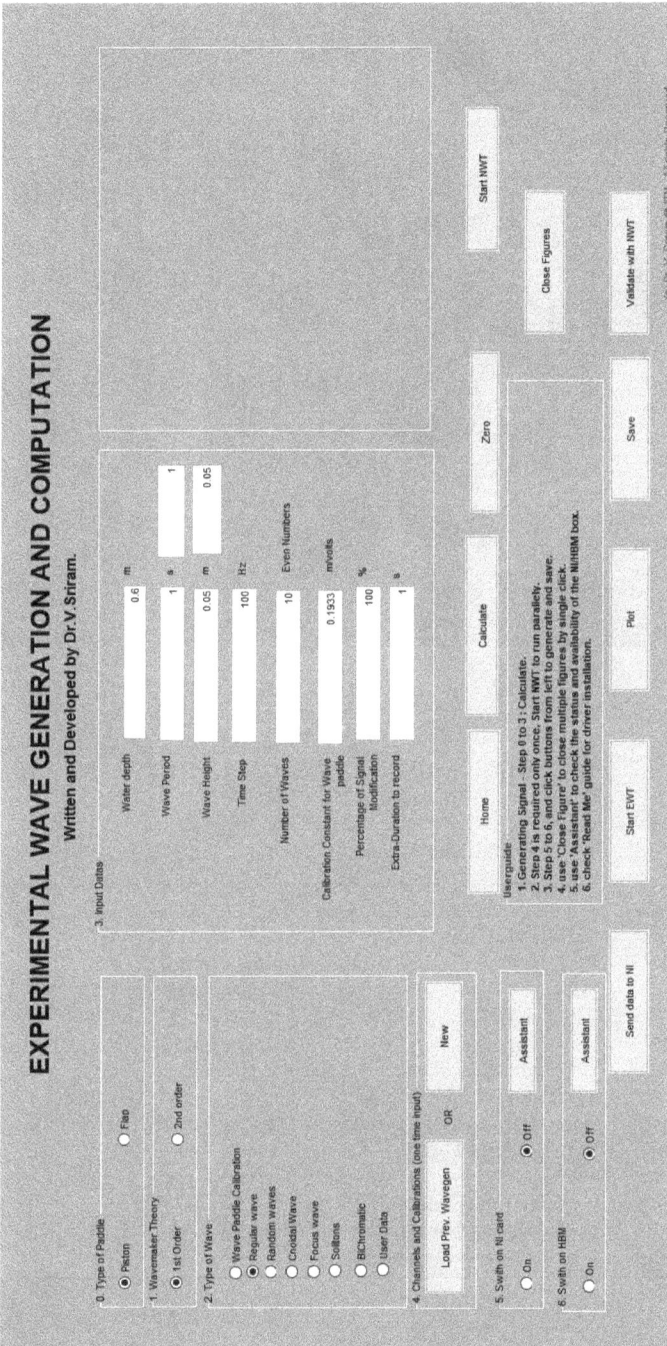

Fig. 10.21   Snapshot view of the WaveGen software.

Fig. 10.22   A view of towing tank in DOE-IITM.

Table 10.3   List of experiments for wave mechanics.

Experiment 1	*Simulation of regular waves using wave maker*
	Calibrate the wave probe
	Data sampling using data acquisition system
	Simulation of regular waves using wave maker
	Plot the FFT of the obtained time history
	Determine the wave height from the record
	Determination of the wavelength
	Determine the wave celerity
	Compare the above results with the analytical solutions from linear theory
	Generate the regular waves in deep water, intermediate, and shallow water depth
	Prepare report as per the guidelines given below and submit it on or before the due date

Table 10.3   (*Continued*)

Experiment 2	*Measurement of bottom pressure, celerity, group celerity*
	Calibrate the pressure transducer
	Generate the regular waves
	Record the bottom pressure using the pressure transducer
	Obtain the wave height using the pressure record from linear wave theory and compare with the experimental wave time history
	Prepare report as per the guidelines given below and submit it on or before the due date
Experiment 3	*Mass transport velocity*
	Measure the mass transport velocity in the flume using high speed camera
	Compare with the linear and higher order theories
	Prepare report as per the guidelines given below and submit it on or before the due date
Experiment 4a	*Measurement of 2D random waves*
	Generate the 2D random waves of the specified wave spectrum
	Obtain the wave time history as well as the pressure time history at a location
	Plot the theoretical and experimental wave spectrum
	Estimate the statistical quantities, $H_s, H_{\mathrm{rms}}, H_{\mathrm{max}}$ and T_{max} from the experiments
	Prepare report as per the guidelines given below and submit it on or before the due date
Experiment 4b	*Dissipated wave energy due to breaking waves*
	Generate plunging and spilling wave breaking
	Determine the dissipated energy of wave due to breaking
	Compare the energy spectrum of the wave before and after breaking
	Prepare report as per the guidelines given below and submit it on or before the due date
Experiment 5	*Estimation of reflection coefficient of the beach*
	Determine the reflection coefficient of a beach using single-probe and two-probe methods
	Measure the wave heights using single-probe and two-probe methods
	Use the theory provided in the class and write a MATLAB code to get the reflection coefficient
	Prepare report as per the guidelines given below and submit it on or before the due date

(NI PXI embedded controller) using the WaveGen Software and are sent to the controller to control the motion of wave paddles. The WaveGen is an in-house software at IITM developed by Sriram *et al.* (2015) and is compatible with various operating systems, with an efficient and user-friendly interface. It can generate the control signal for regular waves, random waves, long waves, Cnoidal waves, solitary waves, focusing waves, and user defined spectrum. A snapshot of the WaveGen software user interface is shown in Fig. 10.21. The wave maker contains five independent wave paddles, which are driven by linear electric actuators.

10.5.7   *Towing tank*

To perform studies related to ships and underwater vehicles, towing tank facility is used. It consists of carriage capable of running at 5 m/s speed and runs over 80 m long rails, to have smooth motion. Four motors are fixed at the hub of the wheels of the carriage to maintain steady speed while testing the models. The test models are fixed to the rigid frames provided in the carriage, towing which the measurements could be obtained. Open water propeller tests were also performed for studying the wakes generated behind the blades and thrust generated by the propeller. A view of the towing tank is shown in Fig. 10.22.

10.5.8   *Physical modeling*

The solution to a variety of problems related to wave structure interaction problems has been solved in the past and is currently in progress through the above-discussed hydrodynamic testing facilities in DOE-IITM. Table 10.3 lists the experiments that are conducted for the students as a training process.

Appendix A

Wave Table

d/L_o	d/L	$2\pi d/L$	tanh $(2\pi d/L)$	sinh $(2\pi d/L)$	cosh $(2\pi d/L)$	H/H'_o	K	$4\pi d/L$	sinh $(4\pi d/L)$	cosh $(4\pi d/L)$	n	C_g/C_o	M
0.0001	0.00398984	0.0250689	0.0250637	0.0250715	1.00031	4.46692	0.999686	0.0501378	0.0501588	1.00126	0.999791	0.0250584	7855.63
0.0002	0.00564308	0.0354565	0.0354417	0.0354639	1.00063	3.75681	0.999372	0.070913	0.0709725	1.00252	0.999581	0.0354268	3928.64
0.0003	0.00691205	0.0434297	0.0434024	0.0434434	1.00094	3.39519	0.999058	0.0868594	0.0869687	1.00377	0.999372	0.0433752	2619.64
0.0004	0.00798219	0.0501536	0.0501116	0.0501746	1.00126	3.16008	0.998744	0.100307	0.100475	1.00503	0.999162	0.0500696	1965.14
0.0005	0.00892529	0.0560793	0.0560206	0.0561087	1.00157	2.98909	0.99843	0.112159	0.112394	1.0063	0.998953	0.0559619	1572.44
0.0006	0.00977819	0.0614382	0.061361	0.0614769	1.00189	2.85635	0.998116	0.122876	0.123186	1.00756	0.998744	0.061284	1310.64
0.0007	0.0105628	0.0663678	0.0662705	0.0664165	1.0022	2.7488	0.997802	0.132736	0.133126	1.00882	0.998535	0.0661734	1123.64
0.0008	0.0112933	0.0709576	0.0708388	0.0710172	1.00252	2.65897	0.997488	0.141915	0.142392	1.01009	0.998326	0.0707202	983.395
0.0009	0.0119796	0.0752698	0.075128	0.0753409	1.00283	2.58222	0.997174	0.15054	0.151109	1.01135	0.998116	0.0749865	874.312
0.001	0.0126289	0.0793497	0.0791835	0.0794329	1.00315	2.51549	0.99686	0.158699	0.159366	1.01262	0.997907	0.0790178	787.046
0.0011	0.0132467	0.0832313	0.0830397	0.0833275	1.00347	2.45665	0.996546	0.166463	0.167233	1.01389	0.997698	0.0828485	715.646
0.0012	0.0138372	0.0869414	0.086723	0.087051	1.00378	2.40416	0.996232	0.173883	0.17476	1.01516	0.997489	0.0865053	656.147
0.0013	0.0144037	0.090501	0.0902547	0.0906246	1.0041	2.3569	0.995919	0.181002	0.181992	1.01643	0.99728	0.0900092	605.801
0.0014	0.014949	0.0939272	0.0936519	0.0940653	1.00441	2.314	0.995605	0.187854	0.188961	1.0177	0.997071	0.0933776	562.648
0.0015	0.0154753	0.0972341	0.0969288	0.0973873	1.00473	2.27479	0.995291	0.194468	0.195696	1.01897	0.996862	0.0966247	525.248
0.0016	0.0159845	0.100433	0.100097	0.100602	1.00505	2.23873	0.994978	0.200867	0.20222	1.02024	0.996653	0.0997621	492.523
0.0017	0.0164782	0.103535	0.103167	0.10372	1.00536	2.2054	0.994664	0.207071	0.208554	1.02152	0.996445	0.1028	463.649
0.0018	0.0169577	0.106548	0.106147	0.10675	1.00568	2.17446	0.99435	0.213096	0.214713	1.02279	0.996236	0.105747	437.982
0.0019	0.0174242	0.109479	0.109044	0.109698	1.006	2.1456	0.994037	0.218959	0.220712	1.02407	0.996027	0.108611	415.018
0.002	0.0178787	0.112335	0.111865	0.112572	1.00632	2.11859	0.993723	0.22467	0.226565	1.02534	0.995818	0.111397	394.35

(*Continued*)

(Continued)

d/L_o	d/L	$2\pi d/L$	tanh $(2\pi d/L)$	sinh $(2\pi d/L)$	cosh $(2\pi d/L)$	H/H'_o	K	$4\pi d/L$	sinh $(4\pi d/L)$	cosh $(4\pi d/L)$	n	C_g/C_o	M
0.0021	0.0183221	0.115121	0.114615	0.115376	1.00663	2.09324	0.99341	0.230243	0.232282	1.02662	0.99561	0.114112	375.65
0.0022	0.0187553	0.117843	0.1173	0.118116	1.00695	2.06936	0.993097	0.235686	0.237874	1.0279	0.995401	0.116761	358.65
0.0023	0.0191788	0.120504	0.119924	0.120796	1.00727	2.04681	0.992783	0.241008	0.243348	1.02918	0.995192	0.119347	343.129
0.0024	0.0195934	0.123109	0.122491	0.12342	1.00759	2.02547	0.99247	0.246217	0.248713	1.03046	0.994984	0.121876	328.901
0.0025	0.0199995	0.12566	0.125003	0.125991	1.00791	2.00522	0.992156	0.251321	0.253975	1.03175	0.994775	0.12435	315.811
0.0026	0.0203977	0.128163	0.127465	0.128514	1.00822	1.98597	0.991843	0.256325	0.259141	1.03303	0.994566	0.126773	303.728
0.0027	0.0207884	0.130618	0.12988	0.130989	1.00854	1.96763	0.99153	0.261235	0.264217	1.03432	0.994358	0.129147	292.541
0.0028	0.0211721	0.133028	0.132249	0.133421	1.00886	1.95013	0.991217	0.266057	0.269207	1.0356	0.99415	0.131476	282.152
0.0029	0.0215492	0.135397	0.134576	0.135811	1.00918	1.9334	0.990903	0.270795	0.274116	1.03689	0.993941	0.133761	272.48
0.003	0.0219199	0.137727	0.136862	0.138162	1.0095	1.91738	0.99059	0.275453	0.27895	1.03818	0.993733	0.136004	263.453
0.0031	0.0222845	0.140018	0.13911	0.140476	1.00982	1.90203	0.990277	0.280036	0.28371	1.03947	0.993524	0.138209	255.008
0.0032	0.0226435	0.142273	0.141321	0.142754	1.01014	1.88729	0.989964	0.284546	0.288402	1.04076	0.993316	0.140376	247.091
0.0033	0.022997	0.144494	0.143497	0.144998	1.01046	1.87312	0.989651	0.288989	0.293028	1.04205	0.993108	0.142508	239.654
0.0034	0.0233453	0.146683	0.14564	0.147209	1.01078	1.85948	0.989338	0.293365	0.297592	1.04334	0.992899	0.144606	232.654
0.0035	0.0236886	0.14884	0.14775	0.14939	1.0111	1.84635	0.989025	0.29768	0.302096	1.04463	0.992691	0.146671	226.054
0.0036	0.0240272	0.150967	0.149831	0.151541	1.01142	1.83368	0.988712	0.301934	0.306543	1.04593	0.992483	0.148704	219.821
0.0037	0.0243611	0.153066	0.151881	0.153664	1.01174	1.82145	0.988399	0.306131	0.310935	1.04723	0.992275	0.150708	213.925
0.0038	0.0246907	0.155137	0.153904	0.15576	1.01206	1.80963	0.988086	0.310273	0.315275	1.04852	0.992067	0.152683	208.339
0.0039	0.0250161	0.157181	0.155899	0.157829	1.01238	1.7982	0.987773	0.314362	0.319566	1.04982	0.991859	0.15463	203.04
0.004	0.0253375	0.1592	0.157869	0.159874	1.0127	1.78714	0.98746	0.3184	0.323808	1.05112	0.991651	0.156551	198.006
0.0041	0.025655	0.161195	0.159813	0.161894	1.01302	1.77642	0.987147	0.32239	0.328004	1.05242	0.991443	0.158446	193.217
0.0042	0.0259687	0.163166	0.161733	0.163891	1.01334	1.76603	0.986834	0.326332	0.332155	1.05372	0.991235	0.160316	188.656
0.0043	0.0262788	0.165114	0.16363	0.165866	1.01366	1.75594	0.986522	0.330229	0.336264	1.05502	0.991027	0.162162	184.308
0.0044	0.0265854	0.167041	0.165504	0.167819	1.01398	1.74616	0.986209	0.334082	0.340331	1.05633	0.990819	0.163985	180.157
0.0045	0.0268886	0.168946	0.167357	0.169751	1.01431	1.73665	0.985896	0.337893	0.344359	1.05763	0.990611	0.165786	176.19
0.0046	0.0271886	0.170831	0.169188	0.171663	1.01463	1.7274	0.985584	0.341662	0.348348	1.05894	0.990403	0.167565	172.396
0.0047	0.0274854	0.172696	0.171	0.173556	1.01495	1.71841	0.985271	0.345392	0.352301	1.06024	0.990195	0.169323	168.764
0.0048	0.0277792	0.174542	0.172791	0.17543	1.01527	1.70966	0.984959	0.349084	0.356217	1.06155	0.989988	0.171061	165.283

0.0049	0.0280701	0.176369	0.174563	0.177285	1.01559	1.70114	0.984646	0.352739	0.360099	1.06286	0.98978	0.172779	161.944
0.005	0.028358	0.178179	0.176317	0.179123	1.01592	1.69283	0.984333	0.356358	0.363948	1.06417	0.989572	0.174478	158.738
0.0051	0.0286432	0.179971	0.178053	0.180944	1.01624	1.68474	0.984021	0.359941	0.367764	1.06548	0.989365	0.176159	155.659
0.0052	0.0289257	0.181746	0.179771	0.182748	1.01656	1.67685	0.983709	0.363492	0.371549	1.06679	0.989157	0.177821	152.697
0.0053	0.0292056	0.183504	0.181472	0.184536	1.01688	1.66914	0.983396	0.367009	0.375303	1.06811	0.988949	0.179467	149.848
0.0054	0.029483	0.185247	0.183157	0.186308	1.01721	1.66162	0.983084	0.370494	0.379028	1.06942	0.988742	0.181095	147.104
0.0055	0.0297578	0.186974	0.184825	0.188065	1.01753	1.65428	0.982771	0.373948	0.382725	1.07074	0.988534	0.182706	144.46
0.0056	0.0300303	0.188686	0.186478	0.189808	1.01785	1.6471	0.982459	0.377372	0.386393	1.07205	0.988327	0.184301	141.91
0.0057	0.0303005	0.190383	0.188116	0.191536	1.01818	1.64009	0.982147	0.380767	0.390034	1.07337	0.988119	0.185881	139.45
0.0058	0.0305683	0.192066	0.189739	0.193249	1.0185	1.63323	0.981835	0.384133	0.39365	1.07469	0.987912	0.187445	137.075
0.0059	0.030834	0.193735	0.191347	0.19495	1.01883	1.62652	0.981522	0.387471	0.397239	1.07601	0.987705	0.188995	134.78
0.006	0.0310974	0.195391	0.192942	0.196637	1.01915	1.61996	0.98121	0.390782	0.400804	1.07733	0.987497	0.19053	132.561
0.0061	0.0313588	0.197033	0.194523	0.198311	1.01947	1.61353	0.980898	0.394067	0.404345	1.07865	0.98729	0.19205	130.416
0.0062	0.0316182	0.198663	0.19609	0.199972	1.0198	1.60724	0.980586	0.397326	0.407863	1.07998	0.987083	0.193557	128.339
0.0063	0.0318755	0.20028	0.197644	0.201621	1.02012	1.60108	0.980274	0.400559	0.411357	1.0813	0.986875	0.19505	126.329
0.0064	0.0321309	0.201884	0.199185	0.203258	1.02045	1.59504	0.979962	0.403768	0.414829	1.08263	0.986668	0.19653	124.381
0.0065	0.0323843	0.203477	0.200714	0.204884	1.02077	1.58912	0.97965	0.406954	0.41828	1.08395	0.986461	0.197997	122.493
0.0066	0.0326359	0.205058	0.202231	0.206498	1.0211	1.58331	0.979338	0.410115	0.421709	1.08528	0.986254	0.199451	120.663
0.0067	0.0328857	0.206627	0.203736	0.208101	1.02142	1.57762	0.979026	0.413254	0.425118	1.08661	0.986047	0.200893	118.887
0.0068	0.0331337	0.208185	0.205229	0.209692	1.02175	1.57204	0.978714	0.416371	0.428506	1.08794	0.98584	0.202323	117.164
0.0069	0.03338	0.209733	0.206711	0.211274	1.02207	1.56656	0.978402	0.419465	0.431875	1.08927	0.985633	0.203741	115.49
0.007	0.0336246	0.211269	0.208181	0.212845	1.0224	1.56118	0.97809	0.422539	0.435225	1.09061	0.985426	0.205147	113.864
0.0071	0.0338675	0.212796	0.209641	0.214405	1.02273	1.5559	0.977778	0.425591	0.438556	1.09194	0.985219	0.206542	112.284
0.0072	0.0341087	0.214312	0.21109	0.215956	1.02305	1.55071	0.977467	0.428623	0.441868	1.09327	0.985012	0.207926	110.748
0.0073	0.0343484	0.215817	0.212528	0.217497	1.02338	1.54562	0.977155	0.431635	0.445163	1.09461	0.984805	0.209299	109.254
0.0074	0.0345865	0.217314	0.213956	0.219028	1.02371	1.54061	0.976843	0.434627	0.448441	1.09595	0.984598	0.210661	107.8
0.0075	0.0348231	0.2188	0.215374	0.22055	1.02403	1.53569	0.976532	0.4376	0.451701	1.09728	0.984392	0.212012	106.386
0.0076	0.0350582	0.220277	0.216782	0.222063	1.02436	1.53086	0.97622	0.440554	0.454945	1.09862	0.984185	0.213354	105.008
0.0077	0.0352918	0.221745	0.218181	0.223567	1.02469	1.5261	0.975908	0.44349	0.458172	1.09996	0.983978	0.214685	103.666
0.0078	0.035524	0.223204	0.21957	0.225062	1.02501	1.52143	0.975597	0.446408	0.461383	1.10131	0.983772	0.216006	102.359

(*Continued*)

(Continued)

d/L_o	d/L	$2\pi d/L$	$\tanh (2\pi d/L)$	$\sinh (2\pi d/L)$	$\cosh (2\pi d/L)$	H/H'_o	K	$4\pi d/L$	$\sinh (4\pi d/L)$	$\cosh (4\pi d/L)$	n	C_g/C_o	M
0.0079	0.0357548	0.224654	0.220949	0.226548	1.02534	1.51683	0.975285	0.449308	0.464579	1.10265	0.983565	0.217318	101.084
0.008	0.0359842	0.226095	0.22232	0.228027	1.02567	1.51231	0.974974	0.452191	0.467759	1.10399	0.983358	0.21862	99.842
0.0081	0.0362122	0.227528	0.223681	0.229496	1.026	1.50786	0.974662	0.455056	0.470925	1.10534	0.983152	0.219913	98.6302
0.0082	0.0364389	0.228952	0.225034	0.230958	1.02632	1.50347	0.974351	0.457905	0.474076	1.10668	0.982945	0.221196	97.448
0.0083	0.0366643	0.230369	0.226378	0.232412	1.02665	1.49916	0.974039	0.460737	0.477212	1.10803	0.982739	0.222471	96.2943
0.0084	0.0368884	0.231777	0.227714	0.233858	1.02698	1.49492	0.973728	0.463554	0.480334	1.10938	0.982532	0.223736	95.1681
0.0085	0.0371113	0.233177	0.229041	0.235296	1.02731	1.49074	0.973417	0.466354	0.483443	1.11073	0.982326	0.224993	94.0684
0.0086	0.0373329	0.234569	0.23036	0.236727	1.02764	1.48662	0.973105	0.469139	0.486538	1.11208	0.982119	0.226241	92.9943
0.0087	0.0375533	0.235954	0.231671	0.23815	1.02797	1.48256	0.972794	0.471909	0.48962	1.11343	0.981913	0.227481	91.9448
0.0088	0.0377725	0.237332	0.232974	0.239566	1.0283	1.47857	0.972483	0.474663	0.492689	1.11478	0.981707	0.228712	90.9192
0.0089	0.0379905	0.238702	0.234269	0.240975	1.02862	1.47463	0.972172	0.477403	0.495745	1.11614	0.9815	0.229935	89.9167
0.009	0.0382074	0.240064	0.235556	0.242377	1.02895	1.47075	0.971861	0.480128	0.498789	1.11749	0.981294	0.23115	88.9365
0.0091	0.0384232	0.24142	0.236836	0.243772	1.02928	1.46692	0.97155	0.48284	0.501821	1.11885	0.981088	0.232357	87.9778
0.0092	0.0386378	0.242768	0.238109	0.24516	1.02961	1.46315	0.971238	0.485537	0.50484	1.12021	0.980882	0.233557	87.0399
0.0093	0.0388513	0.24411	0.239374	0.246542	1.02994	1.45943	0.970927	0.48822	0.507848	1.12157	0.980676	0.234748	86.1223
0.0094	0.0390638	0.245445	0.240632	0.247917	1.03027	1.45576	0.970616	0.49089	0.510844	1.12293	0.980469	0.235932	85.2241
0.0095	0.0392752	0.246773	0.241883	0.249285	1.0306	1.45215	0.970305	0.493546	0.513829	1.12429	0.980263	0.237109	84.3449
0.0096	0.0394855	0.248095	0.243127	0.250648	1.03093	1.44858	0.969994	0.49619	0.516803	1.12565	0.980057	0.238278	83.484
0.0097	0.0396949	0.24941	0.244364	0.252004	1.03126	1.44506	0.969684	0.49882	0.519765	1.12701	0.979851	0.239441	82.6409
0.0098	0.0399032	0.250719	0.245594	0.253354	1.03159	1.44159	0.969373	0.501438	0.522718	1.12838	0.979645	0.240596	81.8149
0.0099	0.0401105	0.252022	0.246818	0.254698	1.03193	1.43816	0.969062	0.504043	0.525659	1.12974	0.979439	0.241743	81.0057
0.01	0.0403168	0.253318	0.248035	0.256036	1.03226	1.43478	0.968751	0.506636	0.52859	1.13111	0.979234	0.242885	80.2126
0.011	0.0423296	0.265964	0.259866	0.269111	1.03558	1.40321	0.965645	0.531929	0.557371	1.14484	0.977177	0.253935	73.0755
0.012	0.0442588	0.278086	0.271133	0.281684	1.03892	1.37519	0.962542	0.556172	0.585292	1.15869	0.975124	0.264388	67.1283
0.013	0.046115	0.289749	0.281904	0.293821	1.04227	1.35009	0.959443	0.579498	0.612482	1.17266	0.973074	0.274313	62.0966
0.014	0.0479068	0.301007	0.292234	0.305573	1.04565	1.32741	0.956347	0.602015	0.639043	1.18675	0.971028	0.283768	57.784
0.015	0.0496411	0.311905	0.302169	0.316986	1.04904	1.30678	0.953254	0.623809	0.665061	1.20096	0.968986	0.292797	54.0469

0.016	0.05133239	0.322478	0.311746	0.328096	1.05245	1.2879	0.950166	0.644955	0.690608	1.21529	0.966947	0.301442	50.7773
0.017	0.0529601	0.332758	0.320997	0.338933	1.05588	1.27055	0.94708	0.665516	0.715743	1.22975	0.964913	0.309734	47.8926
0.018	0.0545538	0.342772	0.32995	0.349523	1.05932	1.25451	0.943999	0.685543	0.740517	1.24433	0.962882	0.317702	45.3289
0.019	0.0561088	0.352542	0.338628	0.35989	1.06279	1.23964	0.94092	0.705084	0.764975	1.25904	0.960854	0.325372	43.0352
0.02	0.0576282	0.362089	0.347052	0.370053	1.06627	1.22579	0.937846	0.724177	0.789155	1.27388	0.958831	0.332765	40.9713
0.021	0.0591148	0.371429	0.355241	0.380029	1.06978	1.21286	0.934775	0.742858	0.813091	1.28884	0.956811	0.339899	39.1042
0.022	0.060571	0.380579	0.36321	0.389833	1.0733	1.20075	0.931707	0.761158	0.836814	1.30394	0.954795	0.346791	37.4071
0.023	0.061999	0.389551	0.370973	0.399479	1.07684	1.18937	0.928643	0.779103	0.860349	1.31917	0.952783	0.353457	35.8578
0.024	0.0634008	0.398359	0.378544	0.408979	1.0804	1.17866	0.925583	0.796718	0.883721	1.33453	0.950774	0.35991	34.4379
0.025	0.0647781	0.407013	0.385933	0.418344	1.08398	1.16855	0.922527	0.814025	0.906952	1.35002	0.94877	0.366162	33.1318
0.026	0.0661323	0.415522	0.393151	0.427583	1.08758	1.159	0.919474	0.831044	0.930059	1.36565	0.946769	0.372223	31.9265
0.027	0.0674651	0.423896	0.400207	0.436705	1.0912	1.14995	0.916425	0.847791	0.953062	1.38142	0.944772	0.378105	30.8106
0.028	0.0687775	0.432142	0.40711	0.445718	1.09484	1.14136	0.913379	0.864284	0.975976	1.39733	0.942779	0.383815	29.7747
0.029	0.0700709	0.440268	0.413867	0.45463	1.09849	1.1332	0.910338	0.880536	0.998816	1.41338	0.94079	0.389362	28.8104
0.03	0.0713462	0.448281	0.420485	0.463447	1.10217	1.12544	0.907299	0.896562	1.0216	1.42957	0.938805	0.394753	27.9106
0.031	0.0726044	0.456187	0.426971	0.472175	1.10587	1.11804	0.904265	0.912374	1.04433	1.4459	0.936823	0.399997	27.069
0.032	0.0738465	0.463991	0.433331	0.48082	1.10959	1.11098	0.901235	0.927982	1.06702	1.46238	0.934846	0.405098	26.2802
0.033	0.0750732	0.471699	0.439571	0.489387	1.11333	1.10423	0.898208	0.943398	1.0897	1.479	0.932872	0.410063	25.5395
0.034	0.0762854	0.479315	0.445695	0.49788	1.11709	1.09778	0.895185	0.95863	1.11235	1.49577	0.930902	0.414898	24.8424
0.035	0.0774837	0.486844	0.451708	0.506305	1.12087	1.0916	0.892166	0.973688	1.135	1.51269	0.928937	0.419608	24.1854
0.036	0.0786688	0.49429	0.457615	0.514665	1.12467	1.08568	0.88915	0.988581	1.15766	1.52976	0.926975	0.424198	23.5651
0.037	0.0798413	0.501657	0.46342	0.522965	1.12849	1.08	0.886139	1.00331	1.18032	1.54698	0.925017	0.428671	22.9785
0.038	0.0810017	0.508949	0.469126	0.531207	1.13233	1.07454	0.883131	1.0179	1.20301	1.56436	0.923063	0.433033	22.4229
0.039	0.0821507	0.516168	0.474737	0.539396	1.1362	1.06931	0.880128	1.03234	1.22572	1.5819	0.921114	0.437287	21.8959
0.04	0.0832887	0.523318	0.480257	0.547534	1.14008	1.06427	0.877128	1.04664	1.24847	1.59959	0.919168	0.441437	21.3955
0.041	0.0844162	0.530403	0.485689	0.555624	1.14399	1.05942	0.874132	1.06081	1.27126	1.61744	0.917226	0.445486	20.9196
0.042	0.0855336	0.537423	0.491035	0.56367	1.14792	1.05475	0.87114	1.07485	1.2941	1.63545	0.915288	0.449439	20.4666
0.043	0.0866413	0.544384	0.496299	0.571673	1.15187	1.05025	0.868152	1.08877	1.31699	1.65362	0.913355	0.453297	20.0347
0.044	0.0877398	0.551285	0.501483	0.579637	1.15585	1.04592	0.865167	1.10257	1.33994	1.67196	0.911425	0.457064	19.6226
0.045	0.0888293	0.558131	0.506589	0.587563	1.15984	1.04173	0.862187	1.11626	1.36296	1.69046	0.9095	0.460743	19.229
0.046	0.0899103	0.564923	0.511621	0.595454	1.16386	1.03769	0.859211	1.12985	1.38605	1.70913	0.907578	0.464336	18.8527

(Continued)

(Continued)

d/L_o	d/L	$2\pi d/L$	$\tanh$ $(2\pi d/L)$	$\sinh$ $(2\pi d/L)$	$\cosh$ $(2\pi d/L)$	H/H_o'	K	$4\pi d/L$	$\sinh$ $(4\pi d/L)$	$\cosh$ $(4\pi d/L)$	n	C_g/C_o	M
0.047	0.090983	0.571663	0.51658	0.603313	1.1679	1.03379	0.856239	1.14333	1.40922	1.72797	0.905661	0.467846	18.4925
0.048	0.0920479	0.578354	0.521468	0.61114	1.17196	1.03002	0.853271	1.15671	1.43246	1.74698	0.903748	0.471275	18.1474
0.049	0.093105	0.584996	0.526287	0.618938	1.17605	1.02638	0.850307	1.16999	1.4558	1.76617	0.901838	0.474626	17.8166
0.05	0.0941549	0.591592	0.53104	0.626709	1.18015	1.02286	0.847347	1.18318	1.47923	1.78553	0.899934	0.477901	17.4991
0.051	0.0951976	0.598144	0.535728	0.634455	1.18429	1.01945	0.844391	1.19629	1.50275	1.80507	0.898033	0.481101	17.1942
0.052	0.0962335	0.604653	0.540352	0.642177	1.18844	1.01615	0.841439	1.20931	1.52638	1.82478	0.896136	0.484229	16.9011
0.053	0.0972628	0.61112	0.544915	0.649876	1.19262	1.01296	0.838491	1.22224	1.55011	1.84468	0.894244	0.487287	16.6193
0.054	0.0982858	0.617548	0.549418	0.657555	1.19682	1.00987	0.835547	1.2351	1.57395	1.86476	0.892355	0.490276	16.348
0.055	0.0993026	0.623936	0.553863	0.665214	1.20105	1.00687	0.832608	1.24787	1.59791	1.88502	0.890471	0.493199	16.0866
0.056	0.100313	0.630288	0.55825	0.672856	1.20529	1.00397	0.829673	1.26058	1.62198	1.90547	0.888592	0.496057	15.8348
0.057	0.101319	0.636603	0.562582	0.680482	1.20957	1.00115	0.826741	1.27321	1.64618	1.92611	0.886716	0.498851	15.5919
0.058	0.102318	0.642884	0.56686	0.688092	1.21387	0.998421	0.823814	1.28577	1.6705	1.94694	0.884845	0.501583	15.3574
0.059	0.103312	0.64913	0.571084	0.695688	1.21819	0.995772	0.820892	1.29826	1.69496	1.96796	0.882978	0.504254	15.1311
0.06	0.104301	0.655344	0.575257	0.703271	1.22253	0.993203	0.817973	1.31069	1.71955	1.98918	0.881115	0.506867	14.9124
0.061	0.105285	0.661527	0.579378	0.710843	1.22691	0.990709	0.815059	1.32305	1.74427	2.01059	0.879256	0.509422	14.7009
0.062	0.106264	0.667678	0.583451	0.718404	1.2313	0.988288	0.812149	1.33536	1.76914	2.03221	0.877402	0.511921	14.4964
0.063	0.107239	0.673801	0.587474	0.725956	1.23572	0.985938	0.809243	1.3476	1.79416	2.05402	0.875552	0.514364	14.2985
0.064	0.108209	0.679894	0.591451	0.733499	1.24017	0.983655	0.806341	1.35979	1.81933	2.07604	0.873706	0.516754	14.1069
0.065	0.109174	0.68596	0.59538	0.741035	1.24464	0.981439	0.803444	1.37192	1.84465	2.09827	0.871865	0.519091	13.9213
0.066	0.110135	0.691999	0.599265	0.748565	1.24914	0.979285	0.800551	1.384	1.87012	2.1207	0.870028	0.521377	13.7415
0.067	0.111092	0.698011	0.603104	0.756089	1.25366	0.977192	0.797663	1.39602	1.89576	2.14334	0.868196	0.523612	13.567
0.068	0.112045	0.703999	0.6069	0.763609	1.25821	0.975159	0.794778	1.408	1.92156	2.1662	0.866368	0.525798	13.3979
0.069	0.112994	0.709961	0.610653	0.771125	1.26279	0.973183	0.791899	1.41992	1.94753	2.18927	0.864544	0.527936	13.2337
0.07	0.113939	0.7159	0.614363	0.778638	1.26739	0.971262	0.789023	1.4318	1.97368	2.21255	0.862724	0.530026	13.0743
0.071	0.114881	0.721816	0.618033	0.786149	1.27202	0.969394	0.786152	1.44363	1.99999	2.23606	0.860909	0.53207	12.9195
0.072	0.115819	0.727709	0.621662	0.793659	1.27667	0.967579	0.783286	1.45542	2.02649	2.25979	0.859099	0.534069	12.7691
0.073	0.116753	0.733581	0.625251	0.801169	1.28136	0.965813	0.780423	1.46716	2.05317	2.28374	0.857293	0.536023	12.6229
0.074	0.117684	0.739432	0.628802	0.80868	1.28606	0.964096	0.777566	1.47886	2.08003	2.30793	0.855491	0.537934	12.4808

0.075	0.118612	0.745261	0.632313	0.816191	1.2908	0.962427	0.774713	1.49052	2.10708	2.33234	0.853694	0.539802	12.3425
0.076	0.119537	0.751072	0.635788	0.823704	1.29557	0.960803	0.771864	1.50214	2.13433	2.35698	0.851901	0.541628	12.208
0.077	0.120458	0.756862	0.639225	0.831221	1.30036	0.959224	0.76902	1.51372	2.16177	2.38186	0.850113	0.543413	12.0771
0.078	0.121377	0.762634	0.642626	0.83874	1.30518	0.957687	0.76618	1.52527	2.18941	2.40697	0.848329	0.545158	11.9496
0.079	0.122293	0.768388	0.645991	0.846264	1.31002	0.956193	0.763345	1.53678	2.21725	2.43232	0.84655	0.546864	11.8254
0.08	0.123206	0.774124	0.649321	0.853792	1.3149	0.954739	0.760514	1.54825	2.2453	2.45792	0.844775	0.54853	11.7044
0.081	0.124116	0.779843	0.652616	0.861325	1.3198	0.953325	0.757689	1.55969	2.27356	2.48376	0.843005	0.550159	11.5865
0.082	0.125023	0.785545	0.655878	0.868865	1.32474	0.951949	0.754867	1.57109	2.30203	2.50985	0.841239	0.55175	11.4716
0.083	0.125928	0.79123	0.659106	0.876411	1.3297	0.950611	0.75205	1.58246	2.33072	2.53619	0.839478	0.553305	11.3595
0.084	0.126831	0.796901	0.6623	0.883965	1.33469	0.949309	0.749238	1.5938	2.35964	2.56279	0.837722	0.554824	11.2502
0.085	0.127731	0.802555	0.665463	0.891526	1.33971	0.948042	0.746431	1.60511	2.38877	2.58964	0.83597	0.556307	11.1435
0.086	0.128628	0.808195	0.668593	0.899096	1.34476	0.94681	0.743628	1.61639	2.41813	2.61675	0.834223	0.557756	11.0394
0.087	0.129524	0.813821	0.671692	0.906675	1.34984	0.945612	0.74083	1.62764	2.44773	2.64412	0.83248	0.55917	10.9378
0.088	0.130417	0.819432	0.674761	0.914264	1.35495	0.944446	0.738037	1.63886	2.47756	2.67176	0.830742	0.560552	10.8385
0.089	0.131308	0.82503	0.677798	0.921863	1.36009	0.943312	0.735248	1.65006	2.50762	2.69966	0.829008	0.5619	10.7416
0.09	0.132196	0.830614	0.680806	0.929473	1.36525	0.942209	0.732464	1.66123	2.53793	2.72784	0.82728	0.563217	10.6469
0.091	0.133083	0.836186	0.683783	0.937094	1.37045	0.941136	0.729685	1.67237	2.56849	2.75629	0.825555	0.564501	10.5544
0.092	0.133968	0.841745	0.686732	0.944727	1.37568	0.940093	0.726911	1.68349	2.59929	2.78502	0.823836	0.565755	10.4639
0.093	0.134851	0.847292	0.689652	0.952372	1.38095	0.939079	0.724141	1.69458	2.63035	2.81403	0.822121	0.566977	10.3755
0.094	0.135732	0.852827	0.692543	0.960031	1.38624	0.938093	0.721376	1.70565	2.66166	2.84332	0.820411	0.56817	10.2891
0.095	0.136611	0.85835	0.695407	0.967702	1.39156	0.937134	0.718616	1.7167	2.69324	2.8729	0.818706	0.569333	10.2045
0.096	0.137488	0.863863	0.698243	0.975388	1.39692	0.936202	0.715861	1.72773	2.72508	2.90276	0.817005	0.570468	10.1218
0.097	0.138364	0.869365	0.701051	0.983088	1.40231	0.935296	0.713111	1.73873	2.75718	2.93293	0.815309	0.571573	10.0408
0.098	0.139238	0.874856	0.703833	0.990804	1.40773	0.934415	0.710366	1.74971	2.78956	2.96338	0.813618	0.572651	9.96164
0.099	0.14011	0.880337	0.706588	0.998534	1.41318	0.93356	0.707625	1.76067	2.82221	2.99414	0.811931	0.573701	9.8841
0.1	0.140981	0.885808	0.709317	1.00628	1.41866	0.932729	0.70489	1.77162	2.85514	3.0252	0.81025	0.574724	9.8082
0.101	0.14185	0.891269	0.71202	1.01404	1.42418	0.931921	0.702159	1.78254	2.88836	3.05657	0.808573	0.57572	9.73386
0.102	0.142718	0.896722	0.714698	1.02182	1.42973	0.931137	0.699433	1.79344	2.92186	3.08825	0.806901	0.57669	9.66106
0.103	0.143584	0.902165	0.71735	1.02962	1.43531	0.930376	0.696713	1.80433	2.95566	3.12024	0.805233	0.577634	9.58975
0.104	0.144449	0.907599	0.719978	1.03744	1.44093	0.929637	0.693997	1.8152	2.98974	3.15255	0.803571	0.578553	9.51988
0.105	0.145312	0.913025	0.722581	1.04527	1.44658	0.928919	0.691286	1.82605	3.02413	3.18518	0.801913	0.579447	9.45141
0.106	0.146175	0.918443	0.72516	1.05312	1.45226	0.928223	0.688581	1.83689	3.05882	3.21813	0.80026	0.580317	9.38431

(*Continued*)

(Continued)

d/L_o	d/L	$2\pi d/L$	tanh $(2\pi d/L)$	sinh $(2\pi d/L)$	cosh $(2\pi d/L)$	H/H'_o	K	$4\pi d/L$	sinh $(4\pi d/L)$	cosh $(4\pi d/L)$	n	C_g/C_o	M
0.107	0.147036	0.923852	0.727715	1.06099	1.45798	0.927548	0.68588	1.8477	3.09382	3.25142	0.798612	0.581162	9.31853
0.108	0.147895	0.929254	0.730246	1.06889	1.46373	0.926893	0.683184	1.85851	3.12913	3.28503	0.796969	0.581983	9.25404
0.109	0.148754	0.934648	0.732754	1.0768	1.46952	0.926258	0.680494	1.8693	3.16475	3.31898	0.795331	0.582782	9.19081
0.11	0.149611	0.940036	0.735239	1.08473	1.47534	0.925643	0.677808	1.88007	3.2007	3.35328	0.793697	0.583557	9.12879
0.111	0.150468	0.945416	0.737701	1.09268	1.4812	0.925046	0.675128	1.89083	3.23696	3.38791	0.792069	0.58431	9.06796
0.112	0.151323	0.950789	0.74014	1.10066	1.48709	0.924469	0.672453	1.90158	3.27356	3.42289	0.790445	0.58504	9.00828
0.113	0.152177	0.956155	0.742557	1.10865	1.49302	0.923909	0.669783	1.91231	3.31049	3.45823	0.788826	0.585748	8.94973
0.114	0.15303	0.961516	0.744952	1.11667	1.49899	0.923368	0.667118	1.92303	3.34775	3.49391	0.787212	0.586436	8.89228
0.115	0.153882	0.96687	0.747325	1.12471	1.50499	0.922844	0.664458	1.93374	3.38536	3.52996	0.785603	0.587101	8.83589
0.116	0.154733	0.972218	0.749677	1.13278	1.51102	0.922338	0.661804	1.94444	3.42331	3.56638	0.783999	0.587747	8.78054
0.117	0.155584	0.97756	0.752008	1.14087	1.5171	0.921848	0.659154	1.95512	3.46161	3.60316	0.7824	0.588371	8.7262
0.118	0.156433	0.982897	0.754317	1.14898	1.52321	0.921375	0.65651	1.96579	3.50027	3.64031	0.780806	0.588976	8.67285
0.119	0.157281	0.988228	0.756606	1.15712	1.52935	0.920918	0.653871	1.97646	3.53928	3.67784	0.779217	0.58956	8.62046
0.12	0.158129	0.993554	0.758874	1.16528	1.53554	0.920477	0.651238	1.98711	3.57866	3.71575	0.777633	0.590125	8.56901
0.121	0.158976	0.998875	0.761121	1.17347	1.54176	0.920051	0.648609	1.99775	3.6184	3.75405	0.776054	0.590671	8.51847
0.122	0.159822	1.00419	0.763349	1.18168	1.54802	0.919641	0.645986	2.00838	3.65853	3.79273	0.7748	0.591198	8.46883
0.123	0.160667	1.0095	0.765557	1.18992	1.55432	0.919246	0.643369	2.01901	3.69903	3.83181	0.772911	0.591707	8.42006
0.124	0.161512	1.01481	0.767744	1.19818	1.56066	0.918866	0.640756	2.02962	3.73991	3.87129	0.771346	0.592197	8.37214
0.125	0.162356	1.02011	0.769913	1.20648	1.56703	0.918499	0.638149	2.04023	3.78118	3.91118	0.769787	0.592669	8.32505
0.126	0.163199	1.02541	0.772062	1.2148	1.57345	0.918148	0.635548	2.05082	3.82284	3.95147	0.768233	0.593123	8.27877
0.127	0.164042	1.03071	0.774192	1.22315	1.5799	0.917809	0.632951	2.06141	3.8649	3.99217	0.766684	0.59356	8.23328
0.128	0.164884	1.036	0.776303	1.23152	1.58639	0.917485	0.63036	2.072	3.90736	4.03329	0.76514	0.59398	8.18856
0.129	0.165726	1.04128	0.778395	1.23993	1.59293	0.917174	0.627775	2.08257	3.95023	4.07484	0.763601	0.594383	8.1446
0.13	0.166567	1.04657	0.780469	1.24836	1.5995	0.916876	0.625195	2.09314	3.99351	4.11681	0.762067	0.59477	8.10138
0.131	0.167407	1.05185	0.782524	1.25682	1.60612	0.916591	0.62262	2.1037	4.03721	4.15921	0.760539	0.59514	8.05887
0.132	0.168247	1.05713	0.784561	1.26532	1.61277	0.916318	0.620051	2.11425	4.08133	4.20206	0.759015	0.595494	8.01707
0.133	0.169086	1.0624	0.786581	1.27384	1.61947	0.916058	0.617487	2.1248	4.12589	4.24534	0.757496	0.595832	7.97596
0.134	0.169925	1.06767	0.788582	1.2824	1.6262	0.91581	0.614929	2.13534	4.17087	4.28908	0.755983	0.596155	7.93552
0.135	0.170764	1.07294	0.790566	1.29098	1.63298	0.915574	0.612377	2.14588	4.2163	4.33326	0.754474	0.596462	7.89575

0.136	0.171602	1.07821	0.792533	1.2996	1.6398	0.91535	0.609829	2.15641	4.26217	4.37791	0.752971	0.596754	7.85661
0.137	0.172439	1.08347	0.794482	1.30825	1.64667	0.915137	0.607288	2.16694	4.30849	4.42302	0.751473	0.597032	7.81811
0.138	0.173277	1.08873	0.796414	1.31693	1.65357	0.914936	0.604752	2.17746	4.35527	4.4686	0.74998	0.597295	7.78022
0.139	0.174114	1.09399	0.798329	1.32564	1.66052	0.914746	0.602221	2.18798	4.4025	4.51465	0.748492	0.597543	7.74294
0.14	0.17495	1.09924	0.800228	1.33439	1.66751	0.914566	0.599696	2.19849	4.45021	4.56118	0.74701	0.597778	7.70624
0.141	0.175787	1.1045	0.802109	1.34317	1.67454	0.914398	0.597177	2.209	4.49839	4.6082	0.745532	0.597998	7.67013
0.142	0.176622	1.10975	0.803975	1.35198	1.68162	0.914239	0.594664	2.2195	4.54705	4.65571	0.74406	0.598205	7.63458
0.143	0.177458	1.115	0.805824	1.36083	1.68875	0.914092	0.592156	2.23001	4.59619	4.70372	0.742593	0.598399	7.59958
0.144	0.178294	1.12025	0.807657	1.36971	1.69591	0.913954	0.589653	2.2405	4.64583	4.75223	0.741131	0.598579	7.56513
0.145	0.179129	1.1255	0.809473	1.37863	1.70312	0.913826	0.587157	2.251	4.69596	4.80126	0.739674	0.598746	7.53121
0.146	0.179964	1.13075	0.811274	1.38759	1.71038	0.913708	0.584666	2.26149	4.7466	4.85079	0.738222	0.598901	7.49781
0.147	0.180799	1.13599	0.81306	1.39658	1.71768	0.9136	0.58218	2.27198	4.79775	4.90085	0.736776	0.599043	7.46492
0.148	0.181633	1.14123	0.814829	1.4056	1.72503	0.913501	0.579701	2.28247	4.84941	4.95144	0.735335	0.599172	7.43253
0.149	0.182468	1.14648	0.816584	1.41467	1.73242	0.913412	0.577227	2.29295	4.90159	5.00256	0.733899	0.59929	7.40062
0.15	0.183302	1.15172	0.818323	1.42377	1.73986	0.913331	0.574759	2.30344	4.95431	5.05422	0.732468	0.599395	7.3692
0.151	0.184136	1.15696	0.820047	1.4329	1.74735	0.91326	0.572297	2.31392	5.00756	5.10643	0.731043	0.599489	7.33825
0.152	0.18497	1.1622	0.821756	1.44208	1.75488	0.913197	0.56984	2.3244	5.06135	5.15919	0.729622	0.599571	7.30776
0.153	0.185804	1.16744	0.82345	1.4513	1.76246	0.913143	0.567389	2.33488	5.11569	5.21251	0.728208	0.599642	7.27773
0.154	0.186638	1.17268	0.825129	1.46055	1.77009	0.913098	0.564945	2.34536	5.17059	5.2664	0.726798	0.599702	7.24813
0.155	0.187471	1.17792	0.826793	1.46984	1.77776	0.913061	0.562506	2.35583	5.22605	5.32086	0.725393	0.599751	7.21898
0.156	0.188305	1.18315	0.828444	1.47917	1.78548	0.913032	0.560072	2.36631	5.28208	5.3759	0.723994	0.599788	7.19025
0.157	0.189138	1.18839	0.83008	1.48854	1.79326	0.913011	0.557645	2.37678	5.33868	5.43153	0.7226	0.599816	7.16193
0.158	0.189972	1.19363	0.831701	1.49796	1.80108	0.912998	0.555223	2.38726	5.39587	5.48775	0.721212	0.599833	7.13403
0.159	0.190806	1.19887	0.833309	1.50741	1.80895	0.912993	0.552808	2.39773	5.45365	5.54458	0.719828	0.599839	7.10653
0.16	0.191639	1.2041	0.834902	1.51691	1.81687	0.912996	0.550398	2.40821	5.51203	5.60201	0.71845	0.599836	7.07943
0.161	0.192473	1.20934	0.836482	1.52644	1.82484	0.913006	0.547994	2.41868	5.57102	5.66005	0.717077	0.599822	7.05272
0.162	0.193306	1.21458	0.838048	1.53602	1.83286	0.913024	0.545597	2.42916	5.63061	5.71872	0.71571	0.599799	7.02639
0.163	0.19414	1.21982	0.8396	1.54564	1.84093	0.913048	0.543205	2.43964	5.69083	5.77802	0.714348	0.599767	7.00043
0.164	0.194974	1.22506	0.841139	1.55531	1.84905	0.91308	0.540819	2.45011	5.75168	5.83796	0.712991	0.599725	6.97484
0.165	0.195807	1.23029	0.842665	1.56502	1.85722	0.913119	0.538439	2.46059	5.81316	5.89854	0.71164	0.599673	6.94961
0.166	0.196641	1.23553	0.844177	1.57477	1.86545	0.913165	0.536065	2.47107	5.87529	5.95978	0.710293	0.599613	6.92473
0.167	0.197475	1.24077	0.845676	1.58456	1.87372	0.913218	0.533697	2.48155	5.93806	6.02168	0.708952	0.599544	6.9002
0.168	0.198309	1.24601	0.847162	1.5944	1.88205	0.913277	0.531335	2.49203	6.0015	6.08424	0.707617	0.599466	6.87602

(*Continued*)

(*Continued*)

d/L_o	d/L	$2\pi d/L$	tanh $(2\pi d/L)$	sinh $(2\pi d/L)$	cosh $(2\pi d/L)$	H/H'_o	K	$4\pi d/L$	sinh $(4\pi d/L)$	cosh $(4\pi d/L)$	n	C_g/C_o	M
0.169	0.199143	1.25125	0.848635	1.60429	1.89043	0.913343	0.528979	2.50251	6.06561	6.14749	0.706287	0.59938	6.85217
0.17	0.199978	1.2565	0.850095	1.61422	1.89887	0.913416	0.526629	2.51299	6.13039	6.21604	0.704962	0.599285	6.82865
0.171	0.200812	1.26174	0.851543	1.6242	1.90736	0.913494	0.524285	2.52348	6.19586	6.27604	0.703642	0.599182	6.80545
0.172	0.201647	1.26698	0.852978	1.63422	1.9159	0.913579	0.521947	2.53397	6.26202	6.34136	0.702328	0.59907	6.78257
0.173	0.202481	1.27223	0.8544	1.64429	1.9245	0.91367	0.519616	2.54445	6.32888	6.4074	0.701019	0.598951	6.76001
0.174	0.203316	1.27747	0.85581	1.65441	1.93315	0.913767	0.51729	2.55495	6.39645	6.47415	0.699716	0.598824	6.73775
0.175	0.204151	1.28272	0.857208	1.66458	1.94186	0.91387	0.514971	2.56544	6.46474	6.54163	0.698418	0.598689	6.7158
0.176	0.204986	1.28797	0.858593	1.67479	1.95062	0.913978	0.512657	2.57594	6.53376	6.60985	0.697125	0.598547	6.69414
0.177	0.205822	1.29322	0.859967	1.68505	1.95944	0.914093	0.51035	2.58643	6.60352	6.67881	0.695838	0.598397	6.67277
0.178	0.206658	1.29847	0.861328	1.69536	1.96831	0.914213	0.508049	2.59694	6.67402	6.74852	0.694556	0.59824	6.6517
0.179	0.207493	1.30372	0.862678	1.70573	1.97725	0.914338	0.505754	2.60744	6.74528	6.819	0.693279	0.598076	6.6309
0.18	0.20833	1.30897	0.864015	1.71614	1.98623	0.914469	0.503465	2.61795	6.8173	6.89025	0.692008	0.597905	6.61038
0.181	0.209166	1.31423	0.865341	1.7266	1.99528	0.914605	0.501183	2.62846	6.8901	6.96229	0.690742	0.597727	6.59014
0.182	0.210003	1.31949	0.866656	1.73711	2.00438	0.914746	0.498906	2.63897	6.96368	7.03511	0.689481	0.597543	6.57017
0.183	0.21084	1.32474	0.867959	1.74768	2.01355	0.914892	0.496636	2.64949	7.03805	7.10874	0.688226	0.597352	6.55046
0.184	0.211677	1.33	0.86925	1.75829	2.02277	0.915044	0.494372	2.66	7.11322	7.18317	0.686976	0.597154	6.53101
0.185	0.212514	1.33527	0.87053	1.76896	2.03205	0.9152	0.492115	2.67053	7.18921	7.25843	0.685732	0.59695	6.51181
0.186	0.213352	1.34053	0.871799	1.77968	2.04139	0.915361	0.489863	2.68106	7.26603	7.33452	0.684493	0.59674	6.49287
0.187	0.21419	1.34579	0.873057	1.79045	2.05079	0.915527	0.487618	2.69159	7.34367	7.41144	0.683259	0.596524	6.47418
0.188	0.215028	1.35106	0.874304	1.80128	2.06025	0.915697	0.485379	2.70212	7.42216	7.48922	0.682031	0.596302	6.45573
0.189	0.215867	1.35633	0.87554	1.81216	2.06977	0.915872	0.483147	2.71266	7.5015	7.56786	0.680808	0.596074	6.43751
0.19	0.216706	1.3616	0.876764	1.8231	2.07935	0.916051	0.48092	2.72321	7.58171	7.64737	0.679591	0.595841	6.41954
0.191	0.217545	1.36688	0.877978	1.83409	2.08899	0.916235	0.4787	2.73375	7.66279	7.72776	0.678379	0.595602	6.4018
0.192	0.218385	1.37215	0.879182	1.84514	2.0987	0.916423	0.476486	2.74431	7.74475	7.80905	0.677172	0.595357	6.38429
0.193	0.219225	1.37743	0.880375	1.85624	2.10846	0.916616	0.474279	2.75486	7.82762	7.89124	0.675971	0.595107	6.367
0.194	0.220065	1.38271	0.881557	1.8674	2.11829	0.916812	0.472078	2.76542	7.91139	7.97434	0.674775	0.594852	6.34993
0.195	0.220906	1.38799	0.882729	1.87861	2.12819	0.917013	0.469883	2.77599	7.99608	8.05837	0.673584	0.594592	6.33309
0.196	0.221747	1.39328	0.88389	1.88989	2.13815	0.917218	0.467695	2.78656	8.0817	8.14334	0.672399	0.594327	6.31645

0.197	0.222589	1.39857	0.885041	1.90122	2.14817	0.917426	0.465513	2.79713	8.16827	8.22925	0.671219	0.594057	6.30003
0.198	0.22343	1.40385	0.886182	1.91261	2.15825	0.917638	0.463337	2.80771	8.25578	8.31613	0.670045	0.593782	6.28382
0.199	0.224273	1.40915	0.887313	1.92405	2.16841	0.917854	0.461168	2.81829	8.34427	8.40398	0.668876	0.593502	6.26782
0.2	0.225115	1.41444	0.888434	1.93556	2.17862	0.918074	0.459005	2.82888	8.43373	8.49281	0.667712	0.593218	6.25201
0.201	0.225958	1.41974	0.889544	1.94713	2.18891	0.918298	0.456849	2.83948	8.52418	8.58263	0.666554	0.59293	6.23641
0.202	0.226802	1.42504	0.890645	1.95876	2.19926	0.918525	0.454699	2.85008	8.61563	8.67347	0.665401	0.592637	6.221
0.203	0.227646	1.43034	0.891736	1.97045	2.20967	0.918755	0.452555	2.86068	8.7081	8.76533	0.664254	0.59234	6.20579
0.204	0.22849	1.43565	0.892818	1.9822	2.22016	0.918989	0.450418	2.87129	8.80159	8.85822	0.663112	0.592038	6.19076
0.205	0.229335	1.44095	0.893889	1.99401	2.23071	0.919226	0.448287	2.88191	8.89612	8.95215	0.661975	0.591733	6.17593
0.206	0.23018	1.44626	0.894952	2.00588	2.24133	0.919466	0.446163	2.89253	8.99171	9.04714	0.660844	0.591424	6.16128
0.207	0.231026	1.45158	0.896004	2.01782	2.25202	0.91971	0.444045	2.90315	9.08836	9.14321	0.659718	0.59111	6.14681
0.208	0.231872	1.45689	0.897048	2.02982	2.26278	0.919956	0.441934	2.91379	9.18609	9.24036	0.658598	0.590794	6.13252
0.209	0.232718	1.46221	0.898081	2.04189	2.27361	0.920206	0.439829	2.92442	9.28491	9.3386	0.657483	0.590473	6.11841
0.21	0.233565	1.46753	0.899106	2.05402	2.28451	0.920458	0.437731	2.93507	9.38483	9.43796	0.656373	0.590149	6.10447
0.211	0.234413	1.47286	0.900121	2.06621	2.29548	0.920714	0.435639	2.94572	9.48588	9.53844	0.655269	0.589821	6.0907
0.212	0.235261	1.47819	0.901128	2.07847	2.30652	0.920972	0.433554	2.95637	9.58806	9.64007	0.65417	0.58949	6.07711
0.213	0.236109	1.48352	0.902125	2.09079	2.31763	0.921234	0.431475	2.96704	9.69139	9.74284	0.653076	0.589156	6.06368
0.214	0.236958	1.48885	0.903113	2.10319	2.32882	0.921498	0.429402	2.9777	9.79587	9.84678	0.651988	0.588819	6.05042
0.215	0.237808	1.49419	0.904093	2.11564	2.34008	0.921764	0.427337	2.98838	9.90154	9.95191	0.650905	0.588478	6.03732
0.216	0.238657	1.49953	0.905063	2.12817	2.35141	0.922033	0.425277	2.99906	10.0084	10.0582	0.649827	0.588135	6.02438
0.217	0.239508	1.50487	0.906025	2.14077	2.36281	0.922305	0.423225	3.00974	10.1165	10.1658	0.648755	0.587788	6.01159
0.218	0.240359	1.51022	0.906978	2.15343	2.37429	0.922579	0.421178	3.02044	10.2257	10.2745	0.647688	0.587439	5.99897
0.219	0.24121	1.51557	0.907922	2.16616	2.38584	0.922856	0.419139	3.03114	10.3362	10.3845	0.646627	0.587087	5.98649
0.22	0.242062	1.52092	0.908858	2.17896	2.39747	0.923135	0.417106	3.04184	10.448	10.4958	0.64557	0.586732	5.97417
0.221	0.242914	1.52628	0.909785	2.19184	2.40918	0.923416	0.415079	3.05255	10.561	10.6083	0.644519	0.586374	5.962
0.222	0.243767	1.53164	0.910704	2.20478	2.42096	0.9237	0.413059	3.06327	10.6754	10.7221	0.643474	0.586014	5.94998
0.223	0.244621	1.537	0.911615	2.21779	2.43282	0.923986	0.411046	3.074	10.791	10.8372	0.642434	0.585652	5.9381
0.224	0.245475	1.54236	0.912517	2.23088	2.44475	0.924273	0.409039	3.08473	10.9079	10.9536	0.641399	0.585287	5.92636
0.225	0.246329	1.54773	0.913411	2.24404	2.45677	0.924563	0.407039	3.09547	11.0262	11.0714	0.640369	0.58492	5.91477
0.226	0.247185	1.55311	0.914297	2.25727	2.46886	0.924855	0.405045	3.10621	11.1458	11.1905	0.639345	0.584551	5.90331
0.227	0.24804	1.55848	0.915174	2.27058	2.48103	0.925149	0.403058	3.11697	11.2667	11.311	0.638326	0.58418	5.89199
0.228	0.248896	1.56386	0.916044	2.28395	2.49328	0.925445	0.401078	3.12772	11.3891	11.4329	0.637312	0.583806	5.88081

(*Continued*)

(*Continued*)

d/L_o	d/L	$2\pi d/L$	tanh $(2\pi d/L)$	sinh $(2\pi d/L)$	cosh $(2\pi d/L)$	H/H'_o	K	$4\pi d/L$	sinh $(4\pi d/L)$	cosh $(4\pi d/L)$	n	C_g/C_o	M
0.229	0.249753	1.56924	0.916906	2.29741	2.50561	0.925743	0.399104	3.13849	11.5128	11.5562	0.636304	0.583431	5.86976
0.23	0.25061	1.57463	0.917759	2.31094	2.51802	0.926042	0.397137	3.14926	11.638	11.6809	0.635301	0.583053	5.85885
0.231	0.251468	1.58002	0.918605	2.32454	2.53052	0.926344	0.395176	3.16004	11.7646	11.807	0.634303	0.582674	5.84806
0.232	0.252327	1.58541	0.919443	2.33823	2.54309	0.926647	0.393223	3.17083	11.8926	11.9346	0.633311	0.582293	5.83741
0.233	0.253186	1.59081	0.920274	2.35199	2.55575	0.926951	0.391275	3.18162	12.0222	12.0637	0.632323	0.581911	5.82688
0.234	0.254045	1.59621	0.921096	2.36582	2.56849	0.927257	0.389335	3.19242	12.1532	12.1942	0.631341	0.581526	5.81647
0.235	0.254905	1.60162	0.921911	2.37974	2.58131	0.927565	0.387401	3.20323	12.2857	12.3263	0.630365	0.58114	5.80619
0.236	0.255766	1.60702	0.922719	2.39373	2.59421	0.927875	0.385473	3.21405	12.4197	12.4599	0.629393	0.580753	5.79603
0.237	0.256627	1.61244	0.923519	2.4078	2.60721	0.928185	0.383552	3.22487	12.5553	12.595	0.628427	0.580364	5.78599
0.238	0.257489	1.61785	0.924312	2.42196	2.62028	0.928497	0.381638	3.2357	12.6924	12.7318	0.627466	0.579974	5.77607
0.239	0.258351	1.62327	0.925097	2.43619	2.63344	0.928811	0.379731	3.24654	12.8311	12.8701	0.62651	0.579583	5.76627
0.24	0.259214	1.62869	0.925875	2.45051	2.64669	0.929126	0.37783	3.25738	12.9715	13.01	0.625559	0.57919	5.75659
0.241	0.260078	1.63412	0.926646	2.4649	2.66003	0.929442	0.375936	3.26823	13.1134	13.1515	0.624614	0.578796	5.74701
0.242	0.260942	1.63955	0.927409	2.47938	2.67345	0.929759	0.374048	3.27909	13.257	13.2947	0.623674	0.578401	5.73756
0.243	0.261807	1.64498	0.928166	2.49395	2.68696	0.930078	0.372167	3.28996	13.4023	13.4395	0.622739	0.578005	5.72821
0.244	0.262672	1.65042	0.928915	2.50859	2.70056	0.930397	0.370293	3.30083	13.5492	13.5861	0.621809	0.577608	5.71897
0.245	0.263538	1.65586	0.929657	2.52332	2.71425	0.930718	0.368426	3.31172	13.6979	13.7343	0.620884	0.57721	5.70984
0.246	0.264405	1.6613	0.930392	2.53814	2.72803	0.93104	0.366565	3.32261	13.8482	13.8843	0.619965	0.576811	5.70082
0.247	0.265272	1.66675	0.931121	2.55304	2.7419	0.931363	0.364711	3.3335	14.0003	14.036	0.619051	0.576411	5.69191
0.248	0.266139	1.6722	0.931842	2.56803	2.75586	0.931687	0.362863	3.34441	14.1542	14.1895	0.618142	0.57601	5.68309
0.249	0.267008	1.67766	0.932557	2.5831	2.76991	0.932011	0.361022	3.35532	14.3099	14.3448	0.617238	0.575609	5.67439
0.25	0.267877	1.68312	0.933265	2.59826	2.78406	0.932337	0.359188	3.36624	14.4674	14.5019	0.616339	0.575207	5.66578
0.251	0.268746	1.68858	0.933966	2.61351	2.79829	0.932663	0.357361	3.37717	14.6267	14.6609	0.615445	0.574805	5.65727
0.252	0.269616	1.69405	0.934661	2.62885	2.81262	0.93299	0.35554	3.3881	14.7879	14.8217	0.614556	0.574402	5.64887
0.253	0.270487	1.69952	0.935349	2.64428	2.82705	0.933318	0.353726	3.39904	14.951	14.9844	0.613673	0.573998	5.64056
0.254	0.271359	1.705	0.936031	2.6598	2.84157	0.933647	0.351918	3.40999	15.116	15.149	0.612794	0.573594	5.63235
0.255	0.272231	1.71048	0.936706	2.67541	2.85619	0.933976	0.350117	3.42095	15.2829	15.3156	0.611921	0.57319	5.62423
0.256	0.273103	1.71596	0.937374	2.69111	2.8709	0.934306	0.348323	3.43192	15.4518	15.4841	0.611053	0.572785	5.61621

0.257	0.273976	1.72144	0.938037	2.7069	2.88571	0.934637	0.346536	3.44289	15.6226	15.6546	0.610189	0.57238	5.60828
0.258	0.27485	1.72694	0.938693	2.72278	2.90061	0.934968	0.344755	3.45387	15.7955	15.8271	0.609331	0.571975	5.60045
0.259	0.275725	1.73243	0.939342	2.73876	2.91562	0.935299	0.342981	3.46486	15.9704	16.0016	0.608478	0.571569	5.5927
0.26	0.2766	1.73793	0.939986	2.75483	2.93072	0.935631	0.341213	3.47586	16.1473	16.1782	0.60763	0.571163	5.58505
0.261	0.277476	1.74343	0.940623	2.771	2.94592	0.935964	0.339452	3.48686	16.3263	16.3569	0.606787	0.570758	5.57748
0.262	0.278352	1.74894	0.941254	2.78726	2.96122	0.936297	0.337698	3.49787	16.5074	16.5377	0.605949	0.570352	5.57001
0.263	0.279229	1.75445	0.94188	2.80362	2.97663	0.93663	0.335951	3.50889	16.6907	16.7206	0.605115	0.569946	5.56262
0.264	0.280106	1.75996	0.942499	2.82008	2.99213	0.936964	0.33421	3.51992	16.8761	16.9057	0.604287	0.56954	5.55531
0.265	0.280985	1.76548	0.943112	2.83663	3.00774	0.937298	0.332476	3.53096	17.0637	17.093	0.603464	0.569134	5.54809
0.266	0.281864	1.771	0.943719	2.85329	3.02345	0.937632	0.330748	3.542	17.2535	17.2825	0.602646	0.568728	5.54095
0.267	0.282743	1.77653	0.94432	2.87004	3.03926	0.937967	0.329027	3.55305	17.4456	17.4742	0.601833	0.568323	5.5339
0.268	0.283623	1.78206	0.944916	2.88689	3.05518	0.938302	0.327313	3.56411	17.6399	17.6682	0.601024	0.567917	5.52692
0.269	0.284504	1.78759	0.945506	2.90384	3.0712	0.938636	0.325606	3.57518	17.8365	17.8645	0.600221	0.567512	5.52003
0.27	0.285385	1.79313	0.94609	2.92089	3.08733	0.938972	0.323905	3.58626	18.0355	18.0632	0.599422	0.567107	5.51322
0.271	0.286267	1.79867	0.946668	2.93804	3.10356	0.939307	0.32221	3.59734	18.2368	18.2642	0.598629	0.566703	5.50648
0.272	0.28715	1.80422	0.947241	2.9553	3.1199	0.939642	0.320523	3.60843	18.4405	18.4676	0.59784	0.566298	5.49983
0.273	0.288033	1.80976	0.947808	2.97266	3.13635	0.939977	0.318842	3.61953	18.6466	18.6734	0.597056	0.565894	5.49325
0.274	0.288917	1.81532	0.94837	2.99012	3.15291	0.940312	0.317167	3.63064	18.8552	18.8817	0.596277	0.565491	5.48674
0.275	0.289801	1.82088	0.948926	3.00769	3.16958	0.940648	0.3155	3.64175	19.0662	19.0924	0.595503	0.565088	5.48031
0.276	0.290687	1.82644	0.949476	3.02537	3.18635	0.940983	0.313839	3.65287	19.2798	19.3057	0.594733	0.564685	5.47396
0.277	0.291572	1.832	0.950022	3.04315	3.20324	0.941318	0.312184	3.66401	19.4958	19.5215	0.593969	0.564283	5.46768
0.278	0.292459	1.83757	0.950562	3.06103	3.22024	0.941653	0.310536	3.67514	19.7145	19.7399	0.593209	0.563882	5.46147
0.279	0.293346	1.84315	0.951096	3.07903	3.23735	0.941988	0.308895	3.68629	19.9358	19.9608	0.592454	0.563481	5.45533
0.28	0.294233	1.84872	0.951626	3.09713	3.25457	0.942323	0.30726	3.69745	20.1597	20.1845	0.591704	0.563081	5.44926
0.281	0.295122	1.8543	0.95215	3.11535	3.27191	0.942657	0.305632	3.70861	20.3862	20.4108	0.590959	0.562681	5.44326
0.282	0.296011	1.85989	0.952669	3.13367	3.28936	0.942992	0.304011	3.71978	20.6155	20.6398	0.590218	0.562282	5.43734
0.283	0.2969	1.86548	0.953182	3.1521	3.30692	0.943326	0.302396	3.73096	20.8475	20.8715	0.589482	0.561884	5.43147
0.284	0.29779	1.87107	0.953691	3.17065	3.32461	0.94366	0.300787	3.74214	21.0823	21.106	0.588751	0.561486	5.42568
0.285	0.298681	1.87667	0.954195	3.18931	3.34241	0.943993	0.299186	3.75334	21.3199	21.3433	0.588024	0.56109	5.41995
0.286	0.299573	1.88227	0.954694	3.20808	3.36032	0.944326	0.297591	3.76454	21.5603	21.5835	0.587302	0.560694	5.41429
0.287	0.300465	1.88787	0.955187	3.22696	3.37836	0.944659	0.296002	3.77575	21.8036	21.8266	0.586585	0.560299	5.4087
0.288	0.301357	1.89348	0.955676	3.24596	3.39651	0.944992	0.29442	3.78697	22.0499	22.0725	0.585873	0.559905	5.40317
0.289	0.302251	1.8991	0.95616	3.26508	3.41478	0.945324	0.292845	3.79819	22.2991	22.3215	0.585165	0.559511	5.3977

(*Continued*)

(Continued)

d/L_o	d/L	$2\pi d/L$	$\tanh(2\pi d/L)$	$\sinh(2\pi d/L)$	$\cosh(2\pi d/L)$	H/H'_o	K	$4\pi d/L$	$\sinh(4\pi d/L)$	$\cosh(4\pi d/L)$	n	C_g/C_o	M
0.29	0.303145	1.90471	0.956639	3.28431	3.43317	0.945655	0.291276	3.80943	22.5512	22.5734	0.584462	0.559119	5.39229
0.291	0.304039	1.91033	0.957114	3.30366	3.45169	0.945987	0.289713	3.82067	22.8064	22.8283	0.583763	0.558728	5.38695
0.292	0.304934	1.91596	0.957583	3.32313	3.47033	0.946317	0.288157	3.83192	23.0647	23.0863	0.583069	0.558337	5.38167
0.293	0.30583	1.92159	0.958048	3.34271	3.48909	0.946648	0.286608	3.84318	23.326	23.3474	0.58238	0.557948	5.37645
0.294	0.306727	1.92722	0.958508	3.36242	3.50797	0.946977	0.285065	3.85444	23.5905	23.6117	0.581695	0.557559	5.37128
0.295	0.307624	1.93286	0.958964	3.38224	3.52698	0.947306	0.283529	3.86571	23.8582	23.8791	0.581014	0.557172	5.36618
0.296	0.308521	1.9385	0.959415	3.40219	3.54611	0.947635	0.281999	3.87699	24.1291	24.1498	0.580339	0.556785	5.36114
0.297	0.30942	1.94414	0.959861	3.42226	3.56537	0.94796	0.280476	3.88828	24.4033	24.4237	0.579667	0.5564	5.35615
0.298	0.310319	1.94979	0.960303	3.44245	3.58476	0.947963	0.278959	3.89958	24.6807	24.701	0.579001	0.556016	5.35122
0.299	0.311218	1.95544	0.960741	3.46277	3.60427	0.948617	0.277449	3.91088	24.9615	24.9816	0.578338	0.555633	5.34635
0.3	0.312118	1.9611	0.961174	3.48321	3.62392	0.948943	0.275945	3.9222	25.2457	25.2655	0.57768	0.555251	5.34154
0.301	0.313019	1.96676	0.961602	3.50378	3.64369	0.949269	0.274447	3.93352	25.5334	25.5529	0.577027	0.55487	5.33678
0.302	0.313921	1.97242	0.962026	3.52447	3.66359	0.949594	0.272956	3.94484	25.8245	25.8438	0.576378	0.554491	5.33207
0.303	0.314823	1.97809	0.962446	3.54529	3.68363	0.949918	0.271472	3.95618	26.1191	26.1382	0.575734	0.554113	5.32742
0.304	0.315725	1.98376	0.962862	3.56624	3.70379	0.950241	0.269994	3.96752	26.4172	26.4362	0.575093	0.553736	5.32282
0.305	0.316629	1.98944	0.963274	3.58732	3.72409	0.950564	0.268522	3.97887	26.719	26.7377	0.574458	0.55336	5.31827
0.306	0.317533	1.99512	0.963681	3.60853	3.74453	0.950885	0.267057	3.99023	27.0244	27.0429	0.573826	0.552985	5.31378
0.307	0.318437	2.0008	0.964084	3.62987	3.76509	0.951206	0.265598	4.0016	27.3336	27.3519	0.573199	0.552612	5.30933
0.308	0.319342	2.00649	0.964483	3.65134	3.7858	0.951527	0.264145	4.01297	27.6464	27.6645	0.572577	0.55224	5.30494
0.309	0.320248	2.01218	0.964878	3.67294	3.80664	0.951846	0.262699	4.02435	27.9631	27.981	0.571958	0.55187	5.3006
0.31	0.321154	2.01787	0.965269	3.69468	3.82762	0.952165	0.261259	4.03574	28.2836	28.3013	0.571344	0.551501	5.29631
0.311	0.322061	2.02357	0.965655	3.71655	3.84873	0.952482	0.259826	4.04714	28.608	28.6255	0.570734	0.551133	5.29207
0.312	0.322969	2.02927	0.966038	3.73855	3.86999	0.952799	0.258399	4.05854	28.9363	28.9536	0.570129	0.550766	5.28787
0.313	0.323877	2.03498	0.966417	3.7607	3.89138	0.953115	0.256978	4.06995	29.2686	29.2857	0.569528	0.550401	5.28373
0.314	0.324785	2.04069	0.966792	3.78298	3.91292	0.95343	0.255564	4.08137	29.605	29.6218	0.568931	0.550038	5.27963
0.315	0.325695	2.0464	0.967163	3.8054	3.9346	0.953744	0.254156	4.0928	29.9454	29.9621	0.568338	0.549675	5.27558
0.316	0.326605	2.05212	0.967531	3.82795	3.95642	0.954057	0.252754	4.10424	30.29	30.3065	0.567749	0.549315	5.27157
0.317	0.327515	2.05784	0.967894	3.85065	3.97838	0.954369	0.251359	4.11568	30.6387	30.655	0.567165	0.548955	5.26762
0.318	0.328426	2.06356	0.968254	3.87349	4.00049	0.954681	0.249969	4.12713	30.9917	31.0078	0.566584	0.548597	5.2637

0.319	0.329338	2.06929	0.96861	3.89647	4.02274	0.954991	0.248587	4.13858	31.349	31.3649	0.566008	0.548241	5.25984
0.32	0.33025	2.07502	0.968962	3.91959	4.04515	0.9553	0.24721	4.15005	31.7106	31.7264	0.565436	0.547886	5.25601
0.321	0.331163	2.08076	0.969311	3.94286	4.0677	0.955608	0.245839	4.16152	32.0767	32.0923	0.564868	0.547533	5.25223
0.322	0.332077	2.0865	0.969656	3.96656	4.09039	0.955916	0.244475	4.173	32.4472	32.4626	0.564304	0.547181	5.2485
0.323	0.332991	2.09224	0.969997	3.98983	4.11324	0.956222	0.243117	4.18449	32.8222	32.8375	0.563745	0.546831	5.2448
0.324	0.333905	2.09799	0.970335	4.01353	4.13624	0.956527	0.241766	4.19598	33.2019	33.2169	0.563189	0.546482	5.24115
0.325	0.334821	2.10374	0.970669	4.03739	4.15938	0.956831	0.24042	4.20748	33.5861	33.601	0.562637	0.546135	5.23754
0.326	0.335736	2.10949	0.971	4.06139	4.18269	0.957134	0.239081	4.21899	33.975	33.9897	0.56209	0.545789	5.23397
0.327	0.336653	2.11525	0.971327	4.08554	4.20614	0.957436	0.237748	4.2305	34.3687	34.3832	0.561546	0.545445	5.23045
0.328	0.33757	2.12101	0.971651	4.10984	4.22975	0.957737	0.236421	4.24203	34.7672	34.7816	0.561006	0.545102	5.22696
0.329	0.338487	2.12678	0.971971	4.13429	4.25351	0.958036	0.2351	4.25356	35.1705	35.1848	0.56047	0.544761	5.22352
0.33	0.339406	2.13255	0.972288	4.1589	4.27743	0.958335	0.233785	4.2651	35.5788	35.5929	0.559939	0.544422	5.22011
0.331	0.340324	2.13832	0.972602	4.18366	4.30151	0.958632	0.232476	4.27664	35.9921	36.006	0.559411	0.544084	5.21674
0.332	0.341243	2.1441	0.972912	4.20857	4.32575	0.958928	0.231174	4.28819	36.4105	36.4242	0.558887	0.543748	5.21341
0.333	0.342163	2.14988	0.97322	4.23364	4.35014	0.959224	0.229877	4.29975	36.8339	36.8475	0.558367	0.543413	5.21012
0.334	0.343084	2.15566	0.973523	4.25887	4.3747	0.959517	0.228587	4.31132	37.2626	37.276	0.55785	0.543081	5.20687
0.335	0.344005	2.16144	0.973824	4.28426	4.39942	0.95981	0.227303	4.32289	37.6965	37.7097	0.557338	0.542749	5.20366
0.336	0.344926	2.16723	0.974122	4.3098	4.4243	0.960102	0.226025	4.33447	38.1357	38.1488	0.55683	0.54242	5.20048
0.337	0.345848	2.17303	0.974416	4.33551	4.44934	0.960392	0.224752	4.34606	38.5803	38.5933	0.556325	0.542092	5.19734
0.338	0.346771	2.17883	0.974707	4.36137	4.47455	0.960681	0.223486	4.35765	39.0304	39.0432	0.555824	0.541765	5.19423
0.339	0.347694	2.18463	0.974995	4.3874	4.49992	0.960969	0.222226	4.36925	39.4859	39.4986	0.555327	0.541441	5.19116
0.34	0.348618	2.19043	0.97528	4.41359	4.52546	0.961256	0.220972	4.38086	39.9471	39.9596	0.554833	0.541118	5.18813
0.341	0.349542	2.19624	0.975562	4.43995	4.55117	0.961541	0.219724	4.39248	40.4139	40.4263	0.554344	0.540797	5.18513
0.342	0.350467	2.20205	0.975841	4.46647	4.57705	0.961826	0.218481	4.4041	40.8865	40.8987	0.553858	0.540477	5.18217
0.343	0.351392	2.20786	0.976117	4.49316	4.6031	0.962109	0.217245	4.41573	41.3649	41.377	0.553375	0.540159	5.17924
0.344	0.352318	2.21368	0.97639	4.52002	4.62931	0.962391	0.216015	4.42736	41.8491	41.8611	0.552897	0.539843	5.17634
0.345	0.353245	2.2195	0.97666	4.54704	4.6557	0.962671	0.21479	4.439	42.3393	42.3512	0.552422	0.539528	5.17348
0.346	0.354172	2.22533	0.976927	4.57424	4.68227	0.96295	0.213572	4.45065	42.8356	42.8473	0.55195	0.539215	5.17065
0.347	0.355099	2.23115	0.977192	4.6016	4.70901	0.963228	0.212359	4.46231	43.3379	43.3495	0.551483	0.538904	5.16785
0.348	0.356027	2.23699	0.977453	4.62914	4.73592	0.963505	0.211152	4.47397	43.8464	43.8578	0.551019	0.538595	5.16509
0.349	0.356956	2.24282	0.977712	4.65685	4.76301	0.96378	0.209951	4.48564	44.3612	44.3725	0.550558	0.538287	5.16236
0.35	0.357885	2.24866	0.977968	4.68474	4.79028	0.964054	0.208756	4.49732	44.8824	44.8935	0.550101	0.537981	5.15966
0.351	0.358815	2.2545	0.978221	4.7128	4.81772	0.964327	0.207567	4.509	45.4099	45.4209	0.549648	0.537677	5.15699

(Continued)

(Continued)

d/L_o	d/L	$2\pi d/L$	$\tanh$ $(2\pi d/L)$	$\sinh$ $(2\pi d/L)$	$\cosh$ $(2\pi d/L)$	H/H'_o	K	$4\pi d/L$	$\sinh$ $(4\pi d/L)$	$\cosh$ $(4\pi d/L)$	n	C_g/C_o	M
0.352	0.359745	2.26034	0.978471	4.74103	4.84535	0.964598	0.206383	4.52069	45.9439	45.9548	0.549198	0.537374	5.15435
0.353	0.360676	2.26619	0.978719	4.76945	4.87316	0.964869	0.205206	4.53238	46.4846	46.4953	0.548751	0.537073	5.15174
0.354	0.361607	2.27204	0.978964	4.79805	4.90115	0.965137	0.204034	4.54409	47.0319	47.0425	0.548309	0.536774	5.14916
0.355	0.362539	2.2779	0.979206	4.82682	4.92932	0.965405	0.202868	4.55579	47.5859	47.5964	0.547869	0.536477	5.14661
0.356	0.363471	2.28375	0.979446	4.85578	4.95768	0.965671	0.201707	4.56751	48.1468	48.1572	0.547433	0.536181	5.14409
0.357	0.364404	2.28962	0.979683	4.88492	4.98622	0.965936	0.200553	4.57923	48.7146	48.7248	0.547001	0.535887	5.1416
0.358	0.365337	2.29548	0.979917	4.91424	5.01495	0.966199	0.19940	4.59096	49.2894	49.2995	0.546571	0.535595	5.13914
0.359	0.366271	2.30135	0.980149	4.94375	5.04387	0.966462	0.19826	4.60269	49.8713	49.8813	0.546146	0.535304	5.13671
0.36	0.367205	2.30722	0.980379	4.97344	5.07298	0.966722	0.197123	4.61443	50.4604	50.4703	0.545723	0.535016	5.13431
0.361	0.36814	2.31309	0.980606	5.00333	5.10228	0.966982	0.195991	4.62618	51.0568	51.0666	0.545304	0.534729	5.13193
0.362	0.369075	2.31897	0.98083	5.0334	5.13177	0.96724	0.194864	4.63793	51.6605	51.6702	0.544889	0.534443	5.12958
0.363	0.370011	2.32485	0.981052	5.06366	5.16146	0.967497	0.193744	4.64969	52.2717	52.2813	0.544476	0.53416	5.12726
0.364	0.370947	2.33073	0.981272	5.09411	5.19134	0.967752	0.192629	4.66146	52.8905	52.8999	0.544067	0.533878	5.12497
0.365	0.371884	2.33662	0.981489	5.12476	5.22141	0.968006	0.191519	4.67323	53.5169	53.5262	0.543661	0.533598	5.1227
0.366	0.372821	2.34251	0.981704	5.15559	5.25168	0.968259	0.190415	4.68501	54.1511	54.1603	0.543259	0.533319	5.12046
0.367	0.373759	2.3484	0.981916	5.18663	5.28215	0.96851	0.189317	4.69679	54.7931	54.8022	0.542859	0.533042	5.11824
0.368	0.374697	2.35429	0.982126	5.21786	5.31282	0.96876	0.188224	4.70859	55.4431	55.4521	0.542463	0.532767	5.11606
0.369	0.375636	2.36019	0.982334	5.24929	5.34369	0.969009	0.187137	4.72038	56.1011	56.11	0.54207	0.532494	5.11389
0.37	0.376575	2.36609	0.982539	5.28091	5.37476	0.969256	0.186055	4.73218	56.7673	56.7761	0.541681	0.532223	5.11175
0.371	0.377515	2.372	0.982743	5.31274	5.40603	0.969502	0.184979	4.74399	57.4417	57.4504	0.541294	0.531953	5.10964
0.372	0.378455	2.3779	0.982944	5.34477	5.43751	0.969746	0.183908	4.75581	58.1244	58.133	0.540911	0.531685	5.10755
0.373	0.379396	2.38381	0.983142	5.377	5.46919	0.969989	0.182842	4.76763	58.8157	58.8242	0.54053	0.531418	5.10549
0.374	0.380337	2.38973	0.983339	5.40943	5.50109	0.970231	0.181782	4.77945	59.5155	59.5239	0.540153	0.531153	5.10345
0.375	0.381278	2.39564	0.983533	5.44207	5.53318	0.970471	0.180728	4.79129	60.224	60.2323	0.539779	0.53089	5.10143
0.376	0.38222	2.40156	0.983725	5.47492	5.56549	0.97071	0.179679	4.80312	60.9412	60.9494	0.539408	0.530629	5.09943
0.377	0.383163	2.40748	0.983915	5.50797	5.59801	0.970947	0.178635	4.81497	61.6674	61.6755	0.53904	0.53037	5.09746
0.378	0.384106	2.41341	0.984103	5.54124	5.63075	0.971183	0.177596	4.82682	62.4026	62.4106	0.538675	0.530112	5.09552
0.379	0.385049	2.41934	0.984289	5.57471	5.66369	0.971418	0.176563	4.83867	63.1469	63.1548	0.538313	0.529856	5.09359
0.38	0.385993	2.42527	0.984473	5.6084	5.69685	0.971652	0.175535	4.85053	63.9005	63.9083	0.537954	0.529601	5.09169

0.381	0.386938	2.4312	0.984655	5.6423	5.73023	0.971884	0.174513	4.8624	64.6633	64.6711	0.537598	0.529348	5.08981
0.382	0.387882	2.43714	0.984835	5.67642	5.76383	0.972114	0.173496	4.87427	65.4357	65.4434	0.537245	0.529097	5.08795
0.383	0.388828	2.44308	0.985012	5.71075	5.79764	0.972343	0.172484	4.88615	66.2177	66.2253	0.536895	0.528848	5.08612
0.384	0.389773	2.44902	0.985188	5.7453	5.83168	0.972571	0.171477	4.89804	67.0094	67.0169	0.536547	0.5286	5.0843
0.385	0.390719	2.45496	0.985362	5.78007	5.86593	0.972798	0.170476	4.90992	67.8109	67.8183	0.536203	0.528354	5.08251
0.386	0.391666	2.46091	0.985534	5.81505	5.90041	0.973023	0.16948	4.92182	68.6224	68.6297	0.535862	0.52811	5.08074
0.387	0.392613	2.46686	0.985704	5.85027	5.93512	0.973246	0.168489	4.93372	69.444	69.4512	0.535523	0.527867	5.07899
0.388	0.39356	2.47281	0.985872	5.8857	5.97005	0.973469	0.167503	4.94563	70.2758	70.283	0.535187	0.527626	5.07726
0.389	0.394508	2.47877	0.986038	5.92136	6.00521	0.973689	0.166522	4.95754	71.118	71.125	0.534854	0.527386	5.07555
0.39	0.395457	2.48473	0.986202	5.95725	6.04059	0.973909	0.165547	4.96945	71.9706	71.9776	0.534524	0.527149	5.07386
0.391	0.396405	2.49069	0.986364	5.99336	6.07621	0.974127	0.164576	4.98138	72.8339	72.8407	0.534197	0.526913	5.07218
0.392	0.397354	2.49665	0.986525	6.0297	6.11206	0.974344	0.163611	4.9933	73.7078	73.7146	0.533872	0.526678	5.07053
0.393	0.398304	2.50262	0.986684	6.06628	6.14815	0.974559	0.162651	5.00523	74.5927	74.5994	0.53355	0.526446	5.0689
0.394	0.399254	2.50859	0.986841	6.10308	6.18447	0.974773	0.161695	5.01717	75.4886	75.4952	0.533231	0.526214	5.06729
0.395	0.400204	2.51456	0.986996	6.14012	6.22102	0.974986	0.160745	5.02912	76.3956	76.4022	0.532915	0.525985	5.0657
0.396	0.401155	2.52053	0.987149	6.1774	6.25781	0.975197	0.1598	5.04106	77.314	77.3204	0.532601	0.525757	5.06412
0.397	0.402106	2.52651	0.987301	6.21491	6.29485	0.975407	0.15886	5.05302	78.2438	78.2502	0.53229	0.525531	5.06256
0.398	0.403058	2.53249	0.987451	6.25266	6.33212	0.975616	0.157925	5.06498	79.1851	79.1915	0.531982	0.525306	5.06103
0.399	0.40401	2.53847	0.987599	6.29065	6.36963	0.975823	0.156995	5.07694	80.1382	80.1445	0.531676	0.525083	5.05951
0.4	0.404962	2.54445	0.987746	6.32888	6.40739	0.976029	0.15607	5.08891	81.1032	81.1094	0.531373	0.524862	5.058
0.401	0.405915	2.55044	0.987891	6.36735	6.4454	0.976233	0.155149	5.10088	82.0802	82.0863	0.531073	0.524642	5.05652
0.402	0.406868	2.55643	0.988034	6.40607	6.48365	0.976436	0.154234	5.11286	83.0694	83.0754	0.530775	0.524424	5.05505
0.403	0.407822	2.56242	0.988176	6.44503	6.52215	0.976638	0.153324	5.12484	84.0709	84.0769	0.530479	0.524207	5.0536
0.404	0.408776	2.56842	0.988316	6.48424	6.5609	0.976838	0.152418	5.13683	85.085	85.0909	0.530186	0.523992	5.05217
0.405	0.409731	2.57441	0.988455	6.52371	6.5999	0.977037	0.151517	5.14883	86.1117	86.1175	0.529896	0.523778	5.05076
0.406	0.410685	2.58041	0.988591	6.56342	6.63916	0.977235	0.150621	5.16082	87.1511	87.1569	0.529608	0.523566	5.04936
0.407	0.41164	2.58641	0.988727	6.60338	6.67867	0.977432	0.14973	5.17283	88.2036	88.2092	0.529323	0.523356	5.04797
0.408	0.412596	2.59242	0.988861	6.6436	6.71844	0.977627	0.148844	5.18483	89.2692	89.2748	0.52904	0.523147	5.04661
0.409	0.413552	2.59842	0.988993	6.68407	6.75846	0.97782	0.147963	5.19685	90.3481	90.3536	0.52876	0.52294	5.04526
0.41	0.414508	2.60443	0.989124	6.7248	6.79875	0.978013	0.147086	5.20887	91.4405	91.4459	0.528482	0.522734	5.04392
0.411	0.415465	2.61044	0.989253	6.76579	6.83929	0.978204	0.146214	5.22089	92.5465	92.5519	0.528207	0.52253	5.04261
0.412	0.416422	2.61646	0.989381	6.80704	6.8801	0.978393	0.145347	5.23291	93.6663	93.6717	0.527934	0.522328	5.0413

(*Continued*)

(*Continued*)

d/L_o	d/L	$2\pi d/L$	$\tanh (2\pi d/L)$	$\sinh (2\pi d/L)$	$\cosh (2\pi d/L)$	H/H'_o	K	$4\pi d/L$	$\sinh (4\pi d/L)$	$\cosh (4\pi d/L)$	n	C_g/C_o	M
0.413	0.41738	2.62247	0.989507	6.84856	6.92118	0.978582	0.144484	5.24495	94.8001	94.8054	0.527663	0.522126	5.04002
0.414	0.418337	2.62849	0.989632	6.89033	6.96252	0.978769	0.143626	5.25698	95.9481	95.9533	0.527395	0.521927	5.03874
0.415	0.419295	2.63451	0.989755	6.93237	7.00413	0.978955	0.142773	5.26902	97.1104	97.1156	0.527129	0.521729	5.03749
0.416	0.420254	2.64053	0.989878	6.97468	7.04601	0.979139	0.141924	5.28107	98.2873	98.2924	0.526865	0.521532	5.03624
0.417	0.421213	2.64656	0.989998	7.01726	7.08815	0.979322	0.14108	5.29312	99.4789	99.4839	0.526604	0.521337	5.03502
0.418	0.422172	2.65259	0.990117	7.06011	7.13058	0.979504	0.140241	5.30517	100.685	100.69	0.526345	0.521144	5.03381
0.419	0.423132	2.65862	0.990235	7.10323	7.17328	0.979685	0.139406	5.31723	101.907	101.912	0.526089	0.520952	5.03261
0.42	0.424092	2.66465	0.990352	7.14663	7.21625	0.979864	0.138576	5.32929	103.144	103.149	0.525834	0.520761	5.03142
0.421	0.425052	2.67068	0.990467	7.1903	7.2595	0.980042	0.13775	5.34136	104.396	104.401	0.525582	0.520572	5.03025
0.422	0.426013	2.67672	0.990581	7.23425	7.30304	0.980219	0.136929	5.35343	105.664	105.669	0.525332	0.520384	5.0291
0.423	0.426974	2.68275	0.990693	7.27848	7.34685	0.980394	0.136113	5.36551	106.948	106.952	0.525085	0.520198	5.02795
0.424	0.427935	2.6888	0.990805	7.32299	7.39095	0.980568	0.135301	5.37759	108.248	108.252	0.524839	0.520013	5.02682
0.425	0.428897	2.69484	0.990915	7.36778	7.43533	0.980741	0.134493	5.38968	109.564	109.568	0.524596	0.51983	5.02571
0.426	0.429859	2.70088	0.991023	7.41286	7.48001	0.980913	0.13369	5.40176	110.896	110.901	0.524355	0.519648	5.02461
0.427	0.430821	2.70693	0.991131	7.45822	7.52497	0.981083	0.132891	5.41386	112.246	112.25	0.524116	0.519468	5.02352
0.428	0.431784	2.71298	0.991237	7.50388	7.57022	0.981252	0.132097	5.42596	113.612	113.616	0.523879	0.519289	5.02244
0.429	0.432747	2.71903	0.991342	7.54982	7.61576	0.98142	0.131307	5.43806	114.995	115	0.523645	0.519111	5.02138
0.43	0.43371	2.72508	0.991446	7.59606	7.6616	0.981587	0.130521	5.45016	116.396	116.4	0.523412	0.518935	5.02033
0.431	0.434674	2.73114	0.991548	7.64259	7.70773	0.981752	0.12974	5.46227	117.814	117.818	0.523182	0.51876	5.01929
0.432	0.435638	2.73719	0.991649	7.68941	7.75417	0.981916	0.128963	5.47439	119.25	119.254	0.522953	0.518586	5.01826
0.433	0.436602	2.74325	0.99175	7.73654	7.8009	0.982079	0.12819	5.4865	120.704	120.708	0.522727	0.518414	5.01725
0.434	0.437567	2.74931	0.991849	7.78396	7.84793	0.982241	0.127422	5.49863	122.176	122.18	0.522503	0.518244	5.01625
0.435	0.438532	2.75538	0.991946	7.83169	7.89527	0.982401	0.126658	5.51075	123.667	123.671	0.522281	0.518075	5.01526
0.436	0.439497	2.76144	0.992043	7.87971	7.94292	0.982561	0.125898	5.52288	125.176	125.18	0.522061	0.517907	5.01428
0.437	0.440463	2.76751	0.992139	7.92805	7.99087	0.982719	0.125143	5.53502	126.704	126.708	0.521842	0.51774	5.01331
0.438	0.441428	2.77358	0.992233	7.97669	8.03913	0.982875	0.124392	5.54715	128.251	128.255	0.521626	0.517575	5.01236
0.439	0.442395	2.77965	0.992327	8.02564	8.0877	0.983031	0.123645	5.5593	129.818	129.822	0.521412	0.517411	5.01142
0.44	0.443361	2.78572	0.992419	8.07491	8.13659	0.983186	0.122902	5.57144	131.404	131.408	0.5212	0.517248	5.01048
0.441	0.444328	2.7918	0.99251	8.12448	8.18579	0.983339	0.122163	5.58359	133.011	133.014	0.520989	0.517087	5.00956

0.442	0.445295	2.79787	0.9926	8.17437	8.23531	0.983491	0.121428	5.59574	134.637	134.641	0.520781	0.516927	5.00865
0.443	0.446262	2.803395	0.992689	8.22458	8.28515	0.983642	0.120698	5.6079	136.284	136.287	0.520574	0.516769	5.00776
0.444	0.44723	2.81003	0.992777	8.27511	8.33531	0.983791	0.119972	5.62006	137.951	137.955	0.52037	0.516611	5.00687
0.445	0.448198	2.81611	0.992864	8.32596	8.3858	0.98394	0.119249	5.63222	139.64	139.643	0.520167	0.516455	5.00599
0.446	0.449166	2.8222	0.99295	8.37713	8.43661	0.984087	0.118531	5.64439	141.349	141.353	0.519966	0.516301	5.00512
0.447	0.450135	2.82828	0.993035	8.42863	8.48774	0.984234	0.117817	5.65656	143.08	143.084	0.519767	0.516147	5.00427
0.448	0.451104	2.83437	0.993119	8.48045	8.53921	0.984379	0.117107	5.66874	144.833	144.836	0.51957	0.515995	5.00342
0.449	0.452073	2.84046	0.993202	8.53261	8.59101	0.984523	0.116401	5.68092	146.607	146.611	0.519375	0.515844	5.00258
0.45	0.453042	2.84655	0.993284	8.5851	8.64314	0.984666	0.115699	5.6931	148.404	148.408	0.519181	0.515694	5.00176
0.451	0.454012	2.85264	0.993365	8.63792	8.69561	0.984807	0.115001	5.70529	150.224	150.227	0.518989	0.515546	5.00094
0.452	0.454982	2.85874	0.993446	8.69108	8.74842	0.984948	0.114306	5.71747	152.066	152.07	0.518799	0.515399	5.00013
0.453	0.455952	2.86483	0.993525	8.74457	8.80156	0.985087	0.113616	5.72967	153.932	153.935	0.518611	0.515253	4.99934
0.454	0.456923	2.87093	0.993603	8.79841	8.85505	0.985226	0.11293	5.74186	155.821	155.824	0.518425	0.515108	4.99855
0.455	0.457894	2.87703	0.99368	8.85258	8.90888	0.985363	0.112248	5.75406	157.733	157.736	0.51824	0.514965	4.99777
0.456	0.458865	2.88313	0.993757	8.90711	8.96307	0.985499	0.111569	5.76627	159.67	159.673	0.518057	0.514822	4.997
0.457	0.459836	2.88924	0.993832	8.96197	9.01759	0.985634	0.110894	5.77847	161.631	161.634	0.517875	0.514681	4.99624
0.458	0.460808	2.89534	0.993907	9.01719	9.07247	0.985768	0.110224	5.79068	163.616	163.619	0.517696	0.514542	4.99549
0.459	0.46178	2.90145	0.993981	9.07276	9.12771	0.985901	0.109557	5.80289	165.627	165.63	0.517518	0.514403	4.99475
0.46	0.462752	2.90756	0.994053	9.12868	9.18329	0.986033	0.108893	5.81511	167.663	167.666	0.517342	0.514265	4.99402
0.461	0.463724	2.91367	0.994125	9.18496	9.23924	0.986164	0.108234	5.82733	169.724	169.727	0.517167	0.514129	4.9933
0.462	0.464697	2.91978	0.994197	9.2416	9.29554	0.986293	0.107578	5.83955	171.811	171.814	0.516994	0.513994	4.99258
0.463	0.46567	2.92589	0.994267	9.29859	9.35221	0.986422	0.106927	5.85178	173.925	173.928	0.516823	0.51386	4.99188
0.464	0.466643	2.932	0.994336	9.35595	9.40924	0.986549	0.106278	5.86401	176.065	176.068	0.516653	0.513727	4.99118
0.465	0.467616	2.93812	0.994405	9.41368	9.46664	0.986676	0.105634	5.87624	178.232	178.235	0.516485	0.513595	4.99049
0.466	0.46859	2.94424	0.994473	9.47176	9.52441	0.986801	0.104993	5.88847	180.426	180.429	0.516318	0.513465	4.98981
0.467	0.469564	2.95036	0.99454	9.53022	9.58255	0.986926	0.104356	5.90071	182.648	182.65	0.516153	0.513335	4.98914
0.468	0.470538	2.95648	0.994606	9.58906	9.64106	0.987049	0.103723	5.91295	184.897	184.9	0.51599	0.513207	4.98847
0.469	0.471512	2.9626	0.994672	9.64826	9.69995	0.987172	0.103093	5.9252	187.175	187.178	0.515828	0.513079	4.98781
0.47	0.472487	2.96872	0.994736	9.70784	9.75921	0.987293	0.102467	5.93745	189.482	189.484	0.515668	0.512953	4.98717
0.471	0.473462	2.97485	0.9948	9.7678	9.81886	0.987413	0.101845	5.9497	191.817	191.82	0.515509	0.512828	4.98652
0.472	0.474437	2.98098	0.994863	9.82814	9.87889	0.987533	0.101226	5.96195	194.182	194.185	0.515351	0.512704	4.98589
0.473	0.475412	2.9871	0.994926	9.88887	9.9393	0.987651	0.100611	5.97421	196.577	196.579	0.515196	0.512581	4.98527

(Continued)

(*Continued*)

d/L_o	d/L	$2\pi d/L$	tanh $(2\pi d/L)$	sinh $(2\pi d/L)$	cosh $(2\pi d/L)$	H/H'_o	K	$4\pi d/L$	sinh $(4\pi d/L)$	cosh $(4\pi d/L)$	n	C_g/C_o	M
0.474	0.476388	2.99323	0.994988	9.94998	10.0001	0.987769	0.0999989	5.98647	199.002	199.004	0.515041	0.51246	4.98465
0.475	0.477364	2.99936	0.995048	10.0115	10.0613	0.987885	0.0993908	5.99873	201.457	201.459	0.514888	0.512339	4.98404
0.476	0.47834	3.0055	0.995109	10.0734	10.1229	0.988	0.0987861	6.01099	203.943	203.946	0.514737	0.512219	4.98343
0.477	0.479316	3.01163	0.995168	10.1357	10.1849	0.988115	0.0981849	6.02326	206.46	206.463	0.514587	0.512101	4.98284
0.478	0.480292	3.01777	0.995227	10.1983	10.2472	0.988228	0.0975872	6.03553	209.01	209.012	0.514438	0.511983	4.98225
0.479	0.481269	3.0239	0.995285	10.2614	10.31	0.988341	0.096993	6.04781	211.591	211.593	0.514291	0.511866	4.98167
0.48	0.482246	3.03004	0.995342	10.3249	10.3732	0.988452	0.0964022	6.06008	214.204	214.207	0.514146	0.511751	4.98109
0.481	0.483223	3.03618	0.995399	10.3888	10.4368	0.988563	0.0958149	6.07236	216.851	216.853	0.514001	0.511636	4.98053
0.482	0.484201	3.04232	0.995455	10.4531	10.5008	0.988672	0.095231	6.08464	219.531	219.533	0.513858	0.511523	4.97997
0.483	0.485178	3.04846	0.995511	10.5178	10.5652	0.988781	0.0946505	6.09693	222.244	222.246	0.513717	0.51141	4.97941
0.484	0.486156	3.05461	0.995565	10.5529	10.63	0.988889	0.0940733	6.10922	224.992	224.994	0.513577	0.511299	4.97886
0.485	0.487134	3.06075	0.995619	10.6484	10.6952	0.988996	0.0934995	6.12151	227.774	227.776	0.513438	0.511188	4.97832
0.486	0.488112	3.0669	0.995673	10.7143	10.7609	0.989102	0.0929291	6.1338	230.591	230.594	0.5133	0.511079	4.97779
0.487	0.489091	3.07305	0.995726	10.7807	10.827	0.989207	0.092362	6.14609	233.444	233.446	0.513164	0.51097	4.97726
0.488	0.490069	3.0792	0.995778	10.8475	10.8935	0.989311	0.0917982	6.15839	236.333	236.335	0.513029	0.510863	4.97674
0.489	0.491048	3.08535	0.995829	10.9147	10.9604	0.989414	0.0912377	6.17069	239.258	239.26	0.512896	0.510756	4.97623
0.49	0.492027	3.0915	0.99588	10.9823	11.0277	0.989517	0.0906805	6.183	242.219	242.221	0.512763	0.510651	4.97572
0.491	0.493006	3.09765	0.99593	11.0504	11.0955	0.989618	0.0901266	6.1953	245.219	245.221	0.512632	0.510546	4.97522
0.492	0.493986	3.1038	0.99598	11.1188	11.1637	0.989719	0.0895759	6.20761	248.255	248.257	0.512502	0.510442	4.97472
0.493	0.494965	3.10996	0.996029	11.1878	11.2324	0.989818	0.0890284	6.21992	251.33	251.332	0.512374	0.510339	4.97423
0.494	0.495945	3.11612	0.996078	11.2571	11.3015	0.989917	0.0884841	6.23223	254.444	254.446	0.512247	0.510238	4.97374
0.495	0.496925	3.12227	0.996126	11.3269	11.371	0.990015	0.087943	6.24455	257.597	257.599	0.512121	0.510137	4.97327
0.496	0.497906	3.12843	0.996173	11.3972	11.441	0.990112	0.0874051	6.25687	260.79	260.792	0.511996	0.510036	4.97279
0.497	0.498886	3.13459	0.99622	11.4679	11.5114	0.990208	0.0868704	6.26919	264.023	264.025	0.511872	0.509937	4.97233
0.498	0.499867	3.14075	0.996266	11.539	11.5823	0.990304	0.0863388	6.28151	267.296	267.298	0.51175	0.509839	4.97186
0.499	0.500847	3.14692	0.996311	11.6106	11.6536	0.990398	0.0858103	6.29383	270.611	270.613	0.511629	0.509742	4.97141
0.5	0.501828	3.15308	0.996357	11.6827	11.7254	0.990492	0.085285	6.30616	273.968	273.97	0.511509	0.509645	4.97096
0.501	0.50281	3.15925	0.996401	11.7552	11.7976	0.990585	0.0847627	6.31849	277.367	277.369	0.51139	0.50955	4.97051
0.502	0.503791	3.16541	0.996445	11.8281	11.8703	0.990677	0.0842436	6.33082	280.808	280.81	0.511272	0.509455	4.97008

0.503	0.504772	3.17158	0.996489	11.9016	11.9435	0.990768	0.0837274	6.34316	284.293	284.295	0.511156	0.509361	4.96964
0.504	0.505754	3.17775	0.996532	11.9755	12.0172	0.990859	0.0832144	6.35549	287.822	287.824	0.511041	0.509268	4.96921
0.505	0.506736	3.18392	0.996574	12.0498	12.0913	0.990948	0.0827043	6.36783	291.396	291.398	0.510926	0.509176	4.96879
0.506	0.507718	3.19009	0.996616	12.1247	12.1659	0.991037	0.0821973	6.38017	295.014	295.016	0.510813	0.509085	4.96837
0.507	0.5087	3.19626	0.996658	12.2	12.2409	0.991125	0.0816932	6.39252	298.678	298.68	0.510701	0.508994	4.96796
0.508	0.509683	3.20243	0.996698	12.2758	12.3165	0.991212	0.0811922	6.40486	302.388	302.39	0.51059	0.508905	4.96755
0.509	0.510665	3.20861	0.996739	12.3521	12.3925	0.991299	0.0806941	6.41721	306.145	306.147	0.510481	0.508816	4.96715
0.51	0.511648	3.21478	0.996779	12.4288	12.469	0.991385	0.080199	6.42956	309.95	309.951	0.510372	0.508728	4.96675
0.511	0.512631	3.22096	0.996818	12.5061	12.546	0.99147	0.0797068	6.44191	313.802	313.803	0.510264	0.508641	4.96635
0.512	0.513614	3.22713	0.996857	12.5838	12.6235	0.991554	0.0792175	6.45427	317.703	317.704	0.510158	0.508554	4.96597
0.513	0.514597	3.23331	0.996896	12.662	12.7015	0.991637	0.0787311	6.46662	321.652	321.654	0.510052	0.508469	4.96558
0.514	0.515581	3.23949	0.996934	12.7408	12.7799	0.99172	0.0782476	6.47898	325.652	325.654	0.509948	0.508384	4.9652
0.515	0.516564	3.24567	0.996972	12.82	12.8589	0.991802	0.077767	6.49134	329.702	329.704	0.509844	0.5083	4.96483
0.516	0.517548	3.25185	0.997009	12.8997	12.9384	0.991883	0.0772893	6.5037	333.803	333.805	0.509742	0.508217	4.96446
0.517	0.518532	3.25803	0.997045	12.9799	13.0184	0.991963	0.0768143	6.51607	337.956	337.958	0.50964	0.508135	4.96409
0.518	0.519516	3.26422	0.997082	13.0607	13.0989	0.992043	0.0763422	6.52843	342.162	342.163	0.50954	0.508053	4.96373
0.519	0.5205	3.2704	0.997118	13.1419	13.1799	0.992122	0.0758729	6.5408	346.42	346.421	0.509441	0.507972	4.96338
0.52	0.521485	3.27659	0.997153	13.2237	13.2615	0.9922	0.0754064	6.55317	350.732	350.733	0.509342	0.507892	4.96302
0.521	0.522469	3.28277	0.997188	13.306	13.3435	0.992278	0.0749427	6.56554	355.098	355.099	0.509245	0.507813	4.96267
0.522	0.523454	3.28896	0.997222	13.3888	13.4261	0.992355	0.0744818	6.57792	359.519	359.521	0.509148	0.507734	4.96233
0.523	0.524439	3.29515	0.997257	13.4721	13.5092	0.992431	0.0740236	6.59029	363.996	363.998	0.509053	0.507656	4.96199
0.524	0.525424	3.30134	0.99729	13.556	13.5928	0.992506	0.0735681	6.60267	368.53	368.531	0.508958	0.507579	4.96166
0.525	0.526409	3.30752	0.997324	13.6404	13.677	0.992581	0.0731153	6.61505	373.12	373.122	0.508865	0.507503	4.96132
0.526	0.527394	3.31372	0.997356	13.7253	13.7617	0.992655	0.0726653	6.62743	377.769	377.77	0.508772	0.507427	4.961
0.527	0.52838	3.31991	0.997389	13.8108	13.847	0.992728	0.0722179	6.63981	382.476	382.477	0.50868	0.507352	4.96067
0.528	0.529365	3.3261	0.997421	13.8968	13.9328	0.992801	0.0717733	6.6522	387.243	387.244	0.508589	0.507278	4.96036
0.529	0.530351	3.33229	0.997453	13.9834	14.0191	0.992873	0.0713312	6.66459	392.069	392.07	0.508499	0.507204	4.96004
0.53	0.531337	3.33849	0.997484	14.0705	14.106	0.992944	0.0708919	6.67698	396.957	396.958	0.50841	0.507131	4.95973
0.531	0.532323	3.34468	0.997515	14.1582	14.1934	0.993015	0.0704551	6.68937	401.906	401.907	0.508322	0.507059	4.95942
0.532	0.533309	3.35088	0.997546	14.2464	14.2814	0.993085	0.070021	6.70176	406.917	406.919	0.508235	0.506987	4.95912
0.533	0.534295	3.35708	0.997576	14.3352	14.37	0.993154	0.0695895	6.71415	411.992	411.993	0.508148	0.506916	4.95882
0.534	0.535282	3.36327	0.997606	14.4245	14.4591	0.993223	0.0691605	6.72655	417.131	417.132	0.508063	0.506846	4.95852
0.535	0.536268	3.36947	0.997635	14.5144	14.5488	0.993291	0.0687342	6.73895	422.334	422.336	0.507978	0.506777	4.95823

(*Continued*)

(Continued)

d/L_o	d/L	$2\pi d/L$	tanh $(2\pi d/L)$	sinh $(2\pi d/L)$	cosh $(2\pi d/L)$	H/H'_o	K	$4\pi d/L$	sinh $(4\pi d/L)$	cosh $(4\pi d/L)$	n	C_g/C_o	M
0.536	0.537255	3.37567	0.997664	14.6049	14.6391	0.993359	0.0683104	6.75135	427.604	427.605	0.507894	0.506708	4.95794
0.537	0.538242	3.38187	0.997693	14.6959	14.7299	0.993426	0.0678891	6.76375	432.939	432.94	0.507811	0.50664	4.95765
0.538	0.539229	3.38807	0.997721	14.7876	14.8213	0.993492	0.0674704	6.77615	438.342	438.343	0.507729	0.506572	4.95737
0.539	0.540216	3.39428	0.997749	14.8798	14.9133	0.993557	0.0670541	6.78855	443.813	443.815	0.507648	0.506505	4.95709
0.54	0.541203	3.40048	0.997777	14.9726	15.0059	0.993622	0.0666404	6.80096	449.354	449.355	0.507567	0.506439	4.95682
0.541	0.54219	3.40668	0.997804	15.0659	15.0991	0.993687	0.0662292	6.81337	454.964	454.965	0.507488	0.506374	4.95654
0.542	0.543178	3.41289	0.997831	15.1599	15.1929	0.993751	0.0658204	6.82577	460.644	460.645	0.507409	0.506309	4.95627
0.543	0.544165	3.41909	0.997858	15.2545	15.2872	0.993814	0.0654141	6.83819	466.397	466.398	0.507331	0.506244	4.95601
0.544	0.545153	3.4253	0.997885	15.3496	15.3822	0.993876	0.0650103	6.8506	472.222	472.223	0.507254	0.506181	4.95575
0.545	0.546141	3.43151	0.997911	15.4454	15.4778	0.993938	0.0646089	6.86301	478.121	478.122	0.507177	0.506117	4.95549
0.546	0.547129	3.43771	0.997936	15.5418	15.5739	0.994	0.0642099	6.87543	484.094	484.095	0.507101	0.506055	4.95523
0.547	0.548117	3.44392	0.997962	15.6388	15.6707	0.99406	0.0638133	6.88784	490.142	490.143	0.507026	0.505993	4.95498
0.548	0.549105	3.45013	0.997987	15.7364	15.7681	0.994121	0.0634191	6.90026	496.267	496.268	0.506952	0.505932	4.95473
0.549	0.550094	3.45634	0.998012	15.8346	15.8662	0.99418	0.0630272	6.91268	502.469	502.47	0.506879	0.505871	4.95448
0.55	0.551082	3.46255	0.998036	15.9335	15.9648	0.994239	0.0626378	6.9251	508.749	508.75	0.506806	0.505811	4.95424
0.551	0.552071	3.46876	0.998061	16.0329	16.0641	0.994298	0.0622507	6.93753	515.108	515.109	0.506734	0.505751	4.954
0.552	0.553059	3.47497	0.998084	16.133	16.164	0.994356	0.0618659	6.94995	521.548	521.549	0.506663	0.505692	4.95376
0.553	0.554048	3.48119	0.998108	16.2338	16.2645	0.994413	0.0614835	6.96238	528.069	528.07	0.506592	0.505634	4.95353
0.554	0.555037	3.4874	0.998131	16.3351	16.3657	0.99447	0.0611033	6.9748	534.672	534.673	0.506522	0.505576	4.9533
0.555	0.556026	3.49362	0.998155	16.4372	16.4675	0.994527	0.0607255	6.98723	541.359	541.36	0.506453	0.505519	4.95307
0.556	0.557015	3.49983	0.998177	16.5398	16.57	0.994582	0.06035	6.99966	548.13	548.131	0.506385	0.505462	4.95284
0.557	0.558005	3.50605	0.9982	16.6431	16.6731	0.994638	0.0599767	7.01209	554.987	554.988	0.506317	0.505406	4.95262
0.558	0.558994	3.51226	0.998222	16.7471	16.7769	0.994692	0.0596057	7.02452	561.93	561.931	0.50625	0.50535	4.9524
0.559	0.559983	3.51848	0.998244	16.8517	16.8814	0.994747	0.0592369	7.03696	568.96	568.961	0.506184	0.505295	4.95218
0.56	0.560973	3.5247	0.998266	16.957	16.9865	0.9948	0.0588704	7.04939	576.08	576.081	0.506118	0.505241	4.95196
0.561	0.561963	3.53092	0.998287	17.063	17.0922	0.994853	0.0585061	7.06183	583.289	583.29	0.506053	0.505187	4.95175
0.562	0.562952	3.53713	0.998308	17.1696	17.1987	0.994906	0.0581439	7.07427	590.589	590.59	0.505989	0.505133	4.95154
0.563	0.563942	3.54335	0.998329	17.2769	17.3058	0.994958	0.057784	7.08671	597.982	597.983	0.505926	0.50508	4.95133
0.564	0.564932	3.54957	0.99835	17.3849	17.4136	0.99501	0.0574263	7.09915	605.468	605.468	0.505863	0.505028	4.95113

0.565	0.565922	3.5558	0.99837	17.4935	17.5221	0.995061	0.0570708	7.11159	613.048	613.048	0.5058	0.504976	4.95093
0.566	0.566913	3.56202	0.99839	17.6029	17.6313	0.995112	0.0567174	7.12403	620.724	620.725	0.505738	0.504924	4.95073
0.567	0.567903	3.56824	0.99841	17.7129	17.7412	0.995162	0.0563661	7.13648	628.496	628.497	0.505677	0.504873	4.95053
0.568	0.568893	3.57446	0.99843	17.8237	17.8517	0.995212	0.056017	7.14892	636.367	636.368	0.505617	0.504823	4.95034
0.569	0.569884	3.58069	0.998449	17.9351	17.963	0.995261	0.05567	7.16137	644.338	644.339	0.505557	0.504773	4.95014
0.57	0.570874	3.58691	0.998468	18.0473	18.075	0.995309	0.0553251	7.17382	652.409	652.41	0.505498	0.504724	4.94995
0.571	0.571865	3.59313	0.998487	18.1602	18.1877	0.995358	0.0549823	7.18627	660.582	660.582	0.505439	0.504675	4.94977
0.572	0.572856	3.59936	0.998506	18.2737	18.3011	0.995406	0.0546416	7.19872	668.858	668.858	0.505381	0.504626	4.94958
0.573	0.573847	3.60559	0.998524	18.388	18.4152	0.995453	0.054303	7.21117	677.238	677.239	0.505324	0.504578	4.9494
0.574	0.574838	3.61181	0.998543	18.503	18.53	0.9955	0.0539664	7.22362	685.725	685.725	0.505267	0.504531	4.94922
0.575	0.575829	3.61804	0.998561	18.6188	18.6456	0.995546	0.0536319	7.23608	694.318	694.319	0.505211	0.504484	4.94904
0.576	0.57682	3.62427	0.998579	18.7353	18.7619	0.995592	0.0532994	7.24853	703.02	703.021	0.505155	0.504437	4.94886
0.577	0.577811	3.63049	0.998596	18.8525	18.879	0.995638	0.0529689	7.26099	711.832	711.832	0.5051	0.504391	4.94869
0.578	0.578802	3.63672	0.998614	18.9704	18.9968	0.995683	0.0526405	7.27345	720.755	720.756	0.505046	0.504345	4.94851
0.579	0.579794	3.64295	0.998631	19.0892	19.1153	0.995727	0.052314	7.28591	729.791	729.791	0.504992	0.5043	4.94834
0.58	0.580785	3.64918	0.998648	19.2086	19.2346	0.995771	0.0519896	7.29837	738.941	738.941	0.504938	0.504256	4.94818
0.581	0.581777	3.65541	0.998664	19.3288	19.3547	0.995815	0.0516671	7.31083	748.206	748.207	0.504886	0.504211	4.94801
0.582	0.582769	3.66164	0.998681	19.4498	19.4755	0.995858	0.0513466	7.32329	757.589	757.589	0.504833	0.504167	4.94785
0.583	0.583761	3.66788	0.998697	19.5715	19.5971	0.995901	0.051028	7.33575	767.089	767.09	0.504782	0.504124	4.94769
0.584	0.584752	3.67411	0.998713	19.694	19.7194	0.995944	0.0507114	7.34822	776.71	776.711	0.50473	0.504081	4.94753
0.585	0.585744	3.68034	0.998729	19.8173	19.8426	0.995986	0.0503967	7.36068	786.453	786.454	0.50468	0.504038	4.94737
0.586	0.586736	3.68657	0.998745	19.9414	19.9665	0.996028	0.050084	7.37315	796.318	796.319	0.504629	0.503996	4.94721
0.587	0.587728	3.69281	0.998761	20.0663	20.0912	0.996069	0.0497731	7.38561	806.308	806.309	0.50458	0.503954	4.94706
0.588	0.588721	3.69904	0.998776	20.1919	20.2166	0.99611	0.0494642	7.39808	816.425	816.425	0.504531	0.503913	4.94691
0.589	0.589713	3.70528	0.998791	20.3183	20.3429	0.99615	0.0491571	7.41055	826.669	826.669	0.504482	0.503872	4.94676
0.59	0.590705	3.71151	0.998806	20.4456	20.47	0.99619	0.048852	7.42302	837.042	837.043	0.504434	0.503832	4.94661
0.591	0.591698	3.71775	0.998821	20.5736	20.5979	0.99623	0.0485486	7.43549	847.546	847.547	0.504386	0.503792	4.94646
0.592	0.59269	3.72398	0.998835	20.7025	20.7266	0.996269	0.0482472	7.44797	858.183	858.184	0.504339	0.503752	4.94632
0.593	0.593683	3.73022	0.99885	20.8321	20.8561	0.996308	0.0479476	7.46044	868.955	868.955	0.504293	0.503713	4.94617
0.594	0.594675	3.73646	0.998864	20.9626	20.9865	0.996346	0.0476498	7.47291	879.862	879.863	0.504247	0.503674	4.94603
0.595	0.595668	3.74269	0.998878	21.0939	21.1176	0.996384	0.0473538	7.48539	890.907	890.908	0.504201	0.503635	4.94589
0.596	0.596661	3.74893	0.998892	21.2261	21.2496	0.996422	0.0470597	7.49786	902.092	902.093	0.504156	0.503597	4.94576

(*Continued*)

(*Continued*)

d/L_o	d/L	$2\pi d/L$	$\tanh$ $(2\pi d/L)$	$\sinh$ $(2\pi d/L)$	$\cosh$ $(2\pi d/L)$	H/H'_o	K	$4\pi d/L$	$\sinh$ $(4\pi d/L)$	$\cosh$ $(4\pi d/L)$	n	C_g/C_o	M
0.597	0.597654	3.75517	0.998906	21.3591	21.3825	0.996459	0.0467673	7.51034	913.418	913.419	0.504111	0.50356	4.94562
0.598	0.598647	3.76141	0.998919	21.4929	21.5161	0.996496	0.0464768	7.52282	924.887	924.888	0.504067	0.503522	4.94549
0.6	0.600633	3.77389	0.998946	21.786	21.786	0.996569	0.045901	7.54778	948.262	948.262	0.50398	0.503449	4.94522
0.61	0.610568	3.83631	0.99907	23.1663	23.1879	0.996912	0.0431259	7.67262	1074.36	1074.36	0.503571	0.503102	4.944
0.62	0.62051	3.89878	0.999179	24.6609	24.6812	0.997222	0.0405167	7.79755	1217.32	1217.32	0.503203	0.50279	4.94292
0.63	0.630457	3.96128	0.999275	26.2527	26.2717	0.997502	0.0380638	7.92255	1379.41	1379.41	0.502872	0.502507	4.94196
0.64	0.64041	4.02381	0.999361	27.948	27.9659	0.997755	0.0357579	8.04762	1563.18	1563.18	0.502574	0.502253	4.94112
0.65	0.650367	4.08638	0.999436	29.7535	29.7703	0.997983	0.0335905	8.17275	1771.54	1771.54	0.502307	0.502023	4.94038
0.66	0.660329	4.14897	0.999502	31.6764	31.6922	0.998188	0.0315535	8.29794	2007.79	2007.79	0.502066	0.501816	4.93972
0.67	0.670294	4.21158	0.999561	33.7243	33.7391	0.998374	0.0296392	8.42317	2275.65	2275.65	0.501851	0.50163	4.93914
0.68	0.680264	4.27422	0.999612	35.9052	35.9191	0.998541	0.0278403	8.54845	2579.36	2579.36	0.501657	0.501463	4.93863
0.69	0.690236	4.33688	0.999658	38.2278	38.2409	0.998691	0.02615	8.67376	2923.73	2923.73	0.501483	0.501312	4.93818
0.7	0.700211	4.39956	0.999698	40.7013	40.7135	0.998826	0.0245619	8.79911	3314.19	3314.19	0.501328	0.501176	4.93778
0.71	0.710189	4.46225	0.999734	43.3354	43.3469	0.998947	0.0230697	8.9245	3756.91	3756.91	0.501188	0.501054	4.93743
0.72	0.720169	4.52496	0.999765	46.1405	46.1513	0.999057	0.0216678	9.04991	4258.89	4258.89	0.501062	0.500945	4.93712
0.73	0.730151	4.58768	0.999793	49.1278	49.138	0.999155	0.0203509	9.17535	4828.08	4828.08	0.50095	0.500846	4.93685
0.74	0.740135	4.65041	0.999817	52.309	52.3186	0.999243	0.0191137	9.30081	5473.46	5473.46	0.50085	0.500758	4.93661
0.75	0.750121	4.71315	0.999839	55.6967	55.7057	0.999322	0.0179515	9.4263	6205.24	6205.24	0.50076	0.500679	4.93639
0.76	0.760108	4.7759	0.999858	59.3043	59.3127	0.999393	0.0168598	9.5518	7035	7035	0.500679	0.500608	4.93621
0.77	0.770097	4.83866	0.999875	63.146	63.1539	0.999457	0.0158343	9.67732	7975.83	7975.83	0.500607	0.500544	4.93604
0.78	0.780086	4.90143	0.999889	67.237	67.2445	0.999514	0.0148711	9.80285	9042.64	9042.64	0.500542	0.500487	4.93589
0.79	0.790077	4.9642	0.999902	71.5935	71.6005	0.999565	0.0139664	9.9284	10252.3	10252.3	0.500484	0.500435	4.93577
0.8	0.800069	5.02698	0.999914	76.2327	76.2393	0.999611	0.0131166	10.054	11623.9	11623.9	0.500432	0.500389	4.93565
0.81	0.810061	5.08977	0.999924	81.1729	81.1791	0.999652	0.0123184	10.1795	13179.1	13179.1	0.500386	0.500348	4.93555
0.82	0.820055	5.15256	0.999933	86.4336	86.4394	0.999689	0.0115688	10.3051	14942.5	14942.5	0.500345	0.500311	4.93546
0.83	0.830049	5.21535	0.999941	92.0355	92.041	0.999722	0.0108647	10.4307	16942.1	16942.1	0.500308	0.500278	4.93539
0.84	0.840044	5.27815	0.999948	98.001	98.0061	0.999751	0.0102034	10.5563	19209.4	19209.4	0.500275	0.500249	4.93532
0.85	0.850039	5.34095	0.999954	104.353	104.358	0.999778	0.00958239	10.6819	21780.2	21780.2	0.500245	0.500222	4.93526

0.86	0.860035	5.40376	0.99996	111.118	111.122	0.999802	0.00899909	10.8075	24695.3	24695.3	0.500219	0.500199	4.9352
0.87	0.870031	5.46657	0.999964	118.321	118.325	0.999823	0.00845128	10.9331	28000.7	28000.7	0.500195	0.500177	4.93515
0.88	0.880028	5.52938	0.999969	125.992	125.995	0.999842	0.00793679	11.0588	31748.7	31748.7	0.500174	0.500158	4.93511
0.89	0.890025	5.59219	0.999972	134.159	134.163	0.999859	0.00745361	11.1844	35998.5	35998.5	0.500155	0.500141	4.93508
0.9	0.900022	5.65501	0.999976	142.857	142.861	0.999874	0.00699982	11.31	40817.4	40817.4	0.500139	0.500126	4.93504
0.91	0.91002	5.71782	0.999978	152.119	152.123	0.999887	0.00657365	11.4356	46281.5	46281.5	0.500124	0.500113	4.93502
0.92	0.920018	5.78064	0.999981	161.982	161.985	0.999899	0.00617342	11.5613	52477.2	52477.2	0.50011	0.500101	4.93499
0.93	0.930016	5.84346	0.999983	172.484	172.487	0.99991	0.00579754	11.6869	59502.5	59502.5	0.500098	0.50009	4.93497
0.94	0.940014	5.90628	0.999985	183.668	183.67	0.99992	0.00544454	11.8126	67468.5	67468.5	0.500088	0.50008	4.93495
0.95	0.950012	5.9691	0.999987	195.576	195.579	0.999929	0.00511303	11.9382	76501.2	76501.2	0.500078	0.500071	4.93493
0.96	0.960011	6.03193	0.999988	208.257	208.26	0.999936	0.0048017	12.0639	86743.3	86743.3	0.50007	0.500064	4.93492
0.97	0.97001	6.09475	0.99999	221.761	221.763	0.999943	0.00450932	12.1895	98356.8	98356.8	0.500062	0.500057	4.9349
0.98	0.980009	6.15758	0.999991	236.14	236.142	0.999949	0.00423474	12.3152	111525	111525	0.500055	0.500051	4.93489
0.99	0.990008	6.2204	0.999992	251.452	251.454	0.999955	0.00397687	12.4408	126457	126457	0.500049	0.500045	4.93488

Appendix B

Expressions for Stokes Fifth Order Theory

The solution is given by a series for ϕ in ascending power of $Hk/2$. Let $S = d + z$ and

$$\phi = \frac{-C}{k} \sum_{n=1}^{N} a_n \left(\frac{Hk}{2}\right)^n \frac{\cosh nkS}{\sinh nkd} \sin n\theta, \tag{B.1}$$

where a_n are polynomials in $\cosh kd$ and $\sinh kd$.

For Stokes' second-order waves, the surface profile, particle velocities, and accelerations are given by

$$\eta = \frac{H}{2}\cos\theta + \frac{\pi}{2}\frac{H^2}{4L}\frac{\cosh kd(2 + \cosh 2kd)}{\sinh^3 kd}\cos 2\theta, \tag{B.2}$$

$$u = \frac{\pi H}{T}\frac{\cosh kS}{\sinh kd}\cos\theta + \frac{3\pi^2 H^2}{4LT}\frac{\cosh 2kS}{\sinh^4 kd}\cos 2\theta, \tag{B.3}$$

$$w = \frac{\pi H}{T}\frac{\sinh kS}{\sinh kd}\sin\theta + \frac{3\pi^2 H^2}{4LT}\frac{\sinh 2kS}{\sinh^4 kd}\sin 2\theta, \tag{B.4}$$

$$\frac{\partial u}{\partial t} = \frac{2\pi^2 H}{T^2}\frac{\cosh kS}{\sinh kd}\sin\theta + \frac{3\pi^3 H^2}{T^2 L}\frac{\cosh 2kS}{\sinh^4 kd}\sin 2\theta, \tag{B.5}$$

$$\frac{\partial w}{\partial t} = \frac{-2\pi^2 H}{T^2}\frac{\sinh kS}{\sinh kd}\cos\theta - \frac{3\pi^3 H^2}{T^2 L}\frac{\sinh 2kS}{\sinh^4 kd}\cos 2\theta. \tag{B.6}$$

Stokes' third-order wave approximations are given as follows:

$$\eta = a\cos\theta + \frac{\pi a^2}{L}f_2\left(\frac{d}{L}\right)\cos 2\theta + \frac{\pi^2 a^3}{L^2}f_3\left(\frac{d}{L}\right)\cos 3\theta, \tag{B.7}$$

where $f_2(d/L)$ and $f_3(d/L)$ are given by

$$f_2\left(\frac{d}{L}\right) = \frac{(2 + \cosh 2kd)\cosh kd}{2\sinh^3 kd}, \quad f_3\left(\frac{d}{L}\right) = \frac{3}{16}\frac{1 + 8\cosh^3 kd}{\sinh^6 kd}.$$

The wavelength is given by

$$L = \frac{gT^2}{2\pi}\tanh kd\left[1 + \left(\frac{2\pi a}{L}\right)^2 \frac{14 + 4\cosh^2 2kd}{16\sinh^4 kd}\right]. \tag{B.8}$$

The particle velocities and accelerations are given by

$$u = C(F_1 \cosh kS \cos \theta + F_2 \cosh 2kS \cos 2\theta$$
$$+F_3 \cosh 3kS \cos 3\theta), \tag{B.9}$$

$$w = C(F_1 \sinh kS \sin \theta + F_2 \sinh 2kS \sin 2\theta$$
$$+F_3 \sinh 3kS \sin 3\theta), \tag{B.10}$$

$$\partial u/\partial t = 2\pi C/T(F_1 \cosh kS \sin \theta + F_2 \cosh 2kS \sin 2\theta$$
$$+F_3 \cosh 3kS \sin 3\theta), \tag{B.11}$$

$$\partial w/\partial t = -2\pi C/T(F_1 \sinh kS \cos \theta + F_2 \sinh 2kS \cos 2\theta$$
$$+F_3 \sinh 3kS \cos 3\theta), \tag{B.12}$$

where

$$F_1 = \frac{ka}{\sinh kd} - (ka)^2 \frac{\cosh^2 kd \left(1 + 5 \cosh^2 kd\right)}{8 \sinh^5 kd}, \quad F_2 = \frac{3}{4}(ka)^2 \frac{1}{\sinh^4 kd},$$
$$F_3 = \frac{3}{64}(ka)^3 \frac{11 - 2 \cosh 2kd}{\sinh^7 kd}.$$

The Stokes' wave has been extended to a fifth-order approximation by Skjelbreia and Hendrickson (1960). Let $S = d + z$, and the particle velocities are written as

$$u = \frac{\partial \phi}{\partial x} \quad \text{and} \quad w = \frac{\partial \phi}{\partial z} = \frac{\partial \phi}{\partial s}. \tag{B.13a}$$

$$\tag{B.13b}$$

The surface boundary conditions are $p = 0$ and $\partial \eta/\partial x = -w/(C - u)$. This is written as

$$\frac{1}{2}(u^2 + w^2) + g(K + S - d) + \frac{\partial \phi}{\partial t} = 0, \tag{B.14}$$

$$w^2 = (C - u)^2 \left(\frac{\partial \eta}{\partial x}\right)^2, \tag{B.15}$$

where C is the wave velocity and K is the elevation of mean wave height above mean water level. It is noted that

$$\frac{\partial \phi}{\partial t} = -C \frac{\partial \phi}{\partial x}.$$

Equation (B.14) can be written as

$$(u - C)^2 - C^2 + w^2 + 2g(K + z) = 0. \tag{B.16}$$

The boundary condition at the seabed is written as

$$\frac{\partial \phi}{\partial s} = 0 \quad \text{at} \quad S = 0. \tag{B.17}$$

To solve the problem, one has to find a solution for $\nabla^2 \phi = 0$ and satisfy Eqs. (B.15)–(B.17). The following series of solutions are assumed for the fifth-order approximation:

$$k\phi/C = (\lambda A_{11} + \lambda^3 A_{13} + \lambda^5 A_{15}) \cosh kS \sin \theta$$
$$+ (\lambda^2 A_{22} + \lambda^4 A_{24}) \cosh 2kS \sin 2\theta$$
$$+ (\lambda^3 A_{33} + \lambda^5 A_{35}) \cosh 3kS \sin 3\theta$$
$$+ \lambda^4 A_{44} \cosh 4kS \sin 4\theta$$
$$+ \lambda^5 A_{55} \cosh 5kS \sin 5\theta, \tag{B.18}$$

$$k\eta = \lambda \cos \theta + (\lambda^2 B_{22} + \lambda^4 B_{24}) \cos 2\theta$$
$$+ (\lambda^3 B_{33} + \lambda^5 B_{35}) \cos 3\theta + \lambda^4 B_{44} \cos 4\theta$$
$$+ \lambda^5 B_{55} \cos 5\theta, \tag{B.19}$$

$$kK = \lambda^2 C_3 + \lambda^4 C_4, \tag{B.20}$$

$$kC^2 = g \tanh kd(1 + \lambda^2 C_1 + \lambda^4 C_2). \tag{B.21}$$

To determine the coefficients A_{ij}, B_{ij}, and C_i in the above equations, one can evaluate $\partial \phi / \partial x$ and $\partial \phi / \partial S$ from Eq. (B.18) and $\partial \eta / \partial x$ from Eq. (B.19), then substitute into Eq. (B.14). Two sides of Eq. (B.14) will contain terms in λ^n multiplied by terms such as $\sin \theta$, $\sin 2\theta$, $\sin 3\theta$, etc. Similar terms of equal value on either side will yield a series of linear equations in A_{ij}, B_{ij}, etc. The coefficients given by Skjelbreia and Hendrickson are shown as follows:

Let

$$s = \sinh kd,$$
$$C = \cosh kid,$$
$$C_0^2 = g \tanh kd.$$

Then,

$$A_{11} = \frac{1}{s},$$

$$A_{13} = \frac{-c^2(5c^2 + 1)}{8s^5},$$

$$ac = 18 - 249c^2,$$

$$A_{15} = \frac{-(1,184c^{10} - 1,440c^8 - 1,992c^6 + 2,641c^4 + ac)}{1,536s^{11}},$$

$$A_{22} = \frac{3}{8s^4},$$

$$A_{24} = \frac{192c^8 - 424c^6 - 312c^4 + 480c^2 + 17}{768s^{10}},$$

$$A_{33} = \frac{13 - 4c^2}{64s^7},$$

$$ac = 512c^{12} + 4,224c^{10},$$

$$A_{35} = \frac{ac - 6,800c^8 - 12,808c^6 + 16,704c^4 - 3,154c^2 + 107}{4,096s^{13}(6c^2 - 1)},$$

$$A_{44} = \frac{80c^6 - 816c^4 + 1,338c^2 - 197}{1,536s^{10}(6c^2 - 1)},$$

$$ac^5 = 163,470c^2 - 16,245,$$

$$A_{55} = \frac{-(2,880c^{10} - 72,480c^8 + 324,000c^6 - 432,000c^4 + ac^5}{61,440s^{11}(6c^2 - 1)(8c^4 - 11c^2 + 3)},$$

$$B_{22} = \frac{(2c^2 + 1)c}{4s^3},$$

$$B_{24} = \frac{C(272c^8 - 504c^6 - 192c^4 + 322c^2 + 21)}{384s^9},$$

$$B_{33} = \frac{3(8c^6 + 1)}{64s^6},$$

$$bc = 88,128c^{14} - 208,224c^{12} + 70,848c^{10},$$

$$B_{35} = \frac{bc + 54,000c^8 - 21,816c^6 + 6,264c^4 - 54c^2 - 81}{12,288s^{12}(6c^2 - 1)},$$

$$B_{44} = \frac{c(768c^{10} - 448c^8 - 48c^6 + 48c^4 + 106c^2 - 21)}{384s^9(6c^2 - 1)},$$

$$bc^5 = 192,000c^{16} - 262,720c^{14} + 83,680c^{12} + 20,160c^{10},$$

$$B_{35} = \frac{bc^5 - 7,280c^8 + 7,160c^6 - 1,800c^4 - 1,050c^2 + 225}{12,288s^{10}(6c^2 - 1)(8c^4 - 11c^2 + 3)},$$

$$C_1 = \frac{8c^4 - 8c^2 + 9}{8s^4},$$

$$cc = 3,840c^{12} - 4,096c^{10},$$

$$C_2 = \frac{cc + 2{,}592c^8 - 1{,}008c^6 + 5{,}944c^4 - 1{,}830c^2 + 147}{512s^{10}(6c^2 - 1)},$$

$$C_3 = \frac{-1}{4sc},$$

$$C_4 = \frac{12c^8 + 36c^6 - 162c^4 + 141c^2 - 27}{192cs^9}.$$

Tables of A_{ij}, B_{ij}, and C_i values are provided in Skjelbreia and Hendrickson (1960). To determine k and λ, relate the wave height H and wave Profile η by the relation

$$H = \eta_{\theta=0} - \eta_{\theta=\pi}.$$

Using Eq. (B.19) and rearranging Eq. (B.21), it is shown that the following simultaneous equations yield d/L and λ to be solved:

$$\frac{\pi H}{d} = \frac{L}{d}[\lambda + \lambda^3 B_{33} + \lambda^5 (B_{35} + B_{55})], \tag{B.22}$$

$$\frac{d}{L_0} = \frac{d}{L} \tanh\left(2\pi\frac{d}{L}\right)(1 + \lambda^2 C_1 + \lambda^4 C_2), \tag{B.23}$$

where $L_0 = gT^2/(2\pi)$.

For a given wave, substituting the values of H, d, and T into Eqs. (B.22) and (B.23), the simultaneous equations yield the correct values of d/L and λ.

Let

$$F_1 = \lambda_2 A_{11} + \lambda^3 A_{13} + \lambda^5 A_{15},$$
$$F_2 = \lambda^2 A_{22} + \lambda^4 A_{24},$$
$$F_3 = \lambda^3 A_{33} + \lambda^5 A_{35},$$
$$F_4 = \lambda^4 A_{44},$$
$$F_5 = \lambda^5 A_{55}.$$

Using Eq. (B.18), particle velocities are derived from

$$u = C\sum_{n=1}^{5} nF_n \cos n\theta \cosh nkS, \tag{B.24}$$

$$w = C\sum_{n=1}^{5} nF_n \sin n\theta \sinh nkS. \tag{B.25}$$

Differentiation of Eqs. (B.24) and (B.25) with respect to time yields the local water particle accelerations

$$\frac{\partial u}{\partial t} = kc^2 \sum_{n=1}^{5} n^2 F_n \sin n\theta \cosh nkS, \tag{B.26}$$

$$\frac{\partial w}{\partial t} = -kc^2 \sum_{n=1}^{5} n^2 F_n \cos n\theta \sinh nkS. \tag{B.27}$$

The absolute horizontal particle acceleration is given by

$$\frac{\partial u}{\partial t} = \frac{\partial u}{\partial t} + u\frac{\partial u}{\partial x} + w\frac{\partial u}{\partial s}. \tag{B.28}$$

The second and third terms in Eq. (B.28) are called convective acceleration terms, and they are neglected in the linear wave theory. From Eq. (B.24), we get

$$\frac{\partial u}{\partial x} = -kc \sum_{n=1}^{5} n^2 F_n \sin n\theta \cosh nkS,$$

$$\frac{\partial u}{\partial s} = Ck \sum_{n=1}^{5} n^2 F_n \cos n\theta \sinh nkS.$$

It is noted that

$$\frac{\partial u}{\partial x} = \frac{-1}{C}\frac{\partial u}{\partial t},$$

$$\frac{\partial u}{\partial s} = \frac{-1}{C}\frac{\partial w}{\partial t}.$$

The absolute vertical particle acceleration is given by

$$\frac{\partial w}{\partial t} = \frac{\partial w}{\partial t} + w\frac{\partial w}{\partial s} + u\frac{\partial w}{\partial x}.$$

Differentiation of Eq. (B.25) with respect to S and x yields

$$\frac{\partial w}{\partial s} = -kc \sum_{n=1}^{5} n^2 F_n \sin n\theta \cosh nkS,$$

$$\frac{\partial w}{\partial x} = Ck \sum_{n=1}^{5} n^2 F_n \cos n\theta \sinh nkS.$$

Appendix C

Basics of Statistics and Random Process

Suppose we have a set of N data points $\{X_i\}$, then the rth central moment can be given by

$$\mu_r = \frac{1}{N} \sum_{i=1}^{N} (X_i - \bar{X})^r. \tag{C.1}$$

The first moment about zero gives the mean value of the dataset. The second moment about mean gives the variance. Similarly, third and fourth moments about mean give skewness and kurtosis, respectively.

$$\text{Mean: } \mu_1 = \frac{1}{N} \sum_{i=1}^{N} (X_i - \bar{X}), \tag{C.2}$$

Variance (measure of spread of data about the mean value),

$$\mu_2 = \frac{1}{N} \sum_{i=1}^{N} (X_i - \bar{X})^2, \tag{C.3}$$

Skewness (measure of the distribution's symmetry),

$$\mu_3 = \frac{1}{N} \sum_{i=1}^{N} (X_i - \bar{X})^3, \tag{C.4}$$

$$\text{Standard deviation, } \sigma = \sqrt{\text{variance}}. \tag{C.5}$$

Coefficient of variation, $\mathbf{C_v}$, characterizes dispersion relative to the mean facilitating comparison among the variables.

$$\mathbf{C_v} = \sigma / \text{Mean}. \tag{C.6}$$

The **coefficient of skewness**, $\mathbf{C_s}$, measures the skewness of a distribution:

$$\mathbf{C_s} = \mu_3 / \sigma^3. \tag{C.7}$$

The probability density function (pdf) of a measured parameter, say, x_i is given by

$$p(x_i) = \frac{\text{prob}(x_i)}{\Delta x}, \tag{C.8}$$

where $\text{prob}(xi)$ is probability of occurrence of value, x_i, $p(x_i) = \frac{f_i}{N}$.

If $x(t)$ is a continuous random variable, one can obtain the central moments as follows:

$$\bar{x} = E(x) = \underset{T \to \infty}{\text{Lim}} \frac{1}{T} \int_0^T x(t)dt \to \text{first moment (Mean)}, \tag{C.9}$$

$$\sigma_x^2 = \underset{T \to \infty}{\text{Lim}} \frac{1}{T} \int_0^T (x(t) - \bar{x})^2 dt \to \text{second moment (Variance)}, \tag{C.10}$$

$$\mu_3 = \underset{T \to \infty}{\text{Lim}} \frac{1}{T} \int_0^T (x(t) - \bar{x})^3 dt \to \text{skewness}. \tag{C.11}$$

(i) $\int_{-\infty}^{\infty} p(x)dx = 1$.
(ii) For any $c < d$,

$$\text{Prob}[c < x < d] = \int_c^d p(x)dx.$$

The cumulative distribution function, $\text{cdf} = P[X \leq x] = \int_{-\infty}^{x} p(x)dx.$

$$\tag{C.12}$$

In Fig. C.1, the total shaded area gives the probability for $x \leq x_4 = P$.

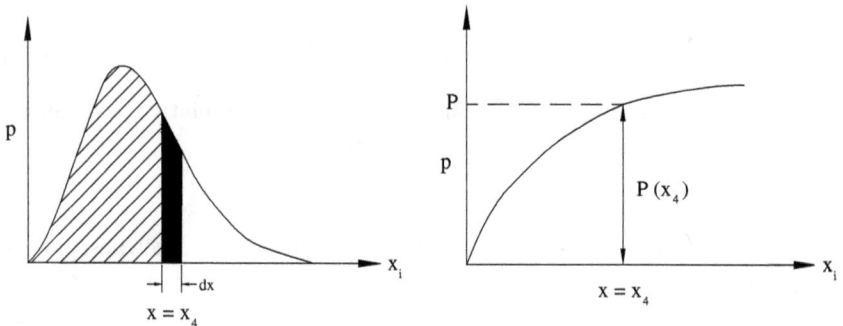

Fig. C.1   Probability distribution function and cumulative distribution curve.

Consider an element of width, dx. If ordinate is P_x, then the total elemental probability would be $p.dx = dP_x$, i.e.,

$$p = \frac{dP_x}{dx}.$$

Probability density is given by change in cumulative probability with respect to x.

The foregoing discussion is for obtaining the pdf of the measured or observed parameter, x_i. It is usual to verify if it follows a standard distribution like the Rayleigh theoretical probability density function given by

$$P(x_i) = \frac{2(x_i)}{x_{rms}^2} \exp^{-\left(\frac{x_i}{x_{rms}}\right)^2}. \tag{C.13}$$

C.1   Classification of Random Data

A single time history representing a random phenomenon (wave elevation for a particular duration of time) is called a sample function or ensemble (or a sample record when observed over a finite time interval). A collection of all possible sample functions which the random phenomenon might have encountered is called a random process or a stochastic process. Ensembles of sample functions form a random process.

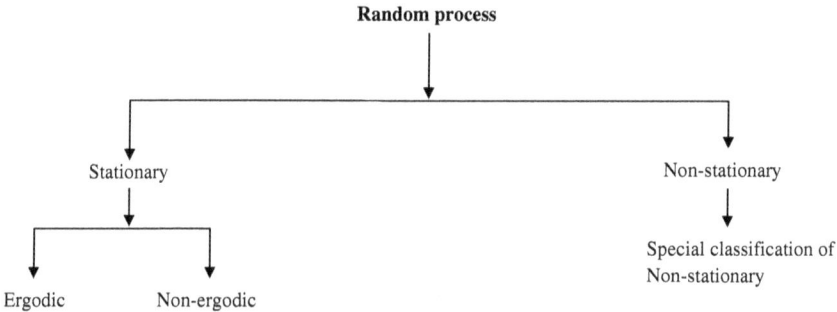

Observations of a random physical phenomenon will result in accumulation of data that cannot be described by means of a mathematical relationship attributed by the uniqueness of the phenomenon.

The mean value calculated at t_1 or first moment of the random process at t_1 is given by

$$\bar{X}_x(t) = \lim_{N \to \infty} \frac{1}{N} \sum_{k=1}^{N} x_k(t_1). \tag{C.14}$$

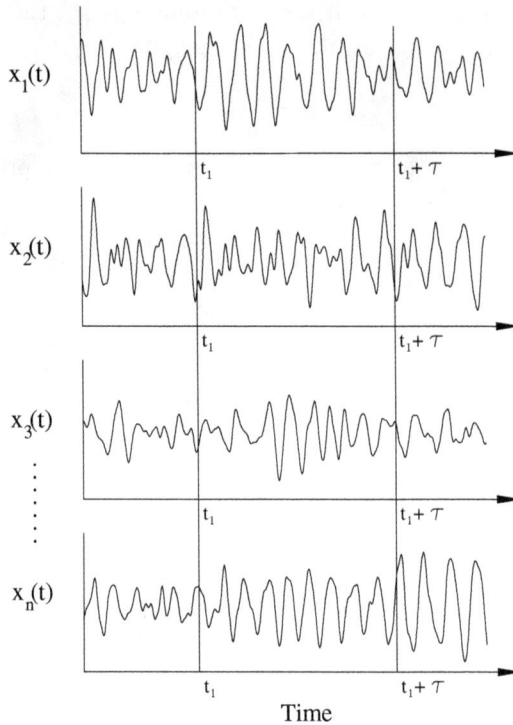

Fig. C.2   Stationary random process/concept of ensemble averaging/calculation of autocorrelation.

Similarly, a correlation (joint moment) between the values of the given random process at two different times (Fig. C.2) called the autocorrelation function can be computed as

$$R_x(t_1, t_1 + \tau) = \underset{N \to \infty}{\text{Lim}} \frac{1}{N} \sum_{k=1}^{N} x_k(t_1) x_k(t_1 + \tau). \qquad (C.15)$$

If $\bar{x}_x(t)$ and $R_x(t_1, t_1 + \tau)$ vary with time, t_1, then the random process $\{x(t)\}$ is said to be *non-stationary*, where the symbol $\{\ \}$ is used to denote an ensemble of sample functions.

If $\bar{x}_x(t)$ and $R_x(t_1, t_1 + \tau)$ do not vary with time, t_1, then the random process is said to be *stationary*. In other words, the probability distribution, mean, mean square, variance of the process is independent of time. Considering the kth sample function of the stationary random process as in the above figure, the $\bar{x}_x(k)$ and $R_x(\tau, k)$ of the kth sample function if found to be almost the same for different sample functions, then the process is *ergodic*. Note that a record obtained from an ergodic random ocean waves will be stationary. But all the stationary process need not be an ergodic process. For an ergodic random process, the time-averaged mean value and autocorrelation function are equal to the corresponding sample averaged mean and the autocorrelation, respectively, i.e.,

$$\bar{X}_x(k) = \bar{X}_x \quad \text{and} \quad R_x(\tau + k) = R_x(\tau). \tag{C.16}$$

Almost all properties of an ergodic random process can be determined by performing time averages over a single sample function.

The autocorrelation function $R_x(\tau)$ of X relates the value of X at time (t) to its value at time $(t+\tau)$ and so provides an indication of the correlation of the signal with itself for various time lags. Autocorrelation at time lag τ can be obtained as the expectation of the lagged product of the data:

$$R(\tau) = E[x(t)x(t + \tau)].$$

or

$$R(\tau) = \operatorname*{Lim}_{T \to \infty} \frac{1}{T} \int_0^T x(t)x(t + \tau)dt. \tag{C.17}$$

For discrete values of data, the autocorrelation can be written as

$$R(k) = \frac{1}{N - k} \sum_{n=1}^{N-k} x_n x_{n+k}, \tag{C.18}$$

where $k = 0, 1, \ldots, m$, in which k is the lag number and m is maximum number of lags. It is usually not desirable to use a greater number of lags than 10% of the total number of data points. If the process does not have zero mean, the mean should be computed and subtracted from each discrete value of the data before calculating the autocorrelation function.

$$R(\tau) = \operatorname*{Lim}_{T \to \infty} \frac{1}{T} \int_0^T x(t)x(t + \tau)dt.$$

When $\tau = 0$,

$$R(0) = E[x^2] \to \text{mean square value.}$$

It is convenient to introduce the normalized autocorrelation function or autocorrelation coefficient as

$$\rho_x(t) = \frac{R_x(\tau)}{R(0)}, \tag{C.19}$$

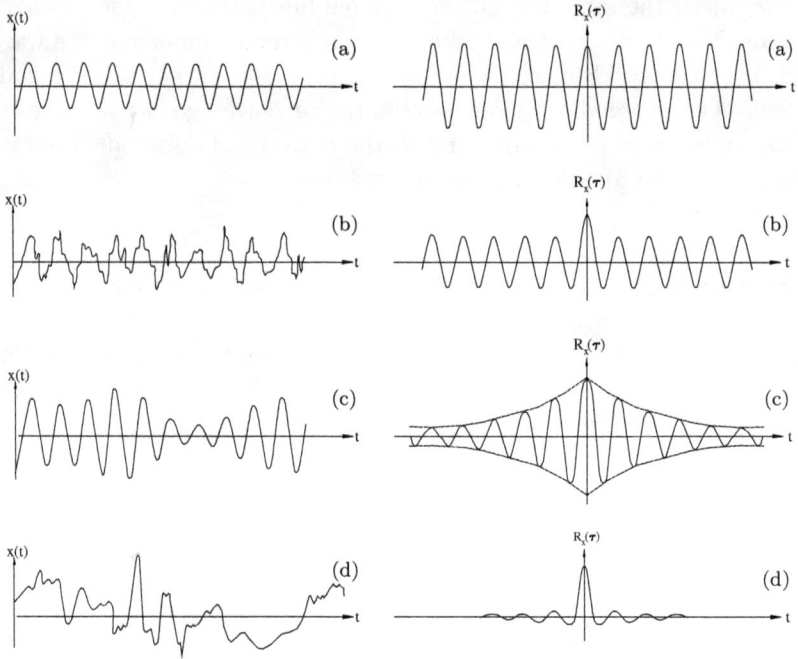

Fig. C.3   Typical time histories and corresponding autocorrelation: (a) periodic wave, (b) periodic wave + random noise, (c) narrowband random noise and (d) broadband random noise (Bendat and Piersol, 1971).

i.e., auto correlation effect is

$$\rho_x(\tau) = \frac{R_x(\tau)}{\langle x^2(t) \rangle} = \frac{\langle x(t)x(t+\tau) \rangle}{\langle x^2(t) \rangle}, \qquad \text{(C.20)}$$

where "$\langle \rangle$" is expected value.

Clearly, a signal is perfectly correlated with itself for zero lag τ, and we have

$$\rho_x(\tau) = 1 \quad \text{at } \tau = 0 \quad \text{and} \quad \rho_x(\tau) = 0 \text{ as } \tau = \infty.$$

It is noted that $R_x(\tau)$ is an even function. i.e.,

$$R_x(\tau) = R_x(-\tau). \qquad \text{(C.21)}$$

The $\rho_x(\tau) = 1$ at $\tau = 0$ would indicate the nature of the timeseries as depicted in Fig. C.3.

C.2   Cross-Correlation Function $R_{XY}(\tau)$

This is useful in relating signals (e.g., velocities or pressures) at two different points in a random flow.

The term cross–covariance function also used for $R_{xy}(\tau)$ in cases where X and Y are both defined to have zero means:

$$R_{xy}(\tau) = \langle x(t)y(t+\tau)\rangle,$$

$$R_{xy}(\tau) = \lim_{T\to\infty} \frac{1}{T} \int_0^T x(t)y(t+\tau)dt. \tag{C.22}$$

Two useful relationships which bound the absolute value of the cross-correlation function are

$$|R_{xy}(\tau)|^2 \le [R_x(0)R_y(0)],$$

$$|R_{xy}(\tau)| \le \frac{1}{2}[R_x(0)R_y(0)].$$

When $R_{xy}(\tau) = 0$, $x(t)$ and $y(t)$ are said to be uncorrelated.

Figure C.4 represents a cross-correlation function. Note that the plot will sometimes display sharp peaks which indicate the existence of correlation between $x(t)$ and $y(t)$ for specific time difference.

C.3   Cross Spectrum

The Wiener–Khintchine relation can be extended to analyze two random wave records. The relation states that the cross-correlation and cross-spectral density functions are a Fourier transform pair. Therefore, the

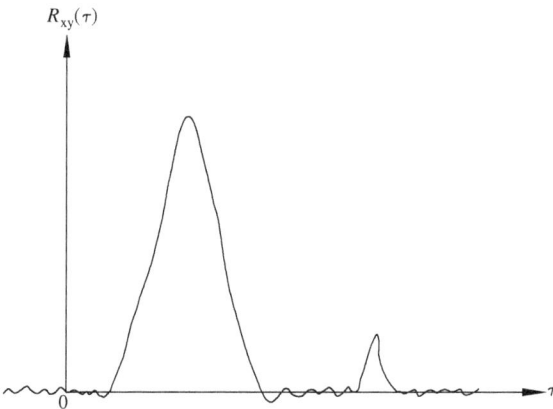

Fig. C.4   Cross-correlation function.

cross-spectral density function can be derived as follows:

$$S_{xy}(\omega) = C_{xy}(\omega) + iQ_{xy}(\omega),$$

where

$$C_{xy}(\omega) = \frac{1}{\pi} \int_0^\infty \left(R_{xy}(\tau) + R_{yx}(\tau) \right) \cos(\omega\tau)d\tau,$$

$$Q_{xy}(\omega) = \frac{1}{\pi} \int_0^\infty \left(-R_{xy}(\tau) + R_{yx}(\tau) \right) \sin(\omega\tau)d\tau.$$

The function $C_{xy}(\omega)$ is known as the *co-spectrum* whereas the other function $Q_{xy}(\omega)$ is known as the *quadrature-spectrum*.

The amplitude of cross spectrum is given by

$$|s_{xy}(\omega)| = \sqrt{(C_{xy}(\omega))^2 + (Q_{xy}(\omega))^2}.$$

We can also observe that the co-spectrum is an even function and the quadrature-spectrum is an odd function.

In order to comprehend the directional characteristics of ocean waves, cross-spectral analysis is needed.

C.4   Coherence Function

Coherence is a measure of the degree of relationship, as a function of frequency, between any two given time series.

Coherence is a statistic measure that can be used to examine the relation between the input and output. It is commonly used to estimate the difference between input and output of a linear system. If the ensemble is ergodic, and the system function is linear, it can be used to *estimate* the cause and effect between the input and output. The coherence function γ^2 is obtained by

$$\gamma^2 = \frac{S_{xy} - S_{xy}^*}{S_{xx} - S_{xy}} = \frac{|S_{xy}|^2}{S_{xx} - S_{xy}}, \tag{C.23}$$

where S_{xy} is the cross spectrum and S_{xx} and S_{yy} are the power spectrum of x and y, respectively. The coherence function, γ^2, is the square of the absolute value of the cross spectrum divided by each power spectrum of the input and output of the system. The coherence function, in its nature, is not meaningful unless averaging is performed.

Bibliography

Abramowitz, M. and Stegun, I.A. (1965). *Handbook of Mathematical Functions*. National Bureau of Standards, Washington, DC.

Airy, G.B. (1845). On tides and waves. In *Encylopaedia Metropolitana*, Vol. 5, Mixed Sciences, London, pp. 241–396.

Balaji, R., Sannasiraj, S.A. and Sundar, V. (2008). Response characteristics of a discus shaped data buoy in nonlinear waves, *Coastal Eng. J.* 50(4), 463–489. 10.1142/S0578563408001892.

Bansal, R.K. (2008). *A Textbook of Fluid Mechanics*, Laxmi Publications (P) Ltd, India.

Bendat, J.S. and Piersol, A.G. (1971). *Random Data: Analysis and Measurement Procedures*, John Wiley and Sons Inc, 1: 75.

Biésel, F. (1951). Les apparails generateurs de houle en laboratorie. *La Houille Blanche* 6(2,4,5).

Biesel, F. and Suquet, F. (1951). Les appareils générateurs de houle en laboratoire, *Houille Blanche*, 6, 147–165, 475–496, 723–737.

Biesel, F. (1952). Study of wave propagation in water of gradually varying depth, in *Gravity Waves*, Circular No. 521, National Bureau of Standards, Washington, DC.

Biesel, F. (1954). Wave Machines, Proc of 1st SNAME Conf on Ships and Waves, Hoboken.

Boccotti, P. (2007). Caisson breakwaters embodying an OWC with a small opening — Part I: Theory, *Ocean Engineering*, 34(5–6), 806–819.

Boccotti, P., Filianoti, P., Fiamma, V. and Arena, F. (2007). Caisson breakwaters embodying an OWC with a small opening — Part II: A small-scale field experiment, *Ocean Engineering*, 34(5–6), 820–841.

Boussinesq, J. (1872). Théorie des ondes et des remous qui se propagent le long d'un canal rectangulaire horizontal, encommuniquant au liquidecontenudansce canal des vitess essensiblement pareilles de la surface au fond. *J. Math. Pureset Appl.* 17, 55–108.

Bowden, K.F. (1983). *Physical Oceanography of Coastal Waters*, Ellis Horwood Limited.

Bretschneider, C.L. (1952). Revised wave forecasting relationship, *Proc. 2nd Conference Coastal Eng.*, Council on Wave Research, University of California, Berkeley, 1–5.

Bretschneider, C.L. (1958). Revisions in wave forecasting: Deep and shallow water, *Proc. 6th Conf. Coastal Eng.*, Council on Wave Research, University of California, Berkeley, pp. 1–18.

Bretschneider, C.L. (1959). Wave variability and wave spectra for wind-generated gravity waves, Technical Memorandum No. 118, Beach Erosion Board, U.S. Army Corps of Engineers, Washington, DC.

Bretschneider, C.L. (1969). Wave forecasting, In *Handbook of Ocean and Underwater Engineering*, in Myers, J.J., Holm, C.H. and McAllister, R.F. (eds.), McGraw-Hill Book Co., New York, Chapter 11.

Brooke, J. (2003). *Wave Energy Conversion*, Elsevier.

Borgman, L.E. (1967). Spectral analysis of ocean wave forces on piling. *Waterways Harbours Division ASCE*, 93(WW2): 129–156.

Borgman, L.E. (1969). Directional spectra models for design use, *Proc. 1st Offshore Tech. Conf.*, Houston, Tx, Paper 1069, pp. 721–746.

Brunn, P. (1981). *Port Engineering*, Gulf Publishing co.

Chan, E.S. and Melville, W.K. (1988). Deep-water plunging wave pressures on a vertical plane wall, *Proceedings of the Royal Society of London. Series A, Mathematical and Physical Sciences* 417(1852), 95–131.

Cartwright, D.E. (1958). On estimating the mean energy of sea waves from the highest waves in record, *Proceedings of the Royal Society of London. Series A, Mathematical and Physical Sciences* 247(1248), 22–48.

Chakrabarti, S.K. (1987). *Hydrodynamics of Offshore Structures*, Springler-Verlag Berlin.

Chakrabarti, S.K. (1994). *Modelling Laws*, Advanced Series on Ocean Engineering.

Cartwright, D.E. and Longuet-Higgins, M.S. (1956). The statistical distribution of the maxima of a random function, *Proc. Royal Soc. London. Ser. A*, 237(1209): 212–232.

Coastal Engineering Manual (2002, 2006). US Army Corps of Engineers. 1110-2-1100, US Army Corps of Engineers, Washington, DC (in 6 volumes).

Collins, J.I. and Weir, W. (1969). Probabilities of wave characteristics in the surf zone, Tetra Tech. Report, TC-149, Pasadena, California, USA.

Cornett, A. (2008). A global wave energy resource assessment, In: *Proceedings of the eighteenth international offshore and polar conference*, 50.

Daugherty, R.L., Franzini, J.B. and Finnemore, E.J. (1985). *Fluid Mechanics with Engineering Applications*, McGraw-Hill.

Dean, R.G. (1965). Stream function representation of nonlinear ocean waves, *J. Geophys. Res.* 70: 18.

Dean, R.G. and Dalrymple, R.A. (1994). *Water Wave Mechanics for Engineers and Scientists*, Prentice-Hall, Inc., Englewood Cliffs, NJ.

Dhinakaran, G., Sundar, V., Sundaravadivelu, R. and Graw, K.U. (2012). Performance of a perforated submerged semicircular breakwater due to non-breaking waves, *J. Eng. Mart. Environ.*, 226(1), 36–50.

DHI User Manual for wave makers at OED IITM (1995).

Fan, S.S. (1968). Diffraction of wind waves, Hydraulic Eng. Lab, University of California, Berkeley, Tech. Rep. HEL 1–10.

Falcão, de O.A.F. (2000). The shoreline OWC wave power plant at the Azores, *Proc. 4th European Wave Energy Conf.*, Aalborg, Denmark, 42–47.

Falcão, de O.A.F. (2002). Control of an oscillating-water-column wave power plant for maximum energy production, *Applied Ocean Research*, 24(2), 73–82.

Fenton, J.D. and Mckee, W.D. (1990). On calculating the lengths of water waves, *Coastal Eng.* 14: 499–513.

Funke, E.R. and Mansard, E.P.D. (1992). On the testing of models in multidirectional seas, *Proc. Coastal Eng.*, Venice, October 4–9.

Funke, E.R. and Mansard, E.P. (1987). A rationale for the use of the deterministic approach to laboratory wave generation, *Proc. Wave Analysis and Generation in Laboratory Basins, 22nd Congress International Association for Hydraulic Research*, pp. 153–195.

Galvin, C.J. (1969). Breaker travel and choice of design wave height, *J. Waterways and Harbors Division, ASCE*, 95(WW2): 175–200.

Garde, R.J. (1989). *Fluid Mechanics through Problems*, Wiley Eastern Limited.

Goda, Y. (1970). A synthesis of breaker indices, *Trans. JSCE*, 2: 227–230.

Goda, Y. (2010). *Random Seas and Design of Maritime Structures*, Advanced Series on Ocean Engineering, World Scientific.

Goda, Y. and Suzuki (1975). Computation of refraction and diffraction of sea waves with Mitsuyasu's directional spectrum, Tech. Note Port and Harbour Res. Inst. No. 230, 1975, p. 45 (in Japanese).

Gourlay, M.R. (1992). Wave set-up, wave run-up and beach water table: Interaction between surf zone hydraulics and groundwater hydraulics, *Coastal Eng.* 17: 93–144.

Guo, J. (2002). Simple and explicit solution of wave dispersion equation, *Coastal Eng.* 45: 71–74.

Hagerman, G. (1995). "Wave power" in *Encyclopedia of Energy Technology and the Environment*, New York: John Wiley & Sons, Inc., pp. 2859–2907, 1995.

Hasselman, K., *et al.* (1973). Measurement of wind-wave growth and swell decay during the Joint North Sea Wave Project (JONSWAP), *Deutschen Hydrographischen Zeit. Erganzunscheft*, 13(A).

Hasselmann, K. (1976). Stochastic climate models, *Part 1. Theory. Tellus*, 28, 473–485.

Haskins, G.L. and Leggett I.M. (1983). Long term forecasting for offshore operations, in *Design of Offshore Structures*. Thomas Telford Ltd.

Havelock, T.H. (1929). Forced surface-waves on water. *Phil. Mag.* 7(8): 569–576.

Heath, T., Whittaker, T.J.T. and Boake, C.B. (2000). The design, construction and operation of the LIMPET wave energy converter (Islay, Scotland), *Proc. 4th European Wave.*

Hiroi, I. (1919). On a method of estimating the force of waves, *J. College Eng. Tokyo Imperial Univ.* 10(1): 1–19.

Horikawa, K. (1978). *Coastal Engineering — An introduction to Ocean Engineering*, University of Tokyo Press.

Hudson, R.Y. (1959). Laboratory investigation of rubble-mound breakwaters, *J. Waterways and Harbors Division, ASCE* 85(WW3), 93–121.

Hughes, S.A. (1993). *Principles of Similitude*, Advanced series on Ocean Engineering.

Hunt, J.N. (1979). Direct solution of wave dispersion equation, *J. Waterways, Ports, Coastal Ocean Division ASCE*, 105(WW4): 457–459.

IEA-OES (2008). International energy agency-Implementing agreement on ocean energy systems, Annual report.

Ippen, A.T. (ed.) (1966). *Estuary and Coastline Hydrodynamics*, McGraw–Hill.

Isobe, M., Kundo, K. and Horikawa, K. (1984). Extension of MLM for estimating directional wave spectrum, *Symp. Description and Modelling of Directional Seas*, Technical University, Denmark, June, A6-1–A6-15.

Iwagaki, Y., Asano, T. and Nagai, F. (1983). Hydrodynamic forces on a circular placed in wave-current co-existing fields, *Mem. Faculty of Eng. Kyoto University*, Japan XLV(1): 11–23.

ITTC (1969). *Recommendations of the 11th International Towing Tank Conference*, Proc. 12th ITTC, Rome.

ITTC (1972). *Recommendations of the 11th International Towing Tank Conference*, Proc. 13th ITTC, Berlin/Hamburg.

ITTC (1996). *Recommendations of the 11th International Towing Tank Conference*, Proc. 11th ITTC, Tokyo.

ITTC (2017). *Recommended Procedures and Guidelines*.

International Ship Structures Congress: ISSC (1964). Proceedings. Reports of the committees 3a-3d, Ship Structures Laboratory. Volume 2.

Karthik Ramnarayan, S., Sannasiraj, S.A. and Sundar, V. (2020). Hydrodynamic characteristics of curved and vertical front face pile-supported breakwaters in regular waves, *Ocean Eng.* in Press.

Keulegan, G.H. (1950). Wave motion, *Engineering Hydraulics*, John Wiley & Sons, New york, Chapter 2.

Keulegan, G.H. and Carpenter, L.H. (1958), Forces on cylinders and plates in an oscillating fluid, *J. Res. Natl. Bur. Stand.*, 60(5): 423–440, doi:10.6028/jres.060.043.

Kobune, K., Sasaki, H. and Hashimoto, N. (1985). Characteristics of ocean waves off Cape Nojima, *Report of the port and Harbour Research Institute, Ministry of Transport*, Nagase, Yokosuka, Japan, Vol. 24(3).

Kobune, K. and Hashimoto, N. (1986). Estimation of directional spectra from the maximum entropy principle, *Proc. 5th Int. Symp. Offshore Mechanics and Artic Engineering*, Tokyo, Japan, Vol. 1, pp. 80–85.

Komar, P.D. and Gaughan, M.K. (1972). Airy wave theory and breaker height prediction, *Proc. 13rd Coastal Eng. Conf.* pp. 405–418.

Kondo, K. and de Koning, J.P.A. (1986). Comparison of directional sea measurements by an orbital following buoy and a combination of a wave gauge and a current meter, *Proc. 5th Int. Symp. Offshore Mechanics and Artic Engineering*, Tokyo, Japan, Vol. 1, pp. 51–58.

Krishnakumar, C. (2009). Hydrodynamic characteristics of slotted screens under uni- and multi-directional waves, PhD thesis, Department of Ocean Engineering, IIT Madras, India.

Kurian, N.P. and Praveen, S.S. (2010). Tsunami wave propagation in the Arabian Sea and its implications of run-up/inundation characteristics along the Kerala coast, *Indian J. Marine Sci.* 39(4): 231–540.

Larson, M. and Kraus, N.C. (1989). SBEACH: Numerical model for simulating storm-induced beach change, Report 1, Tech. Report CERC-89-99, Waterways Experiment Station, US Army Corps of Engineers.

Le Mehaute, B. (1976). *An Introduction to Hydrodynamics and Water Waves*, Springer Science Business Media, New York.

Le Mehaute, B. and Koh, R.C.Y. (1967). On the breaking of waves arriving at an angle to the shore, *J. Hydraulic Res.* 5(1): 67–88.

Longuet-Higgins, M.S. (1952). On the statistical distribution of height of sea waves, *J. Marine Res.* IX(3): 245–268.

MacCamy, R.C. and Fuch, R.A. (1954). Wave forces on piles: a diffraction theory, US Army Corps of Eng., Beach Erosion Board, Tech. Memo No. 69.

Madsen, O.S. (1976). Wave climate of the continental margin: Elements of its mathematical description, Stanley, D.J. and Swift, D.J.P. (eds.), *Marine Sediment Transport in Environmental Management*, New York, Wiley, pp. 65–87.

McCowan, J. (1894). On the highest waves of a permanent type, *Philos. Mag. Edinburgh, Ser.* 5(38): 351–358.

Miche, R. (1944). Mouvements mondulatoires des mersenprofondeurconstante on decroissante, *Ann. Ponts Chaussees*, 114: 131–164, 270–292 and 369–406.

Minikin, R.R. (1963). *Winds, Waves, and Maritime Structures: Studies in Harbor Making and in the Protection of Coasts.* 2nd edition, Griffin, London.

Mitsuyasu, H., Tasai, F., Suhara, T., Mizuno, S., Ohkusu, M., Honda, T. and Rikiishi, K. (1975). Observations of the directional spectrum of ocean waves using a cloverleaf buoy, *J. Phys. Oceanography*, 5: 750–760.

Mobarek, I.E., (1965). Directional Spectra of laboratory wind waves, *J. Waterway Harbours Division, ASCE*, 91(WW3): 91–116.

Modi, P.N. and Seth, S.M. (1985). *Hydraulics and Fluid Mechanics*, Standard Book House.

Mogridge, G.R. and Jamieson, W.W. (1976). Wave loads on square caisson, *Proc. 15th Coastal Eng. Cong*, Honolulu, Hawaii, Vol. III, pp. 2271–2289.

Morison, J.P., OBrien, M.P., Johnson, J.W. and Shaaf, S.A. (1950). The force exerted by surface waves on piles, *Petroleum Trans. AIME* 189.

Munk, W.H. (1949). The solitary wave theory and its application to surf problems, *Ann. N.Y. Acad. Sci.* 51: 376–423.

Narasimha Rao, T.V.S. and Sundar, V. (1982). *Estimation of Wave Power Potential Along the Indian Coastline*, Pergamon Press Ltd.

Nwogu, O.U., Mansard, E.P.D., Miles, M.D. and Isaacson, M. (1987). Estimation of directional wave spectra by the maximum entropy method, *Ocean Eng.* 29(11): 1343–1355.

Ochi, M.K. and Hubble, E.N. (1976). Six parameter wave spectra, *Proc. 15th Coastal Eng. Conf.*, Honolulu, Hawaii, ASCE, pp. 301–328.

Ohneda, H., Igarashi, S., Shinbo, O., Sekihara, S., Suzuki, K., Kubota, H., Ogino, H. and Morita, H. (1991). Construction procedure of a wave power extracting caisson breakwater. In: *Proc. 3rd Symp. Ocean Energy Utilization*, Tokyo, 171–179.

Okawa, Y. and Shuto, N. (1984). Run-up of periodic waves on beaches of non–uniform slope, *Proc. 19th Coastal Eng. Conf.*, ASCE, pp. 328–344.

Ostendorf, D.W. and Madsen, O.S. (1979). An analysis of longshore current and associated sediment transport in the surf zone, Report No. 241, Department of Civil Engineering, MIT.

Parker, M.E. (1977). Wave induced Forces on a partially exposed circular cylinder, M.S. Thesis, Texas A & M University.

Penney, W.G. and Price, A.T. (1952). The diffraction theory of sea waves and the shelter afforded by breakwaters, *Philos. Trans. Roy. Soc. A*, 244(882), 236–253.

Pierson, W.J. and Moskowitz, L. (1964). A proposed spectral form for fully developed wind seas based on the similarity theory of S.A. Kitaigorodskii, *J. Geophys. Res.* 69: 5181–5190.

Pierson, W.J. and Holmes, P. (1965). Irregular wave forces on a pile, *J. of Waterways Harbours Division, ASCE*, 91(WW4), 1–10.

Pilarczyk, K.W. and Zeidler, R.B. (1996). Offshore breakwaters and shore evolution control.

Ploeg, J. and Funke, E.R. (1980). "A Survey of 'Random' Wave Generation Techniques," *Proc. 17th Coastal Engineering Conf. ASCE*, Vol. 1, pp. 135–153.

Ploeg, J. and Funke, E.R. (1985). The generation of laboratory waves: a tool for the understanding of wave dynamics, *Proc. Int. Sym Water Wave Research*, SFB 205, University of Hannover, Germany, pp. 265–283.

Priest, M.S. (1971). Wave forces on exposed pipelines on the ocean bed, *Proc. Offshore Technology Conf.* paper NO: OTC 1383, Houston, Texas, USA.

Rajasekaran, C., Sannasiraj, S.A. and Sundar, V. (2010). Breaking wave impact pressure on a vertical wall. *Int. J. Ocean Climate Syst.* 1(3): 155–165.

Rambabu, M., Narasimha Rao, S. and Sundar, V. (2003). Current induced scour around a vertical pile in cohesive soil, *Ocean Eng. J.* 30: 893–920.

Rattanapitikon, W. and Vivattanasirisak, T. (2000). Examination of breaker height formulas, *Proc. 12th Congr. Asia & Pacific Division of IAHR*, Bangkok, 13–16 November 2000, pp. 425–434.

Ravindran, M. and Koola, P.M. (1991). Energy from sea waves — the Indian wave energy program, *Current Science*, 60: 676–680.

Reddy, D.V. (2013). *Essentials of Offshore Structures: Framed and Gravity Platforms*, CRC Press.

Reid, R.O. (1957). Correlation of water level variations with wave forces on a vertical pile for non-periodic waves, *Proc. 6th Coastal Eng. Conf.*, Florida, pp. 749–786.

Roberts, I. and Shepherd, K. (2009). WaveRush: A New concept for a breakwater wave energy converter, *Proc. Coasts, Marine structures and Breakwaters*, 16–18 Sep.

Rundgren, L. (1958). Water wave forces: a theoretical and laboratory study, *Kungliga Tekniska Hoegskolans handlingar*, Stockholm, Sweden.

Sainflou, M. (1928). Essaisur les digues maritimes verticales, *Ann. des Ponts Chaussees*, 98(11).

Sand, S.E. (1979). Three-dimensional deterministic structure of ocean waves, *Hydrodynamics and Hydraulic Engineering*, Technical University of Denmark, Series Paper No. 24, p. 189.

Sand, S.E. and Mynett, A.E. (1987). Directional wave generation and analysis, *IAHR Seminar*, pp. 209–235.

Sannasiraj, S.A., Sundaravadivelu, R. and Sundar, V. (2000). Diffraction-radiation of multiple floating structures in directional waves, *Ocean Eng. J.* 28: 201–234.

Sarpkaya, T. and Isaacson, M. (1981). *Mechanics of Wave Forces on Offshore Structures*, New York, Van Nostrand Reinhold.

Scott, J.R. (1965). A sea spectrum for model tests and long-term ship prediction, *J. Ship Res.* 9: 145–152.

Seymour, R.J. (1992). *Ocean Energy Recovery: The State of the Art American*, Society of Civil Engineers.

Shore Protection Manual (1984). Published by Dept. of the Army, Waterways Experiment Station, Corps of Engineers, Coastal Engineering Research Center, Vicksburg, Miss, Washington, DC.U.S. Army Coastal Engineering Research Centre.

Sithik, A., Meyer, J., Sriram, V. and Hildebrandt. A. (2020). Experimental investigation of offshore crane load during upending a wind turbine jacket substructure in regular waves, *Marine Structures*, In press.

Silvester, R. (1974). *Coastal Engineering*, Elsevier Scientific Publishing Co.

Singamsetti, S.R. and Sind, H.G. (1980). Characteristics of breaking and shoaling periodic waves normally incident on to plane beaches of constant slope, Report M1371, Delft Hydraulic Lab., Delft, The Netherlands.

Skjelbreia, L. (1959). *Gravity Waves Stokes Third order Approximation, Table of Functions*, American Society of Civil Engineers.

Skjelbreia, L. and Hendrickson, J. (1960). Fifth order gravity wave theory, *Proc. Coastal Engineering Conf.* 184–196.

Skjelbreia, L. and Hendrickson, J. (1960). Fifth order gravity wave theory, *Proc. 7th Coastal Engineering Conf.*, The Hague, The Netherlands, pp. 184–196.

Smith, J.M. and Kraus, N.C. (1990). Laboratory study on macro-features of wave breaking over bars and artificial reefs, Technical Report CERC-90-12, WES, U.S. Army Corps of Engineers.

Sorenson, R.M. (1978). *Basic Coastal Engineering*, A Wiley-Interscience, New York.

Sriram, V., Schlurmann, T. and Schimmels, S. (2015). Focused wave evolution using linear and second order wave maker theory, *Appl. Ocean Res.* 53: 279–296.

Sruthi, C. and Sriram, V. (2017). Wave impact load on jacket structure in intermediate water depth, *Ocean Eng.* 140: 183–194.

Stokes, G.G. (1847). On the theory of oscillatory waves, *Trans. Camb. Phil. Soc.* 8, 441. *Mathematical and Physical Papers*, 1, 197. Cambridge University Press.

Sukanya, R., Sundar, V. and Sannasiraj, S.A. (2016). Stability and settlement analysis of compound geo-synthetic materials in coastal protection measures, *Abstract Submitted to Ninth Int. Conf. Coastal and Port Engineering in Developing Countries* (*PIANC- COPEDEC IX*), 16–21 October, Rio de Janeiro, Brazil.

Sunamura, T. (1980). A laboratory study of offshore transport of sediment and a model for eroding beaches, *Proc. 17th Coastal Eng. Conf. ASCE*, pp. 1051–1070.

Sunamura, T. and Horikawa, K. (1974). Two-dimensional beach transformation due to waves, *Proc. 14th Coastal Eng. Conf. ASCE*, Vol. II, pp. 920–938.

Sundar, V. (1986). Wave characteristics off the south East Coast of India, *Ocean Eng.* 13(4), 327–338.

Sundar, V. Koola, P.M. and Schlenkhoff, A.U. (1999). Dynamic pressures on inclined cylinders due to freak waves, *Ocean Engineering*, 26(9), 841–863, ISSN 0029-8018, https://doi.org/10.1016/S0029-8018(98)00029-8.

Sundar, V., Moan, T. and Hals, J. (2010). January. Conceptual design of OWC wave energy converters combined with breakwater structures. In *International Conference on Offshore Mechanics and Arctic Engineering* (Vol. 49118, pp. 479–489).

Sundar, V., Vengatesan, V., Anandkumar, G. and Schlenkhoff, A. (1998). Hydrodynaamic coefficients for inclined cylinders, *Ocean Engineering*, 25(4–5), 277–294.

Sundar, V., Koola, P. and Schlenkhoff, A. (1999). Dynamic pressures on inclined cylinders due to freak waves, *Ocean Eng.* 26(9), 841–863.

Sundar, V., Sannasiraj, S.A., Murali, K. and Sundaravadivelu, R. (2007). Run-up an inundation along the Indian Peninsula including Andaman Islands due to great Indian ocean tsunami, *J. Waterway, Port, Coastal, Ocean Eng. ASCE*, 133(6): 401–413.

Sundar, V., Vengatesan, V. and Graw, K.U. (2002). Sleeve forces on inclined cylinders due to long and short crested waves, *J. Hydraulic Research, IAHR*, 40(3): 275–287.

Sverdrup, H.U. and Munk, W.H. (1946). *Wind, Sea and Swell: Theory of Relations for Forecasting*, Publication 601, US Navy Hydro Graphic Office, Washington, DC.

Tolman, H.L. (2002). Validation of WAVEWATCH III version 1.15 for a global domain. NOAA/NWS/NCEP/OMB Technical Note Nr. 213, 33 pp.

Tucker, M.J. (1963). Analysis of records of sea waves, *Proc. Institution of Civil Engineers*, London, Vol. 26, pp. 305–316.

Tucker, M.J., Challenor, P.G. and Carter, D.J.T. (1984). Numerical simulation of a random sea, *Appl. Ocean Res.* 6(2): 118–122.

Tung, C.C. and Huang, N.E. (1973). Combined effects of current and waves on fluid force, *Ocean Eng.* 2(4): 183–193.

USACE (U.S. Army Corps of Engineers), (2006). *Coastal Engineering Manual*, Vicksburg, Mississippi: Coastal and Hydraulics Laboratory, Engineer Research and Development Center Report EM 1110-2-1100.

Wiegel, R.L. (1960). A presentation of cnoidal wave theory for practical application, *J. Fluid Mechanics*, 7, 273–286.

Wiegel, R.L. (1964). *Oceanographical Engineering*. Englewood Cliffs, N.J., Prentice-Hall.

Weggel, J.R. (1972). Maximum breaker height, *J. Waterways, Harbors Coastal Eng. Division*, 98(WW4): 529–548.

Weigel, R.L. (1982). *Oceanographical Engineering*, Prentice Hall Inc.

Whittaker, T.J.T. and Stewart, T.P. (1993). An experimental study of nearshore and shorline oscillating water columns with harbours, *Proceedings of the European Wave Energy Symposium*, Edinburgh, Scotland, July, 151–156.

Wilson, B.W. (1963). Deep water wave generation by moving wind systems, *Transactions of the American Society of Civil Engineers* 128(4), 104–133.

You, Z.-Jin. (2008). A close approximation of wave dispersion relation for direct calculation of wavelength in any coastal water depth, *Applied Ocean Research — Appl. Ocean Res.* 30, 113–119. 10.1016/j.apor.2008.08.001

Zhang, H. and Geng, B. (2015). Introduction of the World Largest Wave Flume Constructed by TIWTE. *Procedia Engineering*, 116, 905–911. 10.1016/j.proeng.2015.08.380.

Index